Communications
in Computer and Information Science   **1341**

More information about this series at http://www.springer.com/series/7899

Dana Simian · Laura Florentina Stoica (Eds.)

Modelling and Development of Intelligent Systems

7th International Conference, MDIS 2020
Sibiu, Romania, October 22–24, 2020
Revised Selected Papers

 Springer

Editors
Dana Simian 🄳
Lucian Blaga University of Sibiu
Sibiu, Romania

Laura Florentina Stoica 🄳
Lucian Blaga University of Sibiu
Sibiu, Romania

ISSN 1865-0929   ISSN 1865-0937 (electronic)
Communications in Computer and Information Science
ISBN 978-3-030-68526-3   ISBN 978-3-030-68527-0 (eBook)
https://doi.org/10.1007/978-3-030-68527-0

This Springer imprint is published by the registered company Springer Nature Switzerland AG
The registered company address is: Gewerbestrasse 11, 6330 Cham, Switzerland

Preface

This volume contains selected, peer-reviewed papers presented at the 7th International Conference on Modelling and Development of Intelligent Systems – MDIS 2020, which was held from 22–24 October, 2020, in Sibiu, Romania. The conference was organized by the Research Center in Informatics and Information Technology and the Department of Mathematics and Informatics of the Faculty of Sciences, Lucian Blaga University of Sibiu (LBUS), Romania. Due to the Covid-19 pandemic, this edition of the conference was organized exclusively online.

Intelligent systems modelling and development are currently areas of great interest, making an important contribution to solving the problems that today's society is facing. In this context, the importance of international forums allowing the exchange of ideas in topics connected to artificial intelligence is increasing at the global level.

The MDIS conferences aim to connect scientists, researchers, academics, IT specialists and students who want to share and discuss their original results in the fields related to modelling and development of intelligent systems. The original contributions presented at MDIS 2020 ranged from concepts and theoretical developments to advanced technologies and innovative applications. Specific topics of interest included but were not restricted to: evolutionary computing, genetic algorithms and their applications, swarm intelligence, metaheuristics and applications, data mining, machine learning, intelligent systems for decision support, ontology engineering, adaptive systems, robotics, knowledge-based systems, computational and conceptual models, pattern recognition, e-learning, hybrid computation for artificial vision, modelling and optimization of dynamic systems, multiagent systems and mathematical models for the development of intelligent systems.

Compared to previous editions, this 7th MDIS edition was distinguished by a considerable extension of the geographical area of the authors (Europe, Asia, Africa, North America) and more diverse teams of collaborators.

Four keynote speakers gave presentations, showing the high impact that research in artificial intelligence has or can have in all life areas. The invited keynote speakers and their lectures were:

Frank-Michael Schleif, Learning in Indefinite Proximity Spaces with Applications for Life Science Data
Milan Tuba, Artificial Intelligence for Digital Image Classification
Abdel-Badeeh M. Salem, Artificial Intelligence in Smart Medical Systems
Anca Ralescu, On Explainable Artificial Intelligence

All submitted papers underwent a thorough single-blind peer review. Each paper was reviewed by at least 3 independent reviewers, chosen based on their qualifications and field of expertise.

This volume contains 25 selected papers submitted by authors from 13 countries.

We thank all the participants for their interesting talks and discussions. We also thank the scientific committee members and all of the other reviewers for their help in reviewing the submitted papers and for their contributions to the scientific success of the conference and to the quality of this proceedings volume.

December 2020

Dana Simian
Laura Florentina Stoica

Organization

General Chair

Dana Simian — Lucian Blaga University of Sibiu, Romania

Scientific Committee

Kiril Alexiev	Bulgarian Academy of Sciences, Bulgaria
Nebojsa Bacanin	Singidunum University, Serbia
Arndt Balzer	University of Applied Sciences Würzburg-Schweinfurt, Germany
Alina Bărbulescu	Ovidius University of Constanța, Romania
Lasse Berntzen	Buskerud and Vestfold University College, Norway
Charul Bhatnagar	GLA University, India
Florian Boian	Babeș-Bolyai University, Romania
Peter Braun	University of Applied Sciences Würzburg-Schweinfurt, Germany
Steve Cassidy	Macquarie University, Australia
Nicolae Constantinescu	University of Craiova, Romania
Ovidiu Cosma	Technical University of Cluj-Napoca, Romania
Dan Cristea	Alexandru Ioan Cuza University, Romania
Gabriela Czibula	Babeș-Bolyai University, Romania
Daniela Dănciulescu	University of Craiova, Romania
Thierry Declerck	German Research Center for Artificial Intelligence, Germany
Lyubomyr Demkiv	Lviv Polytechnic National University, Ukraine
Alexiei Dingli	University of Malta, Malta
Oleksandr Dorokhov	Kharkiv National University of Economics, Ukraine
George Eleftherakis	CITY College, International Faculty of the University of Sheffield, Greece
Călin Enăchescu	University of Medicine, Pharmacy, Science and Technology of Târgu Mureș, Romania
Denis Enăchescu	University of Bucharest, Romania
Ralf Fabian	Lucian Blaga University of Sibiu, Romania
Stefka Fidanova	Bulgarian Academy of Sciences, Bulgaria
Ulrich Fiedler	Bern University of Applied Sciences, Switzerland
Martin Fränzle	Carl von Ossietzky University of Oldenburg, Germany
Amir Gandomi	Michigan State University, USA
Dejan Gjorgjevikj	Ss. Cyril and Methodius University, Republic of Macedonia
Teresa Gonçalves	University of Évora, Portugal
Andrina Granić	University of Split, Croatia

Katalina Grigorova	University of Ruse, Bulgaria
Axel Hahn	Carl von Ossietzky University of Oldenburg, Germany
Piroska Haller	University of Medicine, Pharmacy, Science and Technology of Târgu Mureş, Romania
Masafumi Hagiwara	Keio University, Japan
Milena Karova	Technical University of Varna, Bulgaria
Sundaresan Krishnan Iyer	Infosys Limited, India
Raka Jovanovic	Hamad bin Khalifa University, Qatar
Saleema J. S.	Christ University, Bangalore, India
Marcel Kyas	Reykjavik University, Island
Adnan Khashman	European Centre for Research and Academic Affairs, Cyprus
Wolfgang Kössler	Humboldt University of Berlin, Germany
Lixin Liang	Tsinghua University, China
Suzana Loskovska	Ss. Cyril and Methodius University, Republic of Macedonia
Manuel Campos Martínez	University of Murcia, Spain
Ginés García Mateos	University of Murcia, Spain
Gerard de Melo	University of Potsdam, Germany
Matthew Montebello	University of Malta, Malta
Gummadi Jose Moses	Raghu Engineering College Visakhapatnam, India
Eugénio da Costa Oliveira	University of Porto, Portugal
Grażyna Paliwoda-Pękosz	Cracow University of Economics, Poland
Ivaylo Plamenov Penev	Technical University of Varna, Bulgaria
Petrică C. Pop Sitar	Technical University of Cluj-Napoca, Romania
Anca Ralescu	University of Cincinnati, USA
Mohammad Rezai	Sheffield Hallam University, UK
Abdel-Badeeh M. Salem	Ain Shams University, Egypt
Livia Sângeorzan	Transilvania University of Braşov, Romania
Hanumat Sastry	University of Petroleum and Energy Studies, India
Willi Sauerbrei	University of Freiburg, Germany
Vasile-Marian Scuturici	University of Lyon, France
Soraya Sedkaoui	University of Khemis Miliana, Algeria
Francesco Sicurello	University of Milano-Bicocca, Italy
Andreas Siebert	University of Applied Sciences Landshut, Germany
Corina Simian	Whitehead Institute for Biomedical Research, USA
Dana Simian	Lucian Blaga University of Sibiu, Romania
Lior Solomovich	Kaye Academic College of Education, Israel
Srun Sovila	Royal University of Phnom Penh, Cambodia
Ansgar Steland	RWTH Aachen University, Germany
Florin Stoica	Lucian Blaga University of Sibiu, Romania
Detlef Streitferdt	Ilmenau University of Technology, Germany
Grażyna Suchacka	University of Opole, Poland
Ying Tan	Beijing University, China
Jolanta Tańcula	University of Opole, Poland

Mika Tonder	Saimaa University of Applied Sciences, Finland
Claude Touzet	Aix-Marseille University, France
Milan Tuba	Singidunum University, Serbia
Dan Tufiş	Romanian Academy, Romania
Alexander D. Veit	Harvard Medical School, USA
Anca Vasilescu	Transilvania University of Braşov, Romania
Sofia Visa	The College of Wooster, USA
Anca Vitcu	Alexandru Ioan Cuza University, Romania
Badri Vellambi	University of Cincinnati, USA
Xin-She Yang	Middlesex University London, UK

Plenary Lecture 1

Learning in Indefinite Proximity Spaces with Applications for Life Science Data

Frank-Michael Schleif

School of Computer Science

University of Applied Sciences Würzburg-Schweinfurt

frank-michael.schleif@fhws.de

Abstract. Life science data are often encoded in a non-standard way by means of alpha-numeric sequences, graph representations, numerical vectors of variable length or other formats. The majority of more complex data analysis algorithms require fixed-length vectorial input data, asking for substantial pre-processing of non-standard input formats. Domain specific, non-standard proximity measures lead in general to so-called indefinite measures and are widely ignored in favour of simple encodings. These encoding steps are not always easy to perform nor particular effective, with a potential loss of information and interpretability. We present some strategies and concepts of how to employ data-driven similarity measures in the life science context to obtain effective prediction models.

Brief Biography of the Speaker: Frank-Michael Schleif (Dipl.-Inf., University of Leipzig, Ph.D., TU-Clausthal, Germany) was a Marie Curie Senior Research Fellow at the University of Birmingham, Birmingham, UK and a Post-Doctoral Fellow in the group of Theoretical Computer Science (TCS) at Bielefeld University, Bielefeld, Germany, where he also received a *venia legendi* in applied computer science in 2013. He was also a software developer and consultant for the Bruker Corp. Since 2016 he has been with the University of Applied Sciences Würzburg-Schweinfurt, Germany, where he is a Professor for Database Management and Business Intelligence. His current research interests include data management, computational intelligence techniques and machine learning for non-metric models and large-scale problems. Several research stays have taken him to the UK, the Netherlands, Japan and the USA. He is a member of the German chapter of the European Neural Network Society (GNNS), the GI and the IEEE-CIS. He is an editor of Machine Learning Reports and a member of the editorial board of Neural Processing Letters.

Plenary Lecture 2

Artificial Intelligence for Digital Image Classification

Milan Tuba

Singidunum University, Belgrade, Serbia
tuba@ieee.org

Abstract. Artificial intelligence represents one of the leading research fields and the source of major progress in various fields such as medicine, autonomous vehicles, security, agriculture, etc. One of the common problems that is solved by AI methods is classification. A great improvement in solving this task was achieved by convolutional neural networks, a special class of deep neural networks that considers the spatial correlation of input data rather than just plain data. They are used for digital image classification, voice recognition, EEG signal analysis and classification, etc. The results achieved by CNN are significantly better in comparison with the previously existing methods. One of the challenges with CNN is finding the network architecture that has the best performance for the specific application. Numerous hyperparameters such as the number of different layers, number of neurons in each layer, optimization algorithm, activation functions, kernel size, optimization algorithm, etc. have to be tuned. In many cases, the CNN's configuration is set by guessing and estimating (guestimating) better values for the hyperparameters, but recent studies showed promising results when using swarm intelligence algorithms to solve this hard optimization problem. A few examples of using swarm intelligence algorithms for convolutional neural network hyperparameter tuning will be presented.

Brief Biography of the Speaker: Milan Tuba is the Vice Rector for International Relations at Singidunum University, Belgrade, Serbia and was the Head of the Department of Mathematical Sciences at the State University of Novi Pazar and the Dean of the Graduate School of Computer Science at John Naisbitt University. He received B. S. in Mathematics, M. S. in Mathematics, M. S. in Computer Science, M. Ph. in Computer Science, Ph. D. in Computer Science from University of Belgrade and New York University. From 1983 to 1994 he was in the U.S.A., first at Vanderbilt University in Nashville and Courant Institute of Mathematical Sciences, New York University and later as Assistant Professor of Electrical Engineering at Cooper Union School of Engineering, New York. During that time he was the founder and director of Microprocessor Lab and VLSI Lab, leader of scientific projects and theses supervisor. From 1994 he was Assistant Professor of Computer Science and Director of the Computer Center at the University of Belgrade, from 2001 Associate Professor, Faculty of Mathematics, University of Belgrade, from 2004 also a Professor of Computer Science and Dean of the College of Computer Science, Megatrend University, Belgrade. He has taught more than 20 graduate and undergraduate courses, from

VLSI Design and Computer Architecture to Computer Networks, Operating Systems, Image Processing, Calculus and Queuing Theory. His research interests include nature-inspired optimizations applied to computer networks, image processing and combinatorial problems. Prof. Tuba is the author or coauthor of more than 200 scientific papers and coeditor or member of the editorial board or scientific committee of a number of scientific journals and conferences. He has been invited and delivered around 60 keynote and plenary lectures at international conferences. He is a member of the ACM, IEEE, AMS, SIAM, IFNA.

Plenary Lecture 3

Artificial Intelligence in Smart Medical Systems

Abdel-Badeeh M. Salem

Ain Shams University, Cairo, Egypt
abmsalem@yahoo.com, absalem@cis.asu.edu.eg

Abstract. Artificial Intelligence (AI) is devoted to the creation of intelligent computer software and hardware that imitates the human mind. The main goal of AI technology is to make computers smarter by creating software that will allow a computer to mimic some of the functions of the human brain in selected applications. Advances in AI paradigms and smart healthcare systems (SHS) domains highlight the need for ICT systems that aim not only at the improvement of humans' quality of life but at their safety too. SHS are intelligent systems and are based on the concepts, methodologies and theories of many sciences, e.g. artificial intelligence, data science, social science, information science, computer science, cognitive sciences, behavioral science, life sciences and healthcare. The well-known smart healthcare paradigms are: Real-time monitoring devices, Computer-aided surgery devices, Telemedicine devices, Population-based care devices, Personalized medicine from a machine learning perspective, Ubiquitous intelligent computing, Expert decision support systems, and Health 2.0. and Internet of Things (IoT). On the other side, AI can support many tasks and domains, e.g. law, education, healthcare, economy, business, life sciences, environment, energy and military applications. All of these applications employ knowledge base and inferencing techniques to solve problems or help make decisions in specific domains. This talk discusses the potential role of the AI paradigms, computational intelligence and machine learning techniques which are used in developing SHS. The talk focuses on the AI methodologies and their potential usage in recent trends in developing smart healthcare and intelligent systems. The following three paradigms are presented: (a) ontological engineering, (b) case-based reasoning, and (c) data mining. Moreover the talk presents the research results of the author and his colleagues that have been carried out in recent years at Ain Shams University AIKE-Labs, Cairo, Egypt.

Brief Biography of the Speaker: Abdel-Badeeh M. Salem has been a full Professor of Computer Science since 1989 at Ain Shams University, Egypt. His research includes biomedical informatics, big data analytics, intelligent education and learning systems, information mining, knowledge engineering and biometrics. He is founder of the Artificial Intelligence and Knowledge Engineering Research Labs, Ain Sham University, Egypt and Chairman of the Working Group on Bio-Medical Informatics, ISfTeH, Belgium. He has published around 550 papers (105 of them in Scopus). He has been involved in more than 600 international conferences and workshops as a keynote and plenary speaker, member of Program Committees, Workshop/Session organizer, Session and Tutorials Chair. In addition he has been on the Editorial Board of 50 international and national Journals. He is a member of many international scientific societies and associations: an elected member of the Euro Mediterranean Academy of Arts and Sciences, Greece; a member of Alma Mater Europaea of the European Academy of Sciences and Arts, Belgrade; and a member of the European Academy of Sciences and Arts, Austria.

Plenary Lecture 4

On Explainable Artificial Intelligence

Anca Ralescu

University of Cincinnati, Ohio, USA
anca.ralescu@uc.edu

Abstract. In recent years, 'explainable AI' has become one of the phrases most often used in connection with Artificial Intelligence. What does this phrase mean, and what developments in AI made it necessary? This talk aims to discuss these questions through a historical perspective on the field of Artificial Intelligence, and the newer disciplines related to it, Machine Learning and Deep Learning.

Brief Biography of the Speaker: Born and raised in Făgăraş, Romania, Anca Ralescu graduated from the University of Bucharest and Indiana University, Bloomington, where she obtained a PhD in Mathematics. She is currently professor of Computer Science in the EECS Department, University of Cincinnati. Her main interests are in intelligent systems, including Artificial Intelligence, Fuzzy Systems, and Machine Learning, with applications to computer vision and image understanding.

Contents

Machine Learning

Mathematical Models for Development of Intelligent Systems

Modelling and Optimization of Dynamic Systems

Evolutionary Computing

Embedding Human Behavior Using Multidimensional Economic Agents

Florentin Bota[1]([✉])[iD] and Dana Simian[2][iD]

[1] Babes-Bolyai University, Cluj-Napoca, Romania
botaflorentin@cs.ubbcluj.ro
[2] Research Center in Informatics and Information Technology,
Lucian Blaga University, Sibiu, Romania
dana.simian@ulbsibiu.ro

Abstract. This paper contributes to the design of realistic autonomous agents which can simulate the human behavior and respond to multiple factors such as impulse, emotions, social influences, etc.

Therefore, "Multidimensional" Agents (MA), embedding several human social features and evolving in time should replace the perfectly rational agents when modeling complex systems which imply human behavior. The features of a MA are dependent on the system we want to model. In this paper we will refer to the design of Multidimensional Economic Agents (MEA).

We propose a stochastic evolutionary algorithm for the evolution of MEA in a dynamic environment, which takes into account the fairness behavior and the community influence as social features. The fairness concept is an anomaly of the standard "homo economicus" model, where the agents are completely rational and self-interested. We used a bottom-up, data-driven approach to find a new fitness function to develop agents which can play the economics experiment called "Ultimatum Game". In order to simulate the human behavior in agent decision we introduced a neural network component.

We used human experiments and simulations to illustrate the effectiveness of our approach.

Keywords: Multi-agent systems · Evolution strategies · Machine learning

1 Introduction

In a system with linear interactions, the whole is the sum of its parts [14] and such a system is easy to predict, based on several key parameters. This does not apply in complex systems, where multiple factors influence the behavior of the elements and can lead to results which are difficult to predict. Complex systems are systems which present a large number of interacting components, with a behavior that is difficult to model. Such behaviors can be found in economy,

© Springer Nature Switzerland AG 2021
D. Simian and L. F. Stoica (Eds.): MDIS 2020, CCIS 1341, pp. 3–19, 2021.
https://doi.org/10.1007/978-3-030-68527-0_1

where the human decision-making process is very problematic to model. In economy, most of the phenomena are complex and the interactions between the parts can create an emergent behavior, where the whole is qualitatively different than the sum of its parts [14].

The behavior of complex systems is usually studied by means of computational models. Computational models are mathematical concepts, requiring considerable computational resources, used to understand, simulate and predict interactions for existing processes and phenomena [22].

In economy, the standard or neo-classical economic model is used by economists in their analysis of consumer choice. It is based on utility and profit maximization functions by using the rational choice theory [27]. Here, by utility we denoted an universal concept that refers to the capacity of a good or a service to satisfy a need. It can refer to buying a cup of coffee or using a service like Netflix for example. There are several assumptions on which the standard economic model is based:

- People act with full information of external factors
- People have known preferences
- People choose the best available option, in a rational manner

Based on this assumptions it is fairly easy to model economic behavior and the model could often corresponds with actual data [35].

An agent acting according to this model is called *"homo economicus"*, a term with Latin origins which was used for the first time in 1906 by Vilfredo Pareto in his *Manual of Political Economy*. The term describes an economic agent which behaves in a perfect rational way and for which happiness is derived from the consumption of goods and services. The concept continues to influence mainstream economists of the capitalist economy [34].

However there are situations in which the model fails to correspond with real-life data (economic crises, *externalities* like pollution or other social costs). In these cases most economists try to bend the quite flexible neo-classical model to explain the deviations.

As a response, a newer concept, the *behavioral model* tries to explain irrational decisions by examining the human behavior and applying insights from laboratory experiments, psychology, and other social sciences in economics. It states that the actors take decisions based on impulse, emotions or social pressure [4,18].

There are some challenges for the behavioral model, namely it is difficult to make predictions in large-scale environments and there are some alternatives, like *Pareto's Theory of Action*, where we can observe conflicts when logical and non-logical actions collide [7].

From our observations, the standard model better fits to macro-economic system, while the behavioral model works better for micro-decisions.

The structure of the paper is as follows. Section 2 describes our objectives and research questions. In Sect. 3 we briefly present the proposed methodology. Section 4 is dedicated to validation of our theoretical model. We analyse the Ultimatum Game and the proposed representation of the agent strategy. We also present our first results, using some state-of-the-art functions. For model validation we conducted our own experiment with human participants and confirmed the state-of-the-art empirical results. The proposed evolutionary algorithm is described in Sect. 5. Implementation details and results are given in Sect. 6. The conclusions and suggested extensions of our work are reviewed in Sect. 7.

2 Objectives and Research Questions

Our final aim is to design and implement artificial Multidimensional Economic Agents (MEA), based on a computational model able to overcome the disadvantages of the models mentioned above and to simulate and predict real human behavior in a manner which works for both micro and macro economics. We propose four components for the computational model corresponding to four levels or dimensions for MEA:

- Cognitive
- Social
- Emotional
- Rational

Each component has to have a specific weight (importance in the final decision), bias (chance to accelerate a decision by itself – increasing or decreasing the weight, according with a specific magnitude), and a computing function in the final model.

In this paper we focus on the study and evaluation of the following research questions:

> **RQ1:** Can the perceived human behavior be simulated using evolutionary algorithms?

> **RQ2:** How can we integrate neural networks in our MEA concept?

3 Proposed Model and Experiments

Our proposed theory is that the existing human behavior is actually the result of evolution over multiple generations, in a dynamic environment, favoring the emergence of long-term economic winning strategies. These strategies are closely related to human social features, such as fairness and sharing as well as related to environmental restrictions.

As part of the levels for MEA enumerated before, in this article, we propose a solution for modeling the social and cognitive behavior of human actors who play the economics experiment called "Ultimatum Game" (UG) [6,29]. In our work on MEA, the UG was used because it offers us insights regarding agent interactions and the impact of fair and unfair decisions. The development of agents that produce results similar to those of experiments with human subjects when playing UG can only be done taking into account the social level of the used computational model.

We designed a stochastic evolutionary algorithm for the evolution of MEA, which takes into account the fairness behavior and the community influence as social features in finding the winning strategy. To simulate the community influence in agents decision we introduced a community filter. The cognitive experience of human actors is achieved in our experiment using the agents evolving process over several generations. To test the performance of the algorithm we compared the average offer and distribution with state-of-the-art examples from the literature and our own experimental results.

We used a bottom-up, data-driven approach to find a new fitness function for developing and evolving our MEA. To validate the proposed evolutionary algorithm and the effectiveness of our approach we conducted human experiments, implemented 3D tools and compared the results.

The applicability of our algorithm will be tested in further experiments to develop agents that could be used to analyze and predict stock market indices, recessions or even vote intention. The complete algorithm and description are given in Sect. 5.

In this context we also analysed the possibility of using neural networks to simulate human behavior and integrate the results in our MEA concept. The experimental results are promising and will be described in the paper.

4   Model Validation

4.1   Ultimatum Game

Ultimatum Game (UG) is an economical game that inspired many articles from different fields ranging from the game theory, to economic, social, psychological theories [19] and even neuro science [32].

In the first version of UG, name also one shot or standard UG, two players act on the following rules.

A fixed amount of money must be divided between the two players: first, player A makes an offer and player B (second player) can accept the amount or reject the offer in its entirety. If the offer was rejected by player B, no player receives anything. If the second player accepts the proposed amount, the money is split in the manner in which the first player offered. The game ends after the second player formulates his decision.

The standard economic model predicts that, reacting totally rational, both players should try to maximize their profit. This model is based on the classical game theory [17, 21, 23]. The first player should offer the smallest possible amount in order to maximize his payoff and the second player should always accept a non-zero offer. This strategy corresponds to Nash equilibrium subgame perfection point.

The experiments on human subjects, conducted in different conditions, in a variety of studies revealed an "anomaly" of UG when comparing with the standard economic model and the results from the game theory. There were many offers between 40–50% of the sum and the unfair offers were in majority rejected [6,13].

There have been attempts to create models whose outcomes correspond to experimental data using evolutionary game theory [15,17], quantum game theory [23], or golden ratio splits (38.2% minimum percentage for a proposal to be accepted) [30]. However, simple evolutionary models when using a deterministic input, usually favours self-interest and do not comply with the practical experiments with human subjects [26].

On the other hand, many game experiments based on the UG tested various hypothesis concerning the influence of different social, cognitive and emotional factors on the results [11,13,19,31–33]. One of the questions that neuroeconomists have tried to find the answer to is why the subjects reject low offers [13]. Many attempts to explain this behavior have been made in the existing literature. Studies correlated with social dimensions considered factors like punishment of unfair offers, emotions, social orientation of players for explaining different experiment results [3,9]. The existing studies on UG emphasize that human agents are heavily influenced by the payoff and social connections of other players [6,29].

The players strategy changes when many round of the game are played on the same population of players. The experience of playing the game with different roles, led the players to fairer offers [20]. The experience acts here as a cognitive factor. UG was also used to describe the influence of conventional nonconscious mental functions in decisions making [3].

The purpose of our work is to design a computational model that can be used in the development of MAE whose decisions when playing UG are similar to those of human players.

4.2   Related Work

Classical game theory was used to study the existence and uniqueness of Nash equilibrium points [21] in the UG. However it can not explain the results of UG experiments with human players.

New theories link the Nash equilibrium in UC to the golden ratio (38.2% minimum percentage for a proposal to be accepted) [30]. The Rabin model of fairness using a kindness and a personal belief functions [28] tried to fit the simulations results with the real ones.

Agents based simulation were used to better understand the human behavior [2] as well as to study the evolution of fairness in finite small populations [29]. The evolution of fairness in repeated UG on a finite small population of agents was studied in [29] using a stochastic evolution of strategies. UG was repeated assigning randomly the roles (proposer and responder) to agents. The strategy of a proposer agent is defined by the amount, p, offered and the strategy of a responder agent is given by the rejection threshold q. The acceptance condition is $p \geq q$.

In [2] an agents based simulation of one shot UG with a fixed amount of money (100Eu) is made using NetLogo v.5.1.0. The agents are split into two categories selfish and altruistic, defined by the range of their proposal (35–45, respectively 45-55Eu). Each agent has associated a willingness to accept (WTA). The monitored result consists in the average mean cash earn by the selfish and altruist agents and is highly influenced by the WTA. The main purpose is to compare the altruistic and selfish behavior.

4.3   The First Experiment

We aim to prove the hypothesis that the concept of fairness have evolved as an evolutionary advantage during human development. Therefore we built a new agent based model, according to the stochastic evolutionary game theory [29,33]. We proposed a new variant of a stochastic approach, where selection is made by using a stochastic decision function. We observed the evolution of the agents strategies over many generations and we observed that the natural selection favors fairness.

We started from the model proposed in [29]. The agents can have two roles denoted by A (first player who makes the offer) and B (second player having to accept or reject the offer). We model the agents strategies using $\alpha, \beta \in [0,1]$, where α is the percentage of the amount of money offered as player A and β is the lowest percentage from the amount which will be accepted as player B. The pairs of agents reacting in an one shot UG and their roles are chosen randomly.

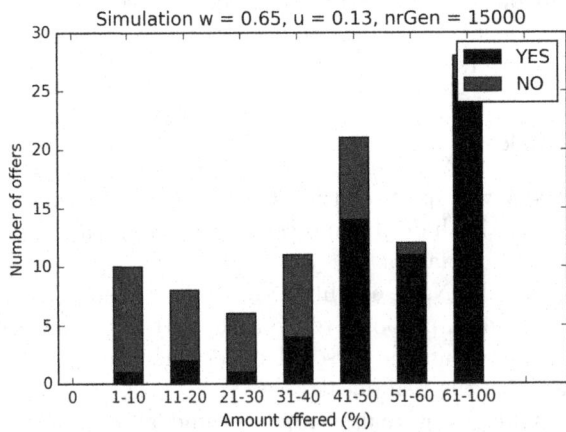

Fig. 1. Intermediate statistical data using the probability function from Eq. (1)

We tested the pairwise comparison process, where a random role model and one individual are chosen from the population, then the focused agent accepts the offer of the role model[33] with probability p.

$$p = \frac{1}{2} + w\frac{\pi_f - \pi_r}{\Delta\pi}. \tag{1}$$

In Eq. (1), $w(0 \leq w \leq 1)$ represents the intensity of selection, $\Delta\pi$ is the maximum payoff difference and π_f and π_r are the payoffs of the selected individuals.

Figure 1 presents the results of this approach. After a close comparison with the experimental data from Fig. 3 we concluded that we cannot simulate human behavior accurately and we decided to improve the process, by using a new equation.

4.4   Second Experiment, with Human Players

We can see in Fig. 3 that most offers are around 50%, representing equal split. This anomaly is very important and we can find the cause in human behavior.

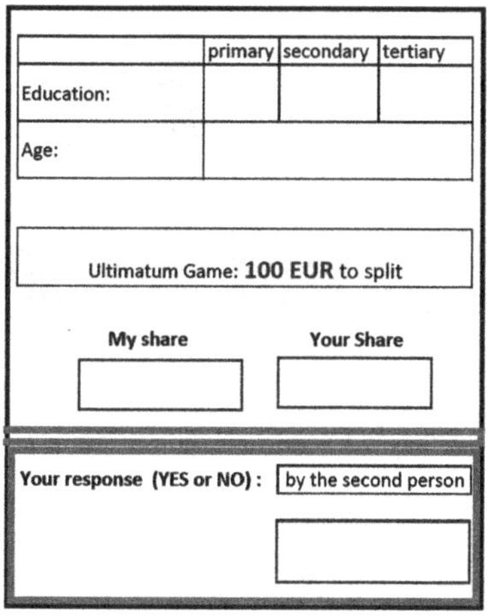

Fig. 2. The card used to play Ultimatum Game with human players

There are numerous experiments regarding UG in the literature, but we used the opportunity to conduct our own analysis. For this purpose we created the playing cards from Fig. 2 and we played the game with 105 people.

Regarding the procedure, 25 of the subjects played the game in Berlin, Germany, during a workshop presentation (SMSA). The other 80 participants were students at the Babes-Bolyai University in Cluj-Napoca, Romania, enrolled in Computer Science studies.

We explained the rules to the participants and they confirmed that they understood them. Each subject was given a paper where they completed their answers.

They played as Player A in the first stage and as Player B in the second stage. The offers were exchanged voluntarily between participants in Berlin (usually the closest person) and they were shuffled and redistributed in Cluj-Napoca.

58% of the participants had completed tertiary education level and the others were in progress. Their average age was 24.49 years and the amount to be split was 100 EUR. Several studies [12,16] confirm that imaginary money can be used instead of real money with no or small deviation of the results. The only differences appear when the amounts of money are very large, but this was not applicable in our case.

After we played the games we analysed the results and the average offer was 43.62%, similar with other experiments (41.04%, [31]).

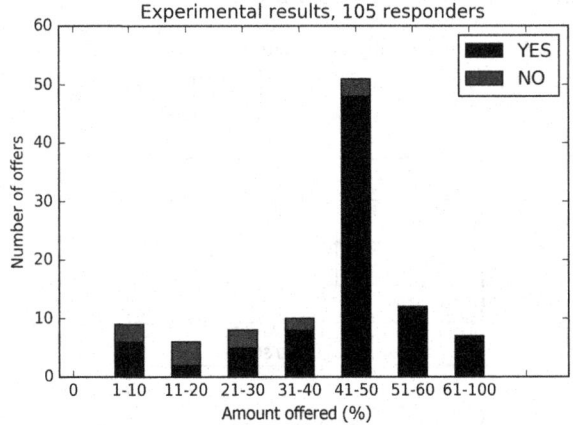

Fig. 3. Experimental results

An anomaly was that the most played strategy was fair split(41.90%) and an unusual amount of offers over 50%. We believe that these anomalies are caused by the sample size (10.0% margin of error) and the community effect (people who know each other are more likely to be fair) for the students.

Our model states that people are inclined to show fairness in interactions with people they already know. Such a correlation was also found during our experiment, where the average offer between the students was 45.02% and the non-student average was 39.2%. Still, another experiment, focused on this aspect is required to determine actual causality.

5   The Proposed Framework

After several experiments we decided to use the function defined in Eq. (2) to determine the cloning probability in the evolution process.

$$p = \frac{1}{3} - w\frac{\pi_r - \pi_f}{\pi_m}. \tag{2}$$

In Eq. (2), $w(0 \leq w \leq 1)$ is the selection's intensity, π_r is the gain of the role model individual, π_f is the payoff of the focused individual and π_m is the amount of money(maximum payoff).

The proposed algorithm is given next.

Algorithm 1. *Ultimatum Game evolution algorithm*

Parameters:

- Selection variable - w;
- Probability of mutation - u;
- Number of generations - $nrGen$;
- Number of agents - $nrAgents$;
- Smallest possible payment amount - $mMin$;
- Largest possible payment amount - $mMax$;
- Communities number - $nrCom$

1: Randomly initialize $nrAgents$ individuals.
2: **for** $nrGen$ generations **do**
3:     $moneyAmount$ = random $(mMin, mMax)$;
4:     reset agents' earnings;
5:     **for** x in $listOfAgents$ **do**
6:         **for** y in $listOfAgents$ **do**
7:             playGame($x, y, moneyAmount$)
8:             playGame($y, x, moneyAmount$)
9:         **end for**
10:     **end for**
11:     average the payoffs from the previous games
12:     $evolve(listOfAgents)$
13: **end for**
14: shuffle $listOfAgents$
15: apply communityFilter($nrCom, listOfAgents$);*
16: play a final game
17: **return** the agents and the last game's results

$^{(*)}$ nrCom=1 in our simulation, each agent pair (x, y) will have $\alpha = \frac{x.\alpha + y.\alpha}{2}$
the function is a simplified version of Rabin's fairness model [28]
$$U_i(a_i, b_j, c_i) = \pi_i(a_i, b_j) + f_j(b_j, c_i) * [1 + f_i(a_i, b_j)]$$
We can use the model to enforce an equilibrium level because the strategies are known when the game is played

---

We begin with a random threshold for offering and receiving money for a list of agents. Each agent has an amount of money in his bank, which increases over time if the other agents agree with his offers.

Over several generations we play the game and use the Eq. (2) to clone or mutate our role agents which were chosen randomly. After a number of generations we play a final game then return the results. The amount of money for each series of games is random and thus varies over the evolution process.

Algorithm 2. *evolve(listOfAgents)*

1: shuffle listOfAgents
2: select *focus*, *role*, two random agents
3: with the probability *p* (given in equation (2)) clone the *role* in the *focus*
4: **if** not cloned **then**
5:     with the probability *u*, mutate *focus*
6: **end if**

6   Implementation and Results

6.1   Ultimatum Game Evolution Algorithm

The results obtained using the Algorithm 1 proved that natural evolution favors fairness, but the distribution is very dependent on the stochastic parameters (see Fig. 4). A further community filter was needed to negotiate offers between the individuals of the same community. In these experiments we used a single community, to better simulate most of our real participants (colleagues).

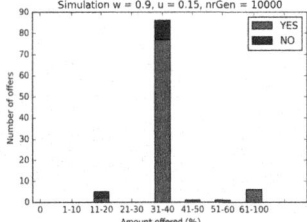

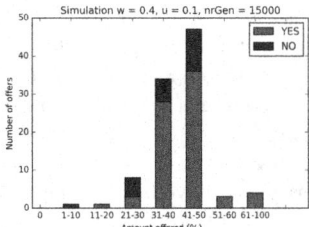

(a) Strong selection, $w = 0.9$ (b) Weak selection, $w = 0.4$

Fig. 4. Evolution of fairness in finite population simulation

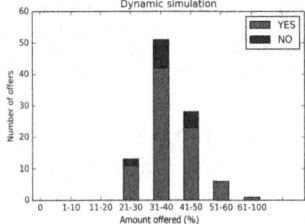
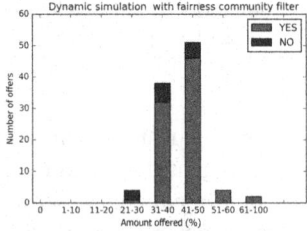

(a) Simple stochastic dynamic population (b) Population with the community filter

Fig. 5. Final results of our agents

The community filter in Algorithm 1 is a simplified version of Rabin's fairness model [28], where the strategies are known in the moment the offers are made. This filter is a new approach and can be used to simulate one or multiple communities and enforce an equilibrium within members of the same community. The purpose of this added operation is to better simulate the social component of our model.

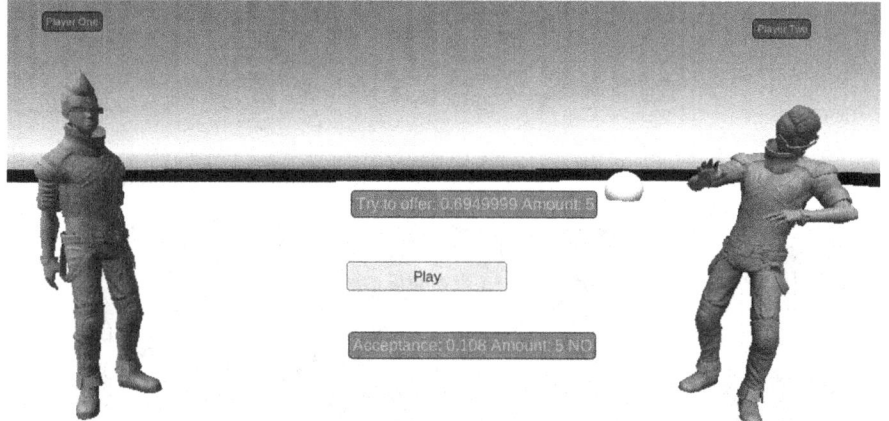

Fig. 6. Ultimatum game simulator, version 0.7

6.2   Ultimatum Game Simulator

Based on the previous work [5] we also implemented several 3D applications, to better illustrate of the concepts involved in the current project. Our first simulation was based on the simplified version of the ultimatum game, where 2 agents play the game as specified in the rules.

Some emotional responses were designed and integrated in our artificial players, as seen in Fig. 6. In this version, the first player gets random amounts of money in the 0–100 range and a *confidence* parameter is calculated based on previous games. If the *confidence* factor is over 0.5 (maximum value is 1 and minimum value is 0), the first player makes the offer to the second player.

The second player also checks his previous games and if the acceptance factor is over 0.5, the offer is accepted. Otherwise, the offer is rejected. In each case, some animations are associated with the characters, so they can represent enjoyment or different levels of discontent.

We played several games on this version. The distribution obtained is represented in Fig. 7. The first results were encouraging, but the low-value offers were more prevalent than in the experiment from Sect. 4.4. To fix this imbalance we added the evolutionary components in our process and ran the experiments again, with the details explained in Sect. 4.

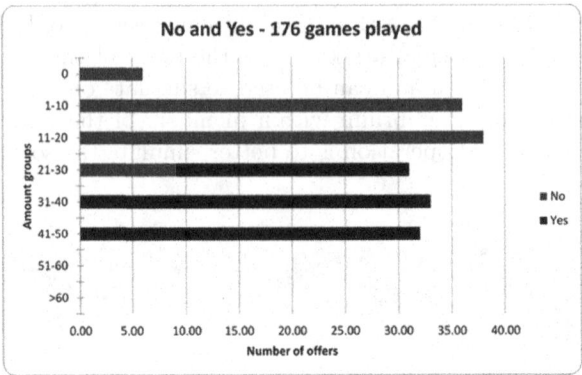

Fig. 7. Results from the first version of the ultimatum game simulation

Fig. 8. Second version of the Ultimatum Game Simulation

For our second version of the simulation, a broader scope was addressed. We used our algorithm from 1 in 3D space and created a platform where multiple agents can interact and evolve based on those relations. In this version, a variable number of agents are created on the platform and play the game with other agents in their proximity. After a round of the game the agents start moving in random directions and repeat the game with other players, depending on the stop position.

Another behavior was discovered in the corners of the map, as seen in Fig. 8, where the agents congregate because of the environment constraints. Based on that behavior, we added the following logic: if multiple agents are in a certain proximity radius (between them), a community is temporary formed, where the offers of the players are affected by the offers of their neighbors.

We added this community filter to better simulate the anomaly in our experiment where the participant from the same university (classes) had a larger than standard share of fair splits (50-50 offers). We used a simplified version of Rabin's fairness model from [28], where the utility of a strategy is dependent on the perceived fairness of the other player. This filter can be described as applying an average between the offers of the same community. This means that if a lot of the members of the same community make offers between a certain value, in time, other members will shift towards that value, as it is perceived normal in that community.

As a simple analogy, in a developing country where the authority level between leaders and the rest of the people are imbalanced, that community will be more inclined to accept unfair offers, because that is probably the normal interaction between them those in a position of power [8]. In a Western country however, the situation is different, and the average of the community will be more inclined towards the fair 50-50 split.

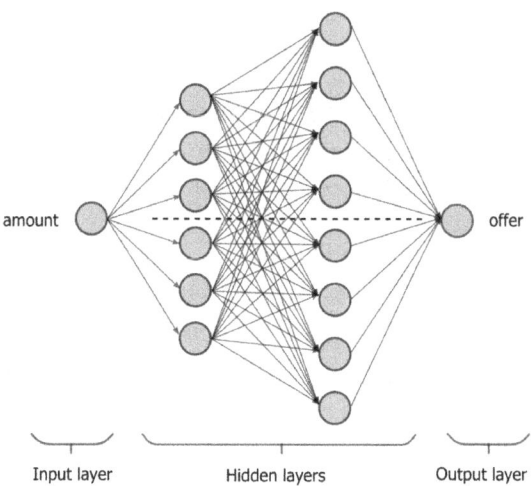

amount   offer

Input layer   Hidden layers   Output layer

Fig. 9. Feed-forward Neural Network

6.3   Using Neural Networks to Play the Ultimatum Game

In order to simulate the Ultimatum Game, we created a simple feed-forward neural network with the following structure (see Fig. 9): one node on the input layer, two hidden layers (with 16 and 32 nodes) and a single node on the output layer. We used the ReLU activation function [24], *ADAM* optimizer and *mean squared error* as a loss function. In this type of structure the information moves in only one direction, from the input layer (amount node), through the hidden layers and then to the output node (offer). The network type was chosen based on several experiments where we determined the optimum size and complexity based on the results and training time.

For the training component we searched for a relevant data set involving Ultimatum Game experiments with actual people. We identified a comprehensive study on fairness in Dictator and Ultimatum Game, conducted by J. Novakova and J. Flegr [25] where they studied a number of 524 people. The set *S1* contains the following variables: identification, dictator_offer, ult_offer, ult_accept, amount, Treatment, logamount and sex_num. The Dataset S1 contains a number of 2619 records from several experiments with variable amounts of money played.

Based on S1 database we created a new ultimatum game dataset, where we played the game between all the unique records from S1. Based on the UG offer and UG minimum acceptance amount we were able to play multiple games and save the result in a larger CSV file, with 140247 records where we stored the offer, the amount played and the accepted value. The actual statistics from the larger dataset are presented in Table 1.

Table 1. Statistics from our generated dataset based on S1 [25]

	Offer	Amount
Count	140247.00	140247.00
Mean	0.41	34081.63
Std	0.18	69533.14
Min	0.00	20.00
25%	0.25	200.00
50%	0.50	2000.00
75%	0.50	20000.00
Max	1.00	200000.00

We split the data set in two parts, 75% for training and 25% for testing, then we ran the training for 50 epochs, given the simple nature of the problem. We used the Keras API [10] over TensorFlow [1], on a local system with a Six-Core AMD processor (3.4 GHz), 32 GB of RAM and a dedicated GPU with 8GB VRAM (Nvidia GeForce RTX 2080 Super). Each epoch took an average of 26 s to process and train.

Figure 10 presents the results, revealing an interesting perspective in our context. The network correctly determined the 40% range based on the training data and offered a limited amount of distribution, similar with the results in our algorithm experiments with strong selection presented in Fig. 4.

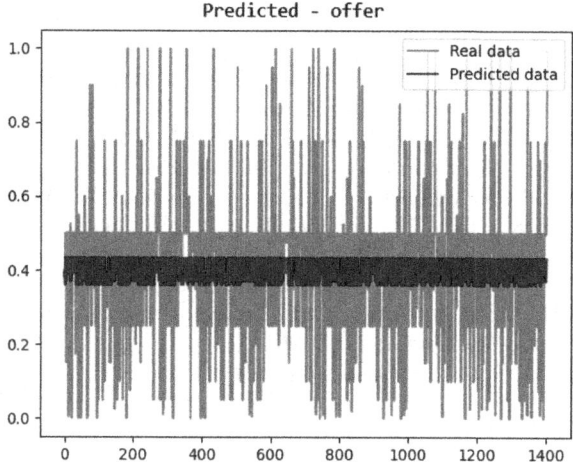

Fig. 10. Results predicted by our neural network in UG

7   Conclusions

In this study we proposed an evolutionary algorithm to demonstrate that the generation of intelligent agents with human-like behavior is possible. We validated the algorithm using many experiments with human and artificial agents. The results (Fig. 5) prove that our algorithm allows us to model the human behavior.

The dynamic environment(variable amounts of money over time during evolution) created a more realistic distribution of fairness in our model. After a number of generations, most of the simulations created fairness strategies, but the results were very sensible to the parameters (Fig. 4).

We were able to simulate the results from the real experiment with the help of stochastic evolutionary game theory (Fig. 5). We also created 3D simulation applications which can be used to describe the concepts involved in this paper. The UG Simulator can be accessed on https://codexworks.com/dev/ug/.

The experiment involving neural networks showed promising results by determining the average offer and we can integrate the component in our MEA model if we can solve the limited distribution problem.

As a future research we plan to study the community(social) effect on economic agents and conduct dedicated experiments. We also aim to use different evolutionary strategies and expand the neural component with other machine learning methods. Other improvements will be to use other economic games or complex interactions to validate our MEA model.

Acknowledgement. Dana Simian was supported from the research action LBUS-RRC-2020-01, financed from Lucian Blaga University of Sibiu & Hasso Plattner Foundation.

References

1. Abadi, M., et al.: Tensorflow: Large-scale machine learning on heterogeneous systems (2015). http://tensorflow.org/. Software available from tensorflow.org
2. Scalco, A., Ceschi, A., Sartori, R., Rubaltelli, E.: Exploring Selfish *versus* altruistic behaviors in the ultimatum game with an agent-based model. In: Bajo, J., et al. (eds.) Trends in Practical Applications of Agents, Multi-Agent Systems and Sustainability. AISC, vol. 372, pp. 199–206. Springer, Cham (2015). https://doi.org/10.1007/978-3-319-19629-9_22
3. Badgaiyan, R.D.: Neuroscience of the Nonconscious Mind (2019)
4. Bank, W.: World Development Report 2015: Mind, Society, and Behavior. World Bank Publications, Washington, D.C. (2014)
5. Bota, F., Dumitrescu, D.: Agent-based computational models implemented in 3d space. In: Modelling and Development of Intelligent Systems, pp. 13–19 (2016)
6. Camerer, C.F.: Behavioral Game Theory: Experiments in Strategic Interaction. Princeton University Press, Princeton (2011)
7. Candela, R., Wagner, R.E.: Vilfredo pareto's theory of action: an alternative to behavioral economics. Il Pensiero Economico Italiano **24**(2), 15–28 (2016)
8. Cardenas, J.C., Carpenter, J.: Behavioural development economics: lessons from field labs in the developing world. J. Dev. Stud. **44**(3), 311–338 (2008)
9. Carolyn Declerck, C.B.: Neuroeconomics of Prosocial BehaviorIndividual Differences in Prosocial Decision Making: Social Values as a Compass. Academic Press, Elsevier, Cambridge (2015)
10. Chollet, F., et al.: Keras (2015). https://github.com/fchollet/keras
11. Camerer, C.F., Ho, T.H.: Behavioral Game Theory Experiments and Modeling (2014)
12. Crump, M.J., McDonnell, J.V., Gureckis, T.M.: Evaluating amazon's mechanical turk as a tool for experimental behavioral research. PloS one **8**(3), e57410 (2013)
13. Daniel Houser, K.M.: Experimental neuroeconomics and non-cooperative games, pp. 47–61 (2009)
14. Farmer, J.D., et al.: Economics needs to treat the economy as a complex system. In: Paper for the INET Conference 'Rethinking Economics and Politics, vol. 14 (2012)
15. Santos, F.P., Santos, F.C., Paiva, A., Pacheco, J.M.: Evolutionary dynamics of group fairness. J. Theor. Biol. **378**, 96–102 (2015)
16. Forsythe, R., Horowitz, J.L., Savin, N.E., Sefton, M.: Fairness in simple bargaining experiments. Games Econ. Behav. **6**(3), 347–369 (1994)
17. Gale, J., Binmore, K.G., Samuelson, L., et al.: Learning to be imperfect: the ultimatum game. Games Econ. Behav. **8**(1), 56–90 (1995)
18. Gilovich, T., Griffin, D., Kahneman, D.: Heuristics and Biases: The Psychology of Intuitive Judgment. Cambridge University Press, New York (2002)
19. Guala, F.: Paradigmatic experiments: the ultimatum game from testing to measurement device. Philos. Sci. **75**(5), 658–669 (2008)
20. Krawczyk, D.: Social cognition. Front. Hum, Neurosci. (2013)
21. Ramsay, K.W., Signorino, C.S.: A statistical model of the ultimatum game (2009)
22. Melnik, R.: Mathematical and Computational Modeling. Wiley Online Library (2015)
23. Mendes, R.V.: The quantum ultimatum game. Quant. Inf. Process. **4**(1), 1–12 (2005). https://doi.org/10.1007/s11128-005-3192-7

24. Nair, V., Hinton, G.E.: Rectified linear units improve restricted boltzmann machines. In: Proceedings of the 27th International Conference on Machine Learning (ICML-10), pp. 807–814 (2010)
25. Novakova, J., Flegr, J.: How much is our fairness worth? the effect of raising stakes on offers by proposers and minimum acceptable offers in dictator and ultimatum games. PloS one **8**(4), e60966 (2013)
26. Nowak, M.A., Page, K.M., Sigmund, K.: Fairness versus reason in the ultimatum game. Science **289**(5485), 1773–1775 (2000)
27. Palgrave, R.H.I.: The New Palgrave: A Dictionary of Economics, vol. 1. Macmillan, New York (1987)
28. Rabin, M.: Incorporating fairness into game theory and economics. Am. Econ. Rev. 1281–1302 (1993)
29. Rand, D.G., Tarnita, C.E., Ohtsuki, H., Nowak, M.A.: Evolution of fairness in the one-shot anonymous ultimatum game. Proc. Nat. Acad. Sci. **110**(7), 2581–2586 (2013)
30. Schuster, S.: A new solution concept for the ultimatum game leading to the golden ratio. Sci. Rep. **7**(5642), 1–11 (2017)
31. Tisserand, J.C.: The ultimatum game, a meta-analysis of 30 years of experimental research. Technical report, Working Paper (2014)
32. Tomasino, B., et al.: Framing the ultimatum game: the contribution of simulation. Front. Hum. Neurosci. **7**, 337 (2013)
33. Traulsen, A., Hauert, C.: Stochastic evolutionary game dynamics. Rev. Nonlinear Dyn. Complex. **2**, 25–61 (2009)
34. Urbina, D.A., Ruiz-Villaverde, A.: A critical review of homo economicus from five approaches. Am. J. Econ. Sociol. **78**(1), 63–93 (2019)
35. Weintraub, R.: Neoclassical economics: the concise encyclopedia of economics (2007). Accessed 26 Sept 2010

Tuning a Mamdani Fuzzy Controller with an Imperialist Competitive Algorithm

Stelian Ciurea[1,2(✉)]

[1] Faculty of Engineering, Department of Computer and Electrical Engineering,
Lucian Blaga University of Sibiu, Sibiu, Romania
stelian.ciurea@ulbsibiu.ro
[2] Research Center in Informatics and Information Technology,
Lucian Blaga University of Sibiu, Sibiu, Romania

Abstract. We have implemented a fuzzy controller with a view to regulating a single-input and single-output second-order linear system. The fuzzy controller was a Mamdami proportional-derivative controller. To determine the parameters of the fuzzy controller we have used an imperialist competitive algorithm. This type of algorithm has a long running time so we implemented also a parallel version of the algorithm that we run on HPC Zamolxes located at the Engineering Faculty of "Lucian Blaga" University from Sibiu. Because we did not have on this computer a version of MATLAB allowing to write parallel algorithms, we implemented the entire application in the C language using the MPI library.

Keywords: Single-input and single-output second-order linear system · Fuzzy controller of type mamdami · Imperialist Competitive Algorithm

1 Introduction

In 2007, Esmaeil Atashpaz Gargari et al. presented a first paper proposing a new heuristic search algorithm for solutions to hard NP-complete problems. This algorithm called Imperialist Competitive Algorithm (hereinafter referred to as ICA) was inspired by socio-political-historical events. Since then, many researchers have applied ICA to solve various problems.

We also have studied how ICA can solve various problems. Thus in [1] we used ICA to determine how this algorithm can solve problems in which the solutions are formed by a set of real values ("continuous" type problems). In [2] we used ICA to solve the problem committed to the Travelling Salesman Problem (so a problem in which the solution is a set of integer values so a "discrete" type problem). In both cases we determined appropriate formulas for the implementation of ICA specific operations. The results obtained and presented in the two articles showed the efficiency of ICA. It is important to mention that in both applications, the implementation of ICA is not very difficult because:

D. Simian and L. F. Stoica (Eds.): MDIS 2020, CCIS 1341, pp. 20–34, 2021.
https://doi.org/10.1007/978-3-030-68527-0_2

- the representation of potential solutions is immediate;
- we have immediate formulas for evaluating the potential solutions (the value of the functions for which the minimum is sought, respectively the cost associated with a permutation that represents the circuit traveled by the travel agent).

Now we set out to use ICA to Determine the Parameters of a Mamdani Fuzzy Controller Used in the Regulation of a SISO (Single-input and Single-output) Second-order Linear System. In order to implement the specific ICA operations, it was necessary:

- An adequate representation of the controllers: the representation we used was a set of integer and real parameters. It is described in Sect. 4.1;
- A function for evaluating the generated solutions. For this it was necessary a mathematical processing of the differential equations that characterize the behaviour of a single-input and single-output second-order linear system of order 2. The formulas used are presented in Sect. 2.
- Implementations of ICA specific operations. To achieve this we used formulas that showed their efficiency in the ICA implemented in [1] for real parameters and in [2] for integer parameters.

The rest of this paper is structured as follows. In the next section, as mentioned before, we present the formulas used in the simulation of the behaviour of the analyzed system. In Sect. 3 we briefly present the designed fuzzy controller. In Sect. 4 we present the ICA we used. Section 5 presents the results obtained by running the ICA with various combinations of its parameters. Section 6 provides some information on the running times of the applications we have implemented. The article ends with some conclusions presented in Sect. 7.

2 The Mathematical Model Used for the Second-Order Linear System

We have implemented an ICA (imperialist competitive algorithm) in order to determine the parameters of a controller so as to enable the best possible unit step response of a second-order automatic system. The function at the output of such a system is given by the following differential equation [3]:

$$a_2 \frac{d^2 y}{dt^2} + a_1 \frac{dy}{dt} + a_0 y(t) = b_2 \frac{d^2 u}{dt^2} + b_1 \frac{du}{dt} + b_0 u(t) \tag{1}$$

where u(t) is the input variable and y(t) the output variable. In case of downtime τ in the transmission of the data, the corresponding differential equation is the following:

$$a_2 \frac{d^2 y}{dt^2} + a_1 \frac{dy}{dt} + a_0 y(t) = b_2 \frac{d^2 u}{dt^2} + b_1 \frac{du}{dt} + b_0 u(t - \tau) \tag{2}$$

Such equations are used in many articles to refer to the operation of systems such as DC motors [4–8]. We have considered the standard structure of an automated system consisting of a second-order system and a fuzzy controller, illustrated in Fig. 1 [3, 9].

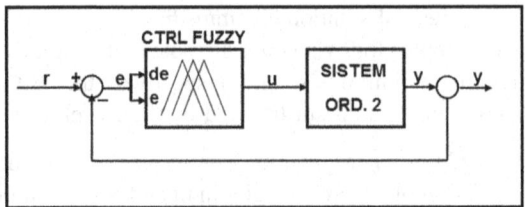

Fig. 1. Automated regulating system

Our aim is for output value y to follow the evolution of reference value r as accurately as possible. To simulate the operation of the system, we have considered its discrete model. The method, described in [3], consists in considering relation (1) at time intervals equal to dt and integrating the terms from that formula twice:

$$\int_{t-dt}^{t} \int_{t-dt}^{t} a_2 \frac{d^2y}{dt^2} + a_1 \frac{dy}{dt} + a_0 y(t) = \int_{t-dt}^{t} \int_{t-dt}^{t} b_2 \frac{d^2y}{dt^2} + b_1 \frac{dy}{dt} + b_0 u(t) \qquad (3)$$

By using the trapezoidal rule for the calculation of integrals, we have obtained the following relation:

$$y_k = \frac{(b_2 + b_1\Delta + b_0\Delta^2)u_k + 2(b_0\Delta^2 - b_2)u_{k-1} + (b_2 - b_1\Delta + b_0\Delta^2)u_{k-2} - 2(a_0\Delta^2 - a_2)y_{k-1} + (a_2 - a_1\Delta + a_0\Delta^2)y_{k-2}}{a_2 + a_1\Delta + a_0\Delta^2}$$
$$(4)$$

where y_k, y_{k-1} and y_{k-2} are output values y at times t_k, t_{k-1} and t_{k-2}, $t_k - t_{k-1} = t_{k-1} - t_{k-2} = dt$; analogous u_k, u_{k-1} and u_{k-2} are values u, and $\Delta = (dt/2)2$.

We have aimed at the operation of this system within the range of [0, tmax] seconds, considering that

- at time $t_0 = 0$, a step signal of an amplitude equal to the unit is applied;
- prior to time t0, the system was idle;
- dt = 0.005 s.

3   The Designed Fuzzy Controller

We have implemented a PD-Mamdani fuzzy controller [10–12] with two inputs and one output. The two entries have been the following:

- "error", marked e, is the percentage difference between the reference value and the system output. We have considered the universe of discourse for this value to be interval [−100, 100]. Its value at time tk is given by the formula:

$$e_k = \frac{r_k - y_k}{r_k} \times 100 \qquad (5)$$

- "error variation", marked de, having universe of discourse $[-5, 5]$ and calculated at time tk by means of the following formula:

$$de_k = \frac{e_k - e_{k-1}}{\Delta t} \tag{6}$$

For controller output u, we have chosen interval $[0, 500]$. We have opted for this range of control because we aimed to track the behaviour of the fuzzy controller system in "hard" conditions (without the controller issuing negative commands).

4 The Designed ICA

ICA are probabilistic algorithms for finding optimal solutions that use heuristic principles [13, 14]. The main idea of this algorithm is the following: a set of possible solutions is randomly generated; these solutions are evaluated using an evaluation function; the best ones (in ICA terminology called metropolises) remain fixed; the others (called colonies) follow (through the operation called "assimilation") the trajectories through which the space of the solutions in the vicinity of the best solutions initially generated is explored. The number of solutions that explore the neighbourhoods of a metropolis depends on its quality and varies throughout the algorithm - so the idea of competition between empires appears. To prevent the "stagnation" of the algorithm, some suboptimal solutions are "thrown" (an operation called "revolution") outside the vicinity of metropolises to travel new paths. These specific operations (competition, assimilation, revolution) are repeated a predetermined number of times. The algorithm stops if all iterations have been executed or the algorithm shuts down because all solutions have become identical. In our application we used ICA in the form presented in [15]:

1. Generate a random set of countries;
2. Evaluate these solutions and the best ones become metropolises;
3. Generate the initial empires by distributing the colonies;
4. Repeat

   a. Assimilation
   b. If a colony has better results than the imperialist country then
      (1) Interchange the colony with the imperialist country
   c. Competition between empires:
      (1) Assess empires
      (2) Distribute the weakest colony of the weakest empire by another empire
      (3) If the weakest empire has no colonies left then
      (4) Remove this empire
5. Until one of the stopping conditions is reached.

We will present the particularities of the algorithm specific to the problem it has to solve.

4.1   Representation of Controllers

The controllers (or "countries" from the point of view of ICA) have been represented in the form of structures having the following makeup presented in detail in [16]:

- 1 integer field for coding fuzzy sets for the three linguistic variables corresponding to the two inputs and to the output: trapezoidal or a combination of two exponential functions;
- 1 integer field representing the number of fuzzy sets in each linguistic variable (seven in our case);
- 1 field stating that the membership functions of fuzzy sets have a symmetry with respect to the central value of the universe of discourse;
- 49 integer fields representing interference matrix as a matrix with 7 rows and 7 columns, where each item can range between 1 and 7 corresponding to the 49 interference rules (1 means ZE, 2 means VS, etc.);
- 3 integer fields that specify the mathematical formula for each of the three logical operations that occur in fuzzy control: AND, OR, implication
- 1 integer field specifying the formula for the aggregation operation;
- 1 field specifying the method used for defuzzification (in our application there are four methods of defuzzification based on weighted averages and three elitist methods);
- 78 real fields for coding the form and locations on the axis of the universe of discourse of the fuzzy sets within each linguistic variable (e, de, and u), with the following implications for any of these variables:
  - 6 values representing the ratio between the larger base of the trapezium and the average value of the universe of discourse – a value within the range [0.25, 2.0];
  - 6 values representing the percentage of the larger base of the trapezium overlapping the larger base of the trapezium placed on its left - a value within the range [0.05, 0.4];
  - 7 values representing the ratio smaller base/larger base – a value within the range [0.01, 0.65];
  - 7 values representing the position of the smaller base for CE – a value within the range [0.05, 0.95] out of the available interval calculated depending on the large basis.

This mode of representation is useful because the specific operations of ICA provide results with values in the same ranges, so correct. But for performing the operations necessary for fuzzy control or for graphical representations, some simple formulas (and related calculations) are needed to determine the classic shape of fuzzy sets.

4.2   Generating the Initial Set of Countries

This has been done randomly, except for the values in the rules matrix. We have used a generation directed so that the values are close to those normally used for PD control. For example, for a rule with antecedents e is NB and de is NM, one of the fuzzy sets ZE, VS and SM of size u has consequently been chosen at random.

4.3   Evaluation of Colonies

To evaluate the colonies in order to implement some of the specific ICA operations (empire determination, colony distribution, competition between empires), we have simulated the behaviour of the system when applying a step signal at time 0. The controller performance was calculated with the formula:

$$performace = \left(\sum_{0}^{2.5}|r_k - y_k|\right) + 10 \times \max\left(y_{max} - y_{fin}, 0\right) \qquad (7)$$

where y_{max} is the maximum that the output value takes in the studied range, and yfin is the final value of the output of the system. This evaluation function was thought of as a penalty: the lower its value, the more efficient the controller. Thus, the value of the first term of this expression is lower the shorter the response time and the lower the steady-state error. The second term is lower the less the system exceeds the steady state value.

4.4   The Assimilation Operation

In the application, we have used different formulae depending on the two types of parameters that make up a colony:

- for integers parameters, we have taken into account that the value of any of these parameters does not condition the value of another parameter. This led to the following formula: the colony parameter receives the value of the corresponding parameter in the metropolis with a given probability, hereinafter referred to as assimilation probability and marked proba;
- for real parameters, we have used the following formula:

$$paramcol_i = paramcol_{i-1} + p \times (1 \pm d) \times (parammet_{i-1} - paramcol_{i-1}) \qquad (8)$$

where $paramcol_i$ is the value of the colony parameter after iteration i, $prammet_i$ is the value of the parameter corresponding to the metropolis of the respective colony after iteration i, and d is a subunit random value, $d \in [-d_a/2; d_a/2]$; d_a is the assimilation deviation.

- for each value in the rules matrix, we have used the following formula:

$$r_{ij} = r_{ij} + \left[p \times \left(m_{ij} - r_{ij}\right)\right] \qquad (9)$$

where r_{ij} and m_{ij} are the denotations of values belonging to the rules matrix of the colony and metropolis, p is the assimilation step, and the square brackets refer to the integer part.

4.5   The Revolution Operation

We have used different formulae depending on the type of parameter whose value is to be changed.

- for each integer parameter, the following formula was applied: with a probability equal to value d_r (revolution deviation)

$$paramcol = nrval - 1 - paramcol \qquad (10)$$

where paramcol is the integer parameter and nrval is the number of values available for that parameter.

- for real parameters, we have used the following formula:

$$paramcol_i = paramcol_{i-1} + p \times (1 \pm d) \times (parammet_{i-1} - paramcol_{i-1}) \qquad (11)$$

$d \in [-d_r/2, d_r/2]$, in this case d_r being the revolution deviation.

- for the fields of the rules matrix, the following formula has been used:

$$r_{ij} = a - 2 + r_{ij} \qquad (12)$$

where a is a random natural number within the range [0, 4].

5 Findings

The following values were chosen for the second-order system parameters:

$$a_0 = 20; \; a_1 = 10; \; a_2 = 1; \; b_0 = 1; \; b_1 = b_2 = 0$$

These values of the coefficients determine, for the system analyzed in open loop, a transfer function with two real poles having the values -7.2361 and -2.7639. The observation interval was [0, 2.5] s. Tests with ten initial sets of countries have been performed. The ICA parameters have been chosen as follows:

- number of countries in a set: 55 and 108;
- initial number of empires in a set: 5 and 8;
- maximum number of iterations in the algorithm: 300;
- share of a colony in the performance of the empire (w): 0.001, fixed;
- approaching step (p): 0.1 and 0.9;
- assimilation and revolution deviations (d_a, d_r): 0.01 and 0.9;
- the probability with which the revolution operation is applied to a country: 10% and 30%.

The results obtained for each of the ten sets of countries (marked with #0, #1,..., #9) are presented in Table 1 and Table 2. The values in the table represent the final performance of each set.

Table 1. The performance of the best controllers obtained for sets of 55 countries

55 countries		p				p			
		0.1				0.9			
		Assimilation deviation d_a				Assimilation deviation d_a			
		0.01		0.01		0.9		0.9	
		$prob_{rev}$		$prob_{rev}$		$prob_{rev}$		$prob_{rev}$	
Set	InitP*	0.1	0.3	0.1	0.3	0.1	0.3	0.1	0.3
#0	244.8	63.66	38.63	63.67	38.15	63.69	33.39	63.69	33.39
#1	131.8	79.34	38.81	60.78	38.79	59.58	38.23	59.58	38.23
#2	197.6	79.77	40.88	80.91	37.68	76.61	30.24	76.61	30.24
#3	110.1	76.43	55.82	70.65	38.49	75.88	41.52	75.88	41.52
#4	149.4	53.72	29.75	53.57	31.22	50.89	29.48	50.89	29.48
#5	214.3	92.04	**22.13**	109.0	38.14	48.35	45.28	48.35	45.28
#6	285.0	86.60	28.66	86.14	28.14	85.72	28.19	85.72	28.90
#7	244.8	101.3	29.71	49.95	36.59	49.92	30.19	49.92	36.9
#8	112.8	42.71	37.55	42.40	39.27	42.31	22.22	42.31	35.65
#9	124.7	42.81	34.20	42.85	34.2	60.31	34.21	60.31	33.86
AvrP*	181.5	71.83	35.61	65.99	36.07	61.32	33.29	61.32	**33.29**
AvrY*		2.52	5.09	2.75	5.03	2.96	5.45	2.96	5.45

*IniP = Initial Performance; AvrP = Average Performance; AvrY = Average Yeld

In the case of the sets of 55 countries, we have noted the following:

- the results are mediocre both as average values of the ten tests and as the values of the best controller: in very few tests, the performance of the best controller drops below 30; the revolution probability has considerable influence: for $prob_{rev}$ = 0.3, both the best performance and the average performance are twice as good as the tests in which $prob_{rev}$ = 0.1;
- a high value of the approaching step leads to an increase of about 6% in the average performance;
- the assimilation deviation does not influence performance, especially when the approaching step is high;
- the best value of the performance function obtained for a controller is 22.13.
- the simulation of the behaviour of the system with the help of this controller is presented in Fig. 2. The system has an override equal to 0.91%, a response time of 0.25 s, and, in steady state, an oscillation around the reference value with an amplitude of 0.0091;
- the yield of the algorithm representing the ratio between the average performance of the best controllers in each set after the first iteration and the average performance after the last iteration had the minimum value of 2.52 and a maximum of 5.45;
- of the ten tests that determined the best average performance, only two had a final performance of less than 30.

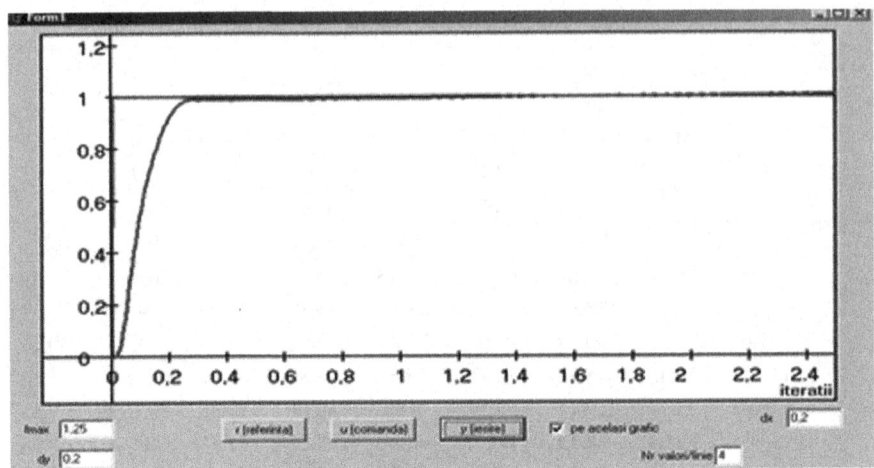

Fig. 2. The behaviour of an automated regulating system with the best controller provided by ICA in a set of 55 countries

Table 2. The performance of the best controllers for sets of 108 countries

108 countries		p				p			
		0.1				0.9			
		Assimilation deviation d_a				Assimilation deviation d_a			
		0.01		0.01		0.9		0.9	
		$prob_{rev}$		$prob_{rev}$		$prob_{rev}$		$prob_{rev}$	
Set	IniP*	0.1	0.3	0.1	0.3	0.1	0.3	0.1	0.3
#0	106.0	39.26	29.6	27.41	26.31	27.41	28.17	27.41	30.47
#1	131.8	50.25	27.06	40.47	28.94	43.57	29.48	43.57	36.42
#2	152.0	36.21	36.35	42.03	37.66	22.94	33.55	22.94	42.39
#3	110.1	29.91	21.50	29.91	21.50	21.33	21.50	21.33	21.50
#4	72.72	24.88	25.34	24.88	23.32	24.88	25.34	24.88	25.34
#5	98.98	24.12	21.26	58.33	17.8	54.28	19.91	54.28	19.91
#6	133.9	45.92	28.98	46.41	27.84	46.41	29.02	46.41	31.15
#7	106.0	75.86	25.96	72.60	30.31	72.6	30.31	72.6	30.31
#8	112.8	27.96	20.73	29.50	20.73	38.13	20.73	38.13	20.73
#9	124.7	39.98	43.46	63.10	43.46	54.27	43.24	54.27	46.83
AvrP*	114.9	39.45	28.02	43.46	**27.78**	40.58	28.12	40.58	30.50
AvrY*		2.91	4.01	2.64	4.13	2.83	4,08	2.83	3,76

*IniP = Initial Performance; AvrP = Average Performance; AvrY = Average Yeld

For the sets made up of 108 countries, the following aspects have been noted:

- the average performance of the best controllers was better by about 45% in the case of a high probability of revolution;
- the assimilation deviation did not actually influence the performance – in most tests the same performance was obtained for $d_a = 0.9$ and $d_a = 0.01$;
- the approaching step had very little influence on performance;
- the efficiency of the algorithm representing the ratio between the average performance of the best controllers in each set after the first iteration and the same value calculated after the last iteration had a minimum value of 6.70 and a maximum one of 10.93;
- in the case of the combination of ICA parameters for which the best average performance was obtained, in eight out of the ten tests, the performance had values less than or approximately equal to 30, which implies a good behaviour of these controllers, even in difficult conditions of operation (limited to non-negative commands only). As a result, in the subsequent experiments, by means of ICA, we have determined the parameters of a number of regulators whose value range for the output signal was symmetrical to the origin, in accordance with the usual engineering practice;
- the best value of the performance function obtained for a controller was equal to 17.80. The simulation of the system operation with this controller is shown in Fig. 3. The system showed an override equal to 3.2%, a response time of 0.185 s and a steady state error of 0.26%.

The types of fuzzy operations for this controller are as follows:

- type of operation AND: prod;
- type of operation OR: max;
- type of implication: min;
- aggregation method: probor (a + b − ab);
- defuzzification method: som.

Table 3 shows the inference matrix of the most powerful controller.

Table 3. Inference matrix of the most efficient controller

u		de						
		NB	NM	NS	ZE	PS	PM	PB
e	NB	ZE	MD	SM	VS	VS	VS	SM
	NM	VS	ZE	ZE	MD	MD	VS	SM
	NS	VS	ZE	ZE	ML	ZE	SM	ML
	ZE	ZE	ZE	ZE	VS	VS	SM	ZE
	PS	VS	ZE	ZE	SM	SM	VL	LA
	PM	ZE	SM	SM	LA	ML	LA	LA
	PB	MD	ML	VL	LA	ML	VL	ML

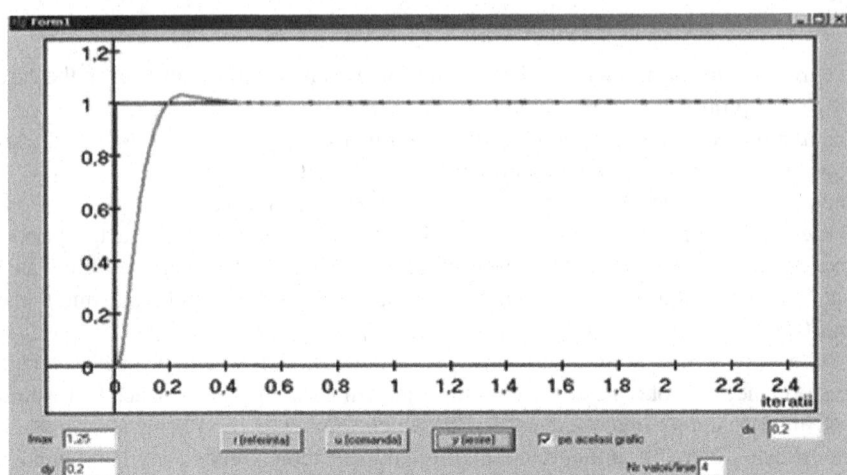

Fig. 3. Behviour of an automated regulating system with the most powerful controller provided by ICA from a set of 108 countries

Figures 4, 5, and 6 illustrate linguistic variables e (error), de (error variation), and u (command). Figure 7 shows the performance variation over the 300 iterations in the test that provided the best controller.

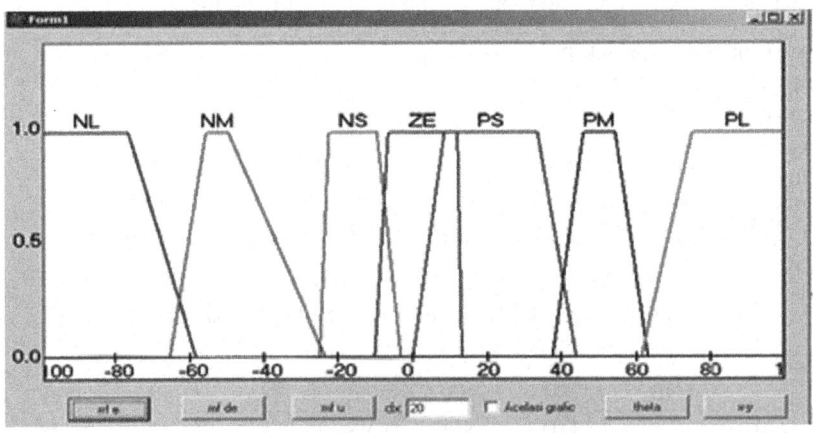

Fig. 4. Membership functions for input *e*

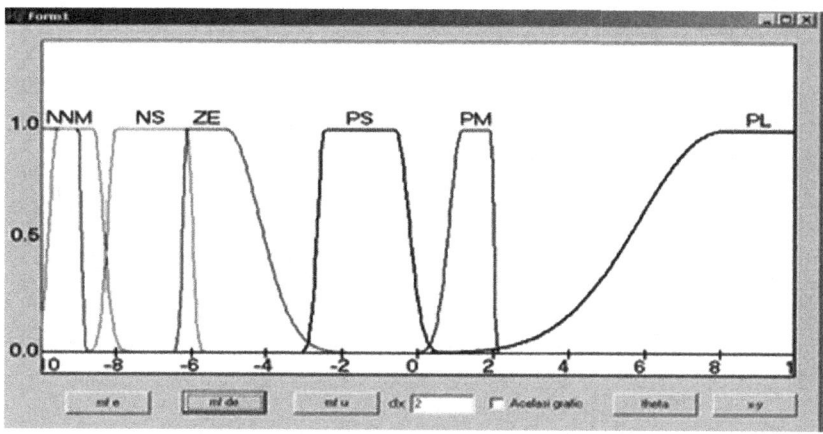

Fig. 5. Membership functions for input *de*

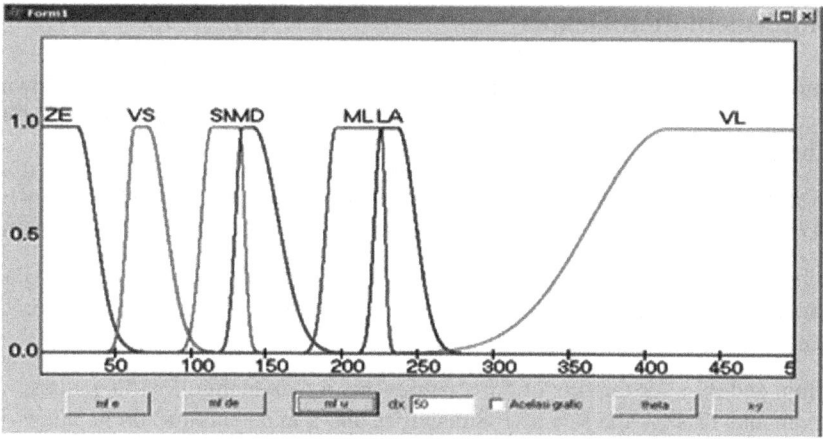

Fig. 6. Membership functions for output *u*

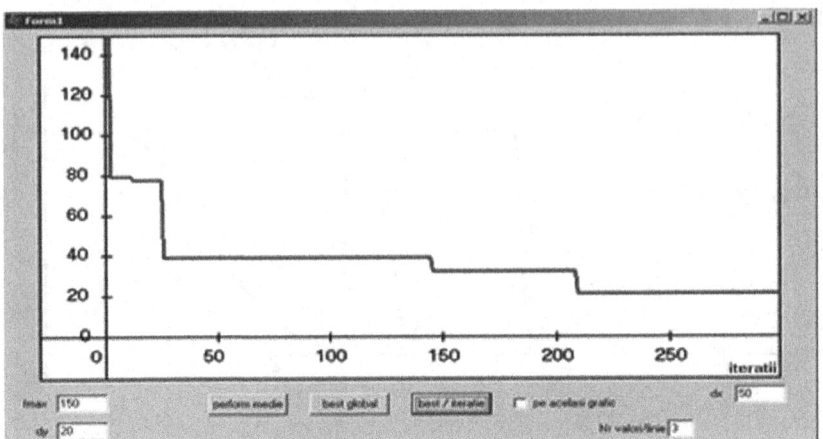

Fig. 7. Performance of the best Mamdani controller vs iteration

6 Implementation Details

The entire application has been implemented in the C language [17]. The following average running times on a computer with an Intel Core i3 microprocessor at 2.93 GHz have been obtained:

- 55 min and 49 s for the sets of 55 countries and a maximum of 300 iterations;
- 117 min and 21 s for the sets of 108 countries and a maximum of 300 iterations.

Running on the Intel HPC System mounted on the premises of the Lucian Blaga University of Sibiu, the parallel ICA, in which a processor was allocated for the evaluation of each controller, had the following average running times:

- 5 min and 39 s for the sets of 55 countries and a maximum of 300 iterations;
- 7 min and 17 s for the sets of 108 countries and a maximum of 300 iterations.

7 Conclusions

The imperialist competitive algorithm has been able to determine parameters for a fuzzy controller to ensure the efficient operation of the second-order automated regulating system, even in difficult conditions (non-negative controls of the controller). Better results have been obtained in the tests in which the number of countries was higher (which was to be expected). Applying a step signal to an automated regulating system, the best controller ensured the operation of the system with an override equal to 3.2%, a response time of 0.185 s, and a steady state error of 0.26%.

Due to the intrinsic parallelism of ICA, running our application on a parallel computer led to a decrease of the running times of more than 10 times.

Acknowledgement. Project financed from Lucian Blaga University of Sibiu & Hasso Plattner Foundation research action LBUS-RRC-2020-0L.

References

1. Ciurea, S.: Imperialist competitive algorithm with variable parameters to determine the global minimum of functions with several arguments. In: Simian, D. (ed.) Proceeding of the 4th International Conference 2015, MDIS, vol. 1, pp. 40–50. Lucian Blaga University Press, Sibiu (2016)
2. Ciurea, S.: An imperialist competitive algorithm optimized to solve the travelling salesman problem. In: Simian, D. (ed.) Proceeding of the 5th International Conference 2017, MDIS, vol. 1, pp. 20–28. Lucian Blaga University Press, Sibiu (2018)
3. Babutia, I., Dragomir, T.L., Mureşan, I., Proştean, O.: Automatic Process Management, 2nd edn. Ed. Facla, Timişoara (1985)
4. Bhartiya, S., Sehra, S., Bhuria, V.: Dc motor speed control using PID- fuzzy logic based controller. Int. J. Artif. Intell. Res. IJAIR 3(5), 311–315 (2013)
5. Bindu, R., Mini Namboothiripad, K.: Tuning of PID controller for DC servo motor using genetic algorithm. Int. J. Emerg. Technol. Adv. Eng. 2(3), 310–314 (2012)
6. Dumitrache, I., Călin, S., Boţan, C., Niţu, C.: Automatizări şi echipamente electronice, 1st edn. Ed. Didactică şi Pedagogică, Bucureşti (1982)
7. Kuma, D., Dhakar, B., Yadav, R.: Tuning a PID controller using evolutionary algorithms for an non-linear inverted pendulum on the cart system. In: Chauhan, B. (ed.) International Conference on Advanced Developments in Engineering and Technology (ICADET 2014) 2014, vol. 4, no. 1, pp. 393–396. Lord Krishna College of Engineering, Ghaziabad (2014)
8. Yazdani, A., Ahmadi, A., Buyamin, S., Rahmat, M., Davoudifar, F., Rahim, H.: Imperialist competitive algorithm-based fuzzy PID control methodology for speed tracking enhancement of stepper motor. Int. J. Smart Sens. Intell. Syst. 5, 717–741 (2012)
9. Ciurea, S.: Determining the parameters of a Sugeno fuzzy controller using a parallel genetic algorithm. In: Dumitrache, I. (ed.) Proceedings of the 9th International Conference on Control Systems and Computer Science 2013, CSCS 2013, vol. 1, pp. 36–43. University Politehnica of Bucharest, Bucharest (2013)
10. Mamdani, E.H.: Application of fuzzy algorithms for the control of a simple dynamic plant. Proc. Inst. Electr. Eng. 121(12), 1585–1588 (1974)
11. Terano, T., Asai, K., Sugeno, M.: Fuzzy Systems Theory and Its Applications, 3rd edn. Academic Press, Cambrige (1992)
12. The MathWorks Inc. https://faculty.petra.ac.id/resmana/private/matlab-help/pdf_doc/fuzzy/fuzzy_tb.pdf. Accessed 10 Sep 2020
13. Pelusi, D.: Optimization of a fuzzy logic controller using genetic algorithms. In: Tang, Y. (ed.) Third International Conference on Intelligent Human-Machine Systems and Cybernetics 2011, IHMSC, vol. 2, pp. 143–146. IEEE Coste Sant'Agostino, Teramo (2011)
14. Michalewicz, Z.: Heuristic methods for evolutionary computation techniques. J. Heuristics 1(2), 177–206 (1995)
15. Atashaz-Gargari, E., Lucas, C.: Imperialist competitive algorithm: an algorithm for optimization inspired by imperialistic competition. In: Tan, K-C., Xu, J.-X. (eds.) IEEE Congress on Evolutionary Computation 2007, CEC, vol. 9, no. 4, pp. 4661–4667. Kluwer Academic Publisher, Norwell (2008)

16. Ciurea, S.: An appropriate representation of a fuzzy controller enabling the determination of its parameters with an imperialist competitive algorithm. "Ovidius" Univ. Ann. Ser. Civ. Eng. **16**, 163–176 (2014)
17. Balaji, P., Bland, W., Gropp, W., Latham, R., Lu, H., Pena, A. J., Raffenetti K., Thakur, R., Zhang, J.: MPICH User's Guide Version 3.1. 1st edn. Mathematics and Computer Science Division Argonne National Laboratory, Argonne (2014)

Optimizing the Integration Area and Performance of VLIW Architectures by Hardware/Software Co-design

Adrian Florea$^{(\boxtimes)}$ and Teodora Vasilas

Computer Science and Electrical Engineering Department, "Lucian Blaga"
University of Sibiu, No. 4, Emil Cioran Street, 550225 Sibiu, Romania
{adrian.florea, teodora.vasilas}@ulbsibiu.ro

Abstract. The cost and the performance are major concerns that the designers of embedded processors shall take into account, especially for market considerations. In order to reduce the cost, embedded systems rely on simple hardware architectures like VLIW (Very Long Instruction Word) processors and they look for compiler support. This paper aims at developing a design space explorer of VLIW architectures from different perspectives like processing performance and integration area. A multi-objective Genetic Algorithm (GA) was used to find the optimum hardware configuration of an embedded system and the optimization rules applied by compiler on the benchmarks code. The first step consisted in representation of the architectural configurations into chromosomes of GA, mapping each architectural parameter or feature into a gene. Each chromosome from a population is a configuration file, and each gene of that chromosome is the value of an architectural parameter (machine and memory hierarchy) or compiler optimization option. The population is composed from a fixed number of such of chromosomes or individuals. The fitness functions of chromosomes (the processing performance - Instructions per Cycle and the integration area of embedded system) used by NSGA-II algorithm for determining the dominated and non-dominated individuals are obtained after the simulations of architectural configurations on different benchmarks of the standard MiBench suite. CACTI tool was used to measure the area of the caches as component of embedded system integration area.

Keywords: Design Space Exploration (DSE) · Area and performance efficiency · Evolutionary algorithms · VLIW architectures · NSGA-II

1 Introduction

Embedded systems are all computing systems that are not used for general purpose [1]. These kinds of systems have nowadays applications in a lot of very different domains. They range from low to very high costs; they can be for industrial or personal use, academic or amusement equipment, weapons, or medical devices, even for automotive or military applications [2]. VLIW (Very Long Instruction Word) architectures are processors that are based on bringing multiple independent RISC instructions into a multiple instruction that they distribute for processing to execution units [3]. They have

© Springer Nature Switzerland AG 2021
D. Simian and L. F. Stoica (Eds.): MDIS 2020, CCIS 1341, pp. 35–51, 2021.
https://doi.org/10.1007/978-3-030-68527-0_3

simple hardware mechanisms because the complexity of these kinds of processors is moved in the compilers' tasks, which render processors faster and which is a good technique for explicit parallel programs [4]. Their logic is so simple because, in addition, they do not use reordering technique or dynamic scheduling like the super-scalar ones. Like RISC architecture, VLIW contains simple instruction sets [5]. However, it is a big challenge for VLIW designers to obtain a good equilibrium between parameters like performance and cost. A tradeoff is needed for the perfor-mance to meet the requirements and the system to have the lowest possible cost, or the system obtains the highest possible performance, but also respects the cost limits [6]. The lack of binary compatibility between successive processors models and the requirement of new software applications for the new models are the main features that relate the embedded processors to VLIW architectures. VLIW architectures became very attractive and useful for a wide variety of embedded applications such as network routing problems or multimedia applications [1].

This paper presents a design space explorer of VLIW architectures from different perspectives like processing performance and integration area. The topic of this paper continues to be relevant even though VLIW architectures have appeared for more than 30 years. Turing award winners J. Hennessy and D. Patterson appreciate that the revolutionary future of Computer Architecture [4] is based on the development of application-specific hardware architectures (DSA), so-called accelerators (Google Tensor Processor Units, Nvidia Deep Learning Accelerator, etc.) and Open Architec-tures that both conceptually follow the VLIW principles. The DSAs are a kind of architectures adapted to a distinctive field; they increase the performance and the efficiency of that specific field. This performance comes from the parallelism that this kind of architectures exploits, from the way they employ the hierarchical memory and by using fewer bits for precision whenever this is possible. These architectures use for the instruction level parallelism VLIW techniques as well [4]. Taking concepts from software development, the open source architectures relies basically on a small instruction set, which runs the full open source software stack, followed by optional standard extensions that designers, depending on their needs, can include or omit. Such example is RISC-V architecture [7], developed at the University of California, Berkeley, which integrates systems ranging from data-center chips to IoT devices.

Because the exploration space is very large and it would be nearly impossible to evaluate all the possibilities of different configurations, recent researches applied all kind of methods to find the best trade-offs. A design space exploration along with a Greedy algorithm and code optimization of VLIW architectures, was used [8] for better cost and performance, by using GSM benchmark from MiBench suite. Same was done in [9] along with different performance models. For a specific application, the best architecture is found in the field of image processing for a VLIW embedded processor by the authors of [10] but these applied Brute Force search algorithm in order to exploit architecture design space. To reduce the execution time, they use the VEX tool [11] for code optimization. Another DSE is used in [12] to find the best configuration for a VLIW architecture memory system, in order to find a trade-off between execution time and energy efficiency. The authors of [13] also try to find the best trade-off between integration area and time. They use different kinds of benchmarks to simulate different architecture and memory configurations using VEX tool.

Next, we briefly present few arguments that show the efficiency of our optimization method. A genetic algorithm will explore many configurations from the design space and it is not stocked in local optima like Greedy algorithm. Furthermore, the GA represents a significantly faster and accurate search method when compared to the brute-force method. In addition, the main differences are that we are treating the problem as a multi-objective one implementing not a simple GA but a multi-objective optimization GA like NSGA-II [14] and our DSE algorithm is automatic. We proved in [15] that manual DSE has not the exploring potential of an automatic DSE.

We implemented the improved version of Non-dominated Sorting Genetic Algorithm (NSGA-II) with two objectives (performance and area) and three objectives (performance, area and energy). Choosing of NSGA-II as multi-objective optimization algorithm was made based on literature review but especially on our previous experience in microarchitectures optimization [15]. NSGA-II has three features: it uses an elitist principle that does not allow an already found Pareto optimal solution to be deleted; it emphasizes non-dominated solutions and it uses an explicit diversity preserving mechanism. Some previous approaches analyze only particular benchmarks like GSM, or Jpeg, or Network. Other researches only aim optimization of the memory subsystem and do not take into account the compiler influence. Our solution overcomes these drawbacks integrating both architectural parameter (machine and memory hierarchy) and compiler optimization rules to determine their influence on the performance metrics but also extends the simulation on different type of embedded benchmarks. Summarizing, the paper's workflow consists in the following stages:

- Define the goal as developing a DSE for VLIW architectures in order to find the best one from perspective of performance and integration area
- Selection, based on literature review, of the most used multi-objective optimization algorithms (NSGA-II) for solving the problem
- The first issue is to map a micro-architectural problem to evolutionary computing ones, mapping the hardware and software configurations to the genes of chromosomes
- Set the objectives and the tools which provide them and apply NSGA-II for a number of generations and extract useful design knowledge from the simulation results

The paper is organized as follows: in Sect. 2 the VEX toolchain is presented and in Sect. 3 the used benchmarks are discussed. The DSE methodology is revealed in Sect. 4, while the experimental results obtained are illustrated in Sect. 5. Finally, Sect. 6 presents conclusions and possible directions for further work.

2 VEX Toolchain

The VEX (VLIW Example) [11] toolchain was created for the VLIW embedded processors for easy simulation and optimization. The environment is very flexible and allows the user to change the value of the simulation parameters and use the compiler in different ways [13]. The VEX system is made up of three elements:

- VEX Instruction Set Architecture (ISA) is a VLIW ISA on 32 bits which contains multiple clusters; it can be customized and scaled in terms of number of execution units, clusters and registers [1]. A cluster is a collection of register files and a tightly coupled set of functional units. Functional units within a cluster directly access only the cluster register files, with only a few exceptions. Partitioning the register file and functional units by clustering represents a solution to register file access bottleneck in case of large issue-width architectures. The read access time of a register file grows approximately linearly with the number of ports producing a negative impact on the overall cycle time.
- VEX C Compiler is a strong ISO/C89 one that uses for its scheduling mechanism a trace scheduling method. It is possible to explore also the architecture by modifying some architectural parameters such as execution units and clusters number, issue width and latencies, without the need of recompilation of the compiler [1]. The compiler is tasked with reorganizing the original program for the purpose of "wrapping" in a single multiple VLIW instruction several primitive and independent RISC instructions, which will be assigned to the execution units in strict accordance with their position in the VLIW instruction.
- The VEX Simulation System is an architecture-level simulator that uses the technology of compiled simulator and includes a simple built-in level 1 Harvard cache simulator [13]. It provides after simulation the Instructions per Cycle (IPC) as a performance metric.

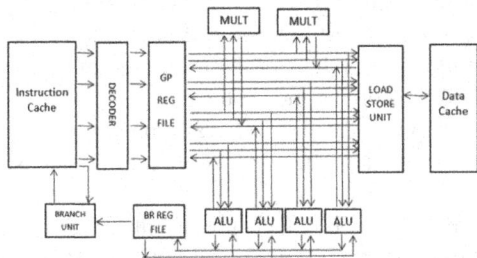

Fig. 1. VEX architecture

The default VEX configuration has one cluster, two register files, one with 64 integer registers for general purpose usage on 32 bits and one containing eight Boolean registers for branches, four ALUs of integer type, two multiplication units having one operand on 16 bits and a second operand on 16 or 32 bits, as well as a data cache port (see Fig. 1). A cluster can issue a maximum number of four operations per instruction.

3 Benchmarks Used

The embedded area is a very large one with a multitude of applications that are hard to be characterized. This means that an embedded benchmark suite shall be diversified enough to cover all of them. MiBench suite [16] is a free set of embedded programs. It contains thirty-five applications, divided in six categories that aim benchmarking. These categories are: Automotive, Security, Consumer Devices, Network, Telecommunications and Office. They are written in C source code and are portable to any framework that has a compiler and they provide small and large embedded applications. The benchmarks described below are those that have been simulated and for which results have been presented.

Bitcount benchmark probes the capacity of manipulating bits by counting them in an integer array, and it is used in **Automotive** industry for airbag controllers, performance monitors of the engine and sensors. This test uses five techniques: a counter of 1 bit per loop, a recursive bit count by nibbles, non-recursive bit count by nibbles using a table look-up, non-recursive bit count by bytes using a table look-up and shift and count bits.

The **Network** category includes the network devices' processors like routers and switches. These benchmarks test if the devices can calculate the shortest path and search in tables or trees. **Dijkstra** algorithm is known for the problems where we search the shortest path. This benchmark creates a large graph using adjacency matrix for representation and computes the shortest path among each pair of nodes.

The **Office automation** benchmarks employ algorithms for text manipulation used by office machines like fax or printers. **Stringsearch** benchmark uses an algorithm for comparison which can search for specific words in certain expressions.

The **Telecommunications** category of benchmarks is created for the small smart devices using Wi-Fi. The algorithms employed here are checksum and frequency analysis, or voice encoding and decoding. **CRC32** benchmark executes a Cyclic Redundancy Check (CRC) of 32 bits on a sound file, in order to find errors in the transmission of the data.

4 Design Space Exploration Methodology

4.1 The Exploration Space

In this paper we made an automatic exploration of the design space of VEX architecture, in order to find the best trade-off between performance and the integration area. The multi-objective optimization problem could be represented in analytical form:

- A vector function f that maps 22 parameters to a 2 objectives
- Minimize the objectives vector $y = f(x) = (f_1(x), f_2(x))$;
- $x = (x_1, x_2, ..., x_{22})$ represents the decision vector (*CacheSize, Sets, LineSize, StrSize, StrSets, StrLineSize, ICacheSize, ICacheSets, ICacheLineSize, #clusters, IssueWidth, MemLoad, MemStore, Memptf, Alu.n, Mpy.n, CopySrc.n, CopyDes.n, Mem.n, Int_REG, Bool_REG, Compiler Optimization levels*)
- $y_1 = f_1(x)$ is obtained after simulation with the VEX tool of MiBench benchmarks on architectural configuration provided by all genes of chromosome;
- $y_2 = f_2(x)$ is obtained after running the CACTI tool [17] on the cache memories configuration provided by first 9 genes of chromosome;
- $y = (y_1, y_2)$ represents the objective vector (Cycles per Instruction – the inverse of IPC, Integration Area)

Table 1. Ranges of memory (left) and machine (middle & right) architectural parameters

Parameter	Value [bytes]	Parameter	Value [bytes]	Parameter	Value [No_of_]
CacheSize	32–67108864	RES: IssueWidth	2–65536	REG: $r0	8/16/32/64
Sets	1–16	RES: MemLoad	2–256	REG: $b0	2/4/8
LineSize	16–67108864	RES: MemStore	2–256	number of clusters	1–4
StrSize	16–67108864	RES: Memptf	2–256		
StrSets	16–67108864	RES: Alu.n	16–65536		
StrLineSize	16–67108864	RES: Mpy.n	2–256		
ICacheSize	4–67108864	RES: CopySrc.n	2–256		
ICacheSets	1–16	RES: CopyDes.n	2–16		
ICacheLineSize	2–256	RES: Mem.n	2–256		

The ISA of VEX allows machines with multiple clusters and has many configurable parameters. The total number of VEX system architectural parameters is 65. From these we selected only 22 genes for encoding into the chromosome (9 memory architecture parameters, 12 machine architecture parameters and one for the Optimization level). Their ranges can be seen in Table 1. We adjusted the right margins in some cases in contrast with the original VEX due to the CACTI, which cannot execute such big caches (>64 MB). The value of one gene differs from another through multiplication by 2. The simulations were done in $\sim$3 weeks, using 5 benchmarks and a local area network with 10 computers. Multiplying all possible values for each parameter, the exploration space has $6 * 10^{19}$ configurations. If 50 configurations took around 31 h uninterruptedly on 5 benchmarks, then all configurations would take $372 * 10^{17}$ h, namely $155 * 10^{16}$ days, which is more than $4 * 10^{15}$ years. This huge design space involves automation and evolutionary algorithms for optimization like NSGA-II.

4.2 Workflow

Figure 2 presents the architecture of the developed application. The application was software implemented in C# using MonoDevelop IDE under Ubuntu Linux.

The second variant of the Non-dominated Sorting Genetic Algorithm [14] was used for the multi-objective optimization part, and the VEX [11] and CACTI [17] tools were used for the evaluation part in the non-dominance check of the NSGA-II (see Fig. 2A). The NSGA-II was implemented based on JMetal.Net[1] library. The chromosomes of the populations are the configurations of the architecture merged with memory system and compiler optimization rules on which the simulation is made. The first part of the

---

[1] https://jmetalnet.sourceforge.net/.

chromosome consists of the memory parameters, like the data or the instruction cache size, the blocks size, and the associativity. The second part of the chromosome comprises the machine parameters. Values like number of clusters, issue width, number of ALUs or number of the registers can be found here. The last genes are the number of clusters and the optimization method (see Fig. 2B).

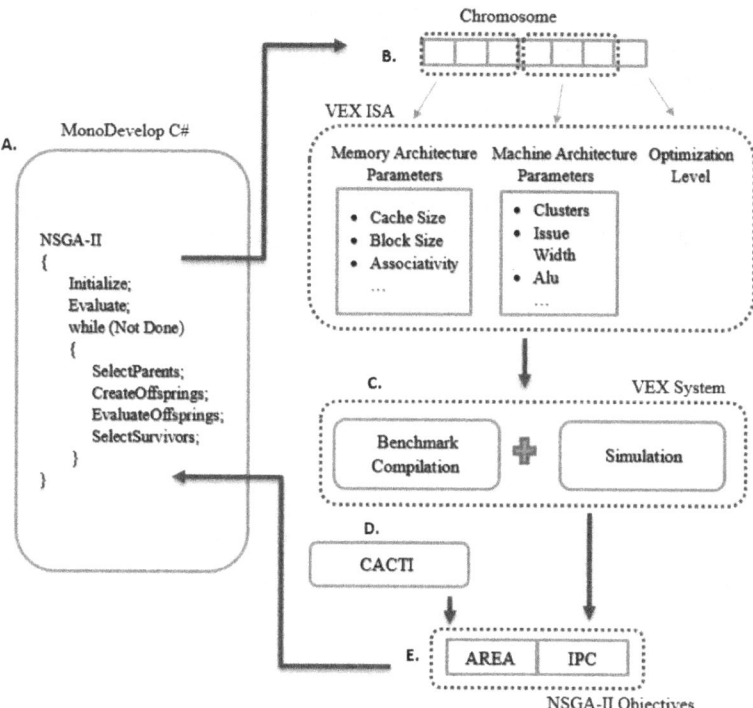

Fig. 2. Application architecture

The developed application is highly parameterized allowing user to configure the number of generations, the selection method, crossover and mutation probabilities, and the benchmark on which the simulation will be run. The NSGA-II algorithm uses the configuration information in order to select the parents and creates the offspring based on the Pareto non-dominance principle. Starting from a randomly initialized population the evaluation is done by creating the configuration (.*cfg* files in VEX) and the machine model (.*mm* files in VEX) files that are needed in the simulation. Then the benchmarks and optimization option are taken, and the simulator command is created. Using a process, the compilation and simulation will be run with the help of the VEX tool (see Fig. 2C). From this simulation will result, as an output, the IPC. The same steps will be taken for calculation of the integration area, namely a command containing the cache parameters will be created and the CACTI tool (see Fig. 2D) will be started using a new process. The integration technology employed was 65 nm. The two processes will output two files, from where the IPC and the integration area will be extracted and set as objectives for the NSGA-II algorithm (see Fig. 2E). This step is applied for the whole population of chromosomes. There are applied genetic operators of crossover and mutation on this population, which will create the first offspring generation. At this point we will have the parent population and the offspring population; both will have the size of the initial population, because from two parents two children will be generated. The two populations will be merged and ranked. The non-dominated sorting and the crowding distance will be calculated in order to find the best solutions.

Once the new population is created, the algorithm will retake these steps, until the number of generations is met.

4.3 Optimization Levels

The compiler applies the optimization level on the benchmark code. It can do the optimizations using different methods. Each level of optimization will only add a new method of optimization over the ones at the lower levels [1]. These methods can be seen in Table 2.

Table 2. Optimization levels

Optimization	Information
-O1	All scalar optimizations
-O2	Minimal loop unrolling and trace scheduling compilation
-O3	Basic loop unrolling and trace scheduling compilation
-O4	Heavy loop unrolling

4.4 Area Estimation

The embedded system cost is associated with the physical part of the system. The total area of the system is calculated adding up the individual areas and ignoring the wiring. After adding all processor components, the remaining area is reserved for the cache.

In their study, Najafi and Salehi [13], estimated the area of the VEX architecture by calculating the area of some parts of the VEX system and reached a total VEX area of 129105 mm^2. We have subtracted the area of the registers from this value and we have obtained the value of 101621 mm^2, as the VEX system area. We used the CACTI tool to calculate the cache area and employed the Formula 1 and 2 proposed in [13], to calculate the area of the registers. CACTI was developed by Norman Jouppi, also developer of VEX and Victim Cache concept, for computer architecture designers to gain a better understanding of different types of caches and their performance. It can calculate: the integrated area, the access time, the dissipated energy and the power consumption.

$$Area = NumberOfRegs * BitsPerReg * Ports^2 \qquad (1)$$

$$Ports = NumberOfReadWritePorts * ALUCluster * IssueWidthCluster \qquad (2)$$

Where for each ALU there is a total of 3 read/write ports (2 read ports and 1 write port), *NumberOfRegs* is the number of registers, *BitsPerReg* is the number of bits of each register and *ALUCluster* and *IssueWidthCluster* are the numbers of issue width and ALUs for each cluster. We took the total area as the sum of the estimated VEX architecture, the measured cache area and the calculated register area.

5   Experimental Results

The first simulation results aim two objectives (performance and integration area) and focus on the features of the multi-objective optimization genetic algorithm, namely:

- Varying the selection method: Binary Tournament, Best Solution Selection and Random Solution Selection
- Varying the crossover probabilities
- Varying the mutation probabilities

The last part considers also energy in analysis. The whole simulation process is time consuming and for this reason was selected only one benchmark per categories of embedded applications from MiBench suite: Automotive, Networks, Telecommunications, Office automation and Security [16].

The crossover operator was a Simulate Binary (SBX) one, and the probabilities used were 0.6 and 0.8. The mutation operator was a Polynomial one, with the probabilities 0.8, 0.1, 0.045 and 0.01. Two approaches were used for the optimization algorithm. The first one used as objectives Area and CPI (cycles per instruction) which is the inversion of performance (CPI = 1/IPC), and the second one evaluated Area, CPI and Energy. All objectives must be minimized.

Next, we present the results categorized on the number of optimization objectives, the selection methods and the mutation values. The results will be graphically represented using the Pareto front between CPI and AREA [mm^2], and the evolution of the algorithm by using the Hypervolume (HV) indicator for each generation. The higher the HV, the more qualitative solutions are obtained.

5.1   Results on the Dijkstra Benchmark

Here are the results for the simulations using Dijkstra benchmark with small and large datasets, where all the selection methods were used, along with crossover probabilities of 0.6 and 0.8 and a mutation probability of 1/number_of_variables, which is approximatively 0.045. Because the whole compilation and simulation process is time consuming, we stopped the algorithm after 50 generations.

For the Binary Tournament Selection method, it can be seen in Fig. 3 (left) that with a crossover probability of 0.6 there are more points on the Pareto front. It can be also seen that the best results are for Dijkstra Small with the 0.6 crossover probability.

This result found in the last generation provided a cache hit rate of 95.67% and it consists in two clusters, with the optimization level "–O3" and 64 KB instruction cache. Figure 3 (right) indicates that the Hypervolume is much higher for the Dijkstra Small benchmark over the generations. However, it can be noticed that the best solutions are obtained after 46 generations.

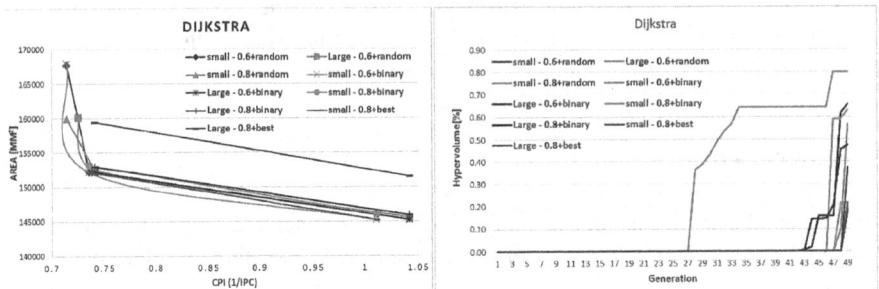

Fig. 3. Dijkstra Pareto Front and Hypervolume

For the Best Solution Selection method, only the 0.8 crossover probability had some notable results but inferior to Binary Tournament Selection method. The simulation on the smaller dataset had a better trade-off between the CPI and the AREA, as shown by Fig. 3 (left). For the point with a smaller CPI value of the Dijkstra Small Pareto Front, the hit rate was 90.69% on two clusters with the optimization level "–O3" and a cache size of 16 KB. The second point had a hit rate of 95.54% on two clusters using an "–O4" optimization level, with a cache size of 8 KB.

For the Random Solution selection technique, the 0.8 crossover probability led to approximately same results as the 0.6 probability. For this technique we have not included in the results the larger dataset, and that on the smaller dataset led to a small number of points. The CPI difference between them is rather insignificant, so for the points with the smaller area the hit rate was 78.35%, on 2 clusters, an "–O4" optimization level and 2 KB instruction cache and 88.41% hit rate, one the default cluster with "–O4" and 512B for the other. From the point of view of the Hypervolume, the results are rather bad since the HV starts to increase only in the last generations. Analyzing the results on Dijkstra benchmark it reveals a general conclusion, namely, as aggressive is the optimization as shortest the code produced, requiring smaller caches.

5.2 Results on the CRC32 Benchmark

In this section are illustrated the results for the simulations using CRC32 benchmark where all the selection methods were used, along with crossover probabilities of 0.6 and 0.8 and a mutation probability of 0.045.

For the Binary Tournament selection method, as Fig. 4 (left) shows, the crossover probability of 0.6 has more points on the Pareto front, than the 0.8 probability. For both of them there is a trade-off between CPI and AREA. In this case, for both crossover probabilities the hit rate was 80%, four clusters were used, using the "–O1"

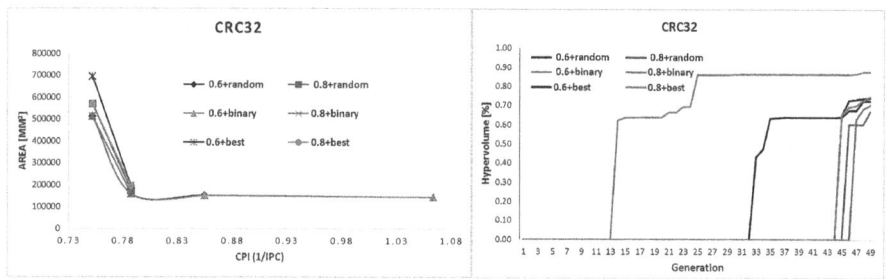

Fig. 4. CRC32 Pareto Front and Hypervolume

optimization level and 4 KB of instruction cache. In Fig. 4 (right) one can also see that the Hypervolume for the 0.6 crossover probability is increasing faster than for the 0.8 probability, but it stagnates after a while.

The results for the Best Solution Selection method were not the best and brought some huge area values, as it can be seen in Fig. 4 (left). For the point with lower area values, the hit rate was 75% and 70% respectively, using no optimization and a number of four clusters, with a cache size of 4 KB and 512 KB respectively. Overall, the 0.8 crossover probability makes a better trade-off between the CPI and the AREA.

The Random Solution selection technique returned small amount of points on the Pareto front. For the middle point of the 0.6 crossover probability, the hit rate was 85% with four clusters, level one of optimization and a 1 MB of instruction cache. Overall, the value of 0.6 crossover probability makes a little better trade-off between the CPI and the AREA. From the Hypervolume plot in Fig. 4 (right), it can be seen that its value increases just in the second half of the number of generations but takes rather big values.

5.3   Results on the Bitcount Benchmark

This section presents the simulations results on Bitcount benchmark with small and large datasets, where all the selection methods were used, along with crossover probabilities of 0.6 and 0.8 and the mutation probability of 0.045.

For the Binary Tournament selection method some better results were obtained for BitCount Small with 0.6 crossover probability and for BitCount Large with 0.8 crossover probability (see the left side of Fig. 5). For both, the results were found in the last generation, using the "–O3" optimization and four clusters. The hit rate of the former was 99.99% on a 32 KB instruction cache and the hit rate for the latter was almost 100% on a 2 KB of instruction cache.

The results for the Best Solution Selection method returned multiple values, as shown in Fig. 5 (left). For the area where the points meet, with a 0.6 crossover probability for the large dataset, and 0.8 for small and large datasets, the hit rate was 100% with an "–O4" optimization level and 4 clusters, with a cache size of 8 KB and 4 KB, and a 93.54% hit rate on the 8 KB cache respectively. Overall, the 0.8 crossover probability makes a better trade-off between the CPI and the AREA. In terms of

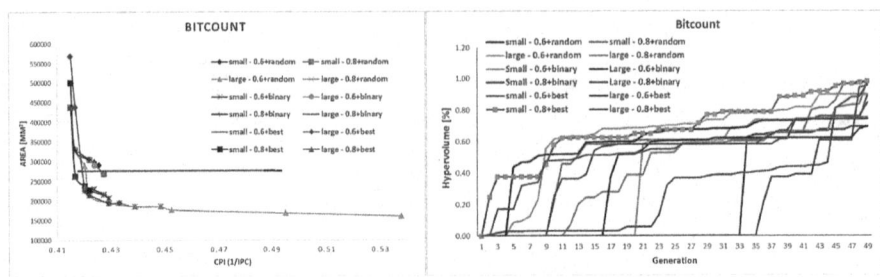

Fig. 5. BitCount Pareto Front and Hypervolume

Hypervolume, BitCount Small with the crossover probability of 0.8 has the highest results a Hypervolume of 0.95, as it can be seen in Fig. 5 (right).

For Random Solution selection technique, the results contain again a lot of points on the Pareto front, as shown in Fig. 5 (left). The best results are for the simulations with the large datasets, and approximately no difference between the distinct crossover probability values. For the middle point of the BitCount large benchmark with 0.8 crossover probability, the hit rate was 100%, on four clusters, with an "–O4" and 8 KB of instruction cache. It can also be seen that HV values increase over the generations, which means the quality of solutions increase.

5.4   Results on the StringSearch Benchmark

This section express the results for the simulations using StringSearch benchmark with small and large datasets, where all the selection methods were used, along with crossover probabilities of 0.6 and 0.8 and the mutation probability of 0.045.

For the Binary Tournament selection method, we have noticed good results only for the 0.8 crossover probability for both the small and the large input datasets. For the middle point of the small dataset (see Fig. 6), the instruction cache hit rate was 81.03% and it through the use of four clusters, with "–O3" optimization level. For the larger dataset, the hit rate was 77.46%, using four clusters and the optimization level "–O4". Both results were found in the last generations with an instruction cache of 8 KB. The Hypervolume for the one with the larger dataset provided better results for the last generations and it reached values as high as 0.8.

For the Best Solution Selection method, we obtained good results, as it can be seen in Fig. 6. For the point of the StringSearch Large which is the nearest to the left corner of the plot, the instruction cache hit rate was 77.1% on four clusters, at "–O4" optimization level and 32 KB instruction cache. For the third point of the simulation made on the large dataset with 0.8 crossover probability, the hit rate was 97.54% on four clusters, the "–O3" optimization level and a 2 KB instruction cache. As to the large dataset, the results were good for both crossover probabilities, unlike the small data set which showed bad results. Both crossover probabilities had good results.

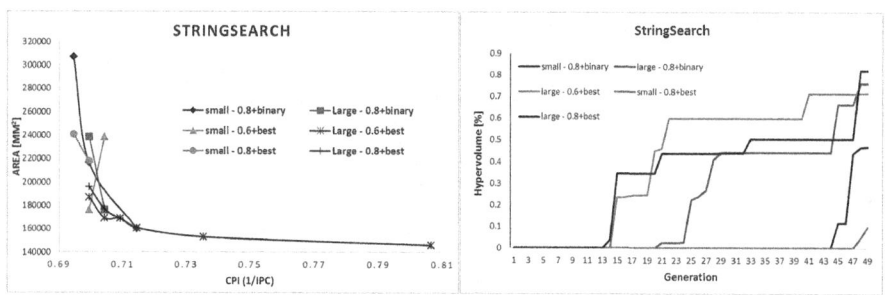

Fig. 6. StringSearch Pareto Front and Hypervolume

5.5 Different Mutation Probabilities

Here are presented the results for the simulations made by using the Binary Tournament selection method, along with a crossover probability of 0.6. We have changed the mutation probabilities with the values 0.8, 0.1, 1/number_of_variables, which is approximatively 0.045 and 0.01. Next, we will present the results for each benchmark.

Dijkstra. From Fig. 7 one can see that for all mutation probability values, the results are quite the same. If we take the point where all lines meet, we have a hit rate of 78.5%, with two clusters and an "–O3" optimization level and a 16 KB of cache. From the Hypervolume, it can be seen that the Hypervolume of the 0.8 mutation value increases along with the first generations, which is a good thing.

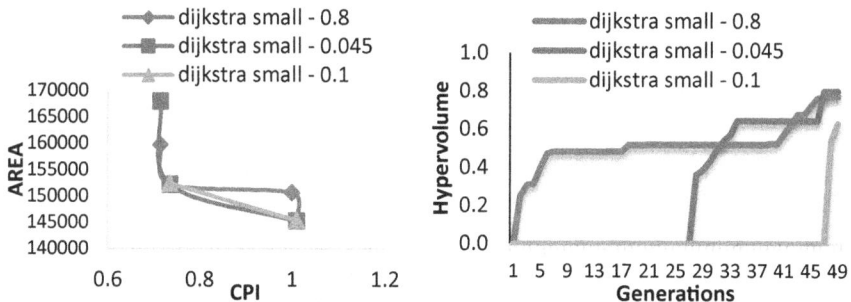

Fig. 7. Dijkstra Pareto Front and Hypervolume for different mutation values

CRC32. In Fig. 8 it can be seen that only the result of the mutation probability of 0.01 differs from the others for the CRC32 benchmark. In the second point of the Pareto Front, we have a 70% hit rate, on two clusters, no optimization and a 512 KB of cache. The Hypervolume plot shows that the evolution stagnates quickly, and all mutation values, except for the 0.01, have higher values.

BitCount. Figure 9 shows that for the BitCount benchmark all values are on the same Pareto front, but the closest values to the initial ones are for the mutation value of 0.8.

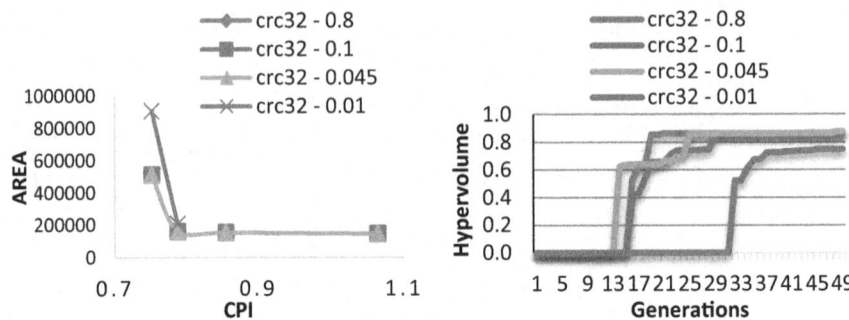

Fig. 8. CRC32 Pareto Front and Hypervolume for different mutation values

On a cache with a size of 16 KB, using four clusters, it resulted an optimization level of "–O3" with a 100% hit rate. A good thing in the Hypervolume plot is that, over generations, the algorithm still evolves, and that, only for the 0.01 mutation, probability evolves more slowly.

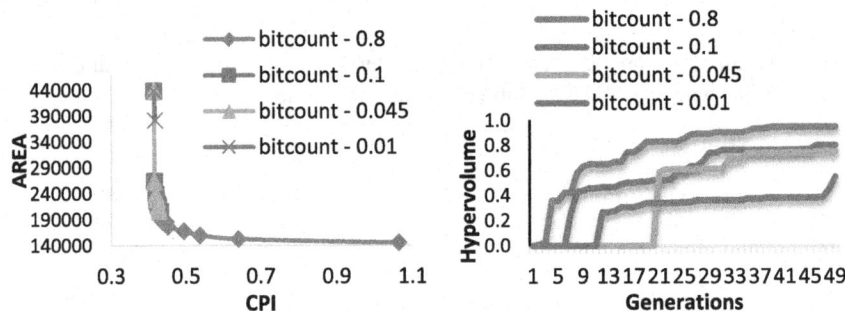

Fig. 9. BitCount Pareto Front and Hypervolume for different mutation values

5.6   Using Energy as the Third Objective

In this section, we present the results for the simulations with Energy introduced as the third objective for optimization. The runs were made by using the Binary Tournament selection method, along with a crossover probability of 0.6 and a mutation probability of 0.045. Next, we will present the results for the BitCount benchmark.

The optimal configuration had a hit rate of 93.02%, using four clusters and an "–O4" optimization level, along with a 2 KB cache size. From the Hypervolume plot (see Fig. 10) it can be seen that in the latter half of the generations, the algorithm evolves, and reaches values close to 100%.

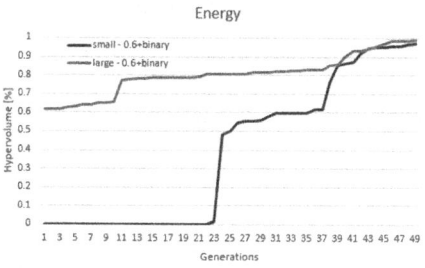

Fig. 10. BitCount small - energy as a third objective

6   Conclusions and Further Work Directions

The future is dedicated to DSA which use the advantage of the VLIW architectures. VLIWs are not performant for general purpose, but they have best performance and efficiency when they are used for specific fields. We have proved this by our simulations on embedded benchmarks. In this paper, we have developed a highly parameterized software application that automatically explores the design space of the VLIW architectures from multiple perspectives. After all simulations, we can conclude that:

- The optimization process, due to the large number of parameters and the necessity to recompile the benchmarks with new optimization flags, is very time consuming. For almost all benchmarks and simulations performed, a large number of generations is required (more than 46 from 50 simulated), so that the optimization algorithm converges to a solution.
- A high mutation rate induces an increase of the performance and an increase of the Hypervolume of the first generations, which means finding good solutions from the very beginning. However, with increasing the number of generations even the low mutation rates reach the same performance. In future we set a varyingly mutation which depends on the generation, so that in the first generation the probability of mutation is high and decreases with the growing number of generations.
- The selection mechanisms proved to be partly efficient. Thus, the Binary Tournament method was effective on the Automotive and Telecommunications benchmarks, whereas the Best Selection Solution method proved optimal on the Office and Network benchmarks.
- For VLIW architectures to be optimal in terms of energy, it is necessary to use small caches and apply aggressive optimizations (O3/O4) rules by the compiler. The BitCount benchmark had a cache hit rate of 99% and the best quality results with almost every run. When the third objective was added, the hit rate decreased by about 10–15%. The best results were found on a configuration of 8 KB cache, along with all four clusters used and heavy loop unrolling optimization.
- The VLIW optimal configurations (integration area and performance) depend on the simulated embedded application.

Further developments would use more types of embedded benchmarks and implement the stop condition of the GA when the HV does not change its value anymore, in order to save time (now, it stops when the number of generations is met). Another interesting development would be integrating the temperature as another objective for the optimization problem, besides the previously mentioned ones in this paper.

References

1. Fisher, J.A., Faraboschi, P., Young, C.: Embedded Computing: A VLIW Approach to Architecture, Compilers and Tools, 1st edn. Morgen Kaufmann, San Francisco (2005)
2. Applications of Embedded System Based Real-Time Projects, Elprocus. https://www.elprocus.com/real-time-applications-of-embedded-systems/. Accessed 11 July 2020
3. Koren, I.: VLIW Architecture Summary, Department of Electrical & Computer engineering, University of Massachusetts. https://www.ecs.umass.edu/ece/koren/architecture/VLIW/1/history.html. Accessed 13 July 2020
4. Hennessy, J.L., Patterson, D.A.: A new golden age for computer architecture. Commun. ACM **62**(2), 48–60 (2019)
5. Kailas, K., Moreno, J.: VLIW Architecture, IBM Research Blog. https://researcher.watson.ibm.com/researcher/view_group.php?id=2831. Accessed 11 July 2020
6. Pestana, S.G., Rijpkema, E., Radulescu, A., Goossens, K., Gangwal, O.P.: Cost-performance trade-offs in networks on chip: a simulation-based approach. In: Proceedings Design, Automation and Test in Europe Conference and Exhibition, vol. 2, pp. 764–769. IEEE (2004)
7. Patterson, D., Waterman, A.: The RISC-V Reader: An Open Architecture Atlas, 1st edn. Strawberry Canyon LLC, San Francisco (2017)
8. Ahuja, S., Brar, G.S.: A VLIW architecture for GSM benchmark & Code Optimization using VEX. https://www.geocities.ws/rushtosumit/report_vliw.pdf. Accessed 11 July 2020
9. Thiele, L., Chakraborty, S., Gries, M., Künzli, S.: Design space exploration of network processor architectures. In: Network Processor Design: Issues and Practices, vol. 1, pp. 55–89 (2003)
10. Saptono, D., Brost, V., Yang, F., Prasetyo, E.: Design space exploration for a custom VLIW architecture: direct photo printer hardware setting using VEX compiler. In: Proceedings International Conference on Signal Image Technology and Internet Based Systems, pp. 416–421. IEEE (2008)
11. VEX Toolchain: Hewlett-Packard Laboratories. https://www.hpl.hp.com/downloads/vex/. Accessed 5 July 2020
12. Jungeblut, T., Sievers, G., Porrmann, M., Rückert, U.: Design space exploration for memory subsystems of VLIW architectures. In: Proceedings Fifth International Conference on Networking, Architecture, and Storage, pp. 377–385. IEEE (2010)
13. Najafi, M.H., Salehi, M.E.: Exploring the design space for area-efficient embedded VLIW packet processing engine. In: Proceedings 21st Iranian Conference on Electrical Engineering, pp. 1–6. IEEE (2013)
14. Deb, K., Pratap, A., Agarwal, S., Meyarivan, T.A.: A fast and elitist multiobjective genetic algorithm: NSGA-II. IEEE Trans. Evol. Comput. **6**(2), 182–197 (2002)
15. Gellert, A., Calborean, H., Vintan, L., Florea, A.: Multi-objective optimisations for a superscalar architecture with selective value prediction. IET Comput. Digital Tech. **6**(4), 205–213 (2012)

16. Guthaus, M., Ringenberg, J., Ernst, D., Austin, T., Mudge T., Brown, R.: MiBench: a free, commercially representative embedded benchmark suite. In: Proceedings Fourth Annual International Workshop on Workload Characterization, pp. 3–14. IEEE (2001)
17. Shivakumar, P., Jouppi N.: CACTI 3.0, HP Labs. https://www.hpl.hp.com/research/cacti/. Accessed 23 June 2020

Feed-Forward Neural Network Training by Hybrid Bat Algorithm

Stefan Milosevic⑩, Timea Bezdan⑩, Miodrag Zivkovic⑩,
Nebojsa Bacanin(✉)⑩, Ivana Strumberger⑩, and Milan Tuba⑩

Singidunum University, Danijelova 32, 11000 Belgrade, Serbia
stefan.milosevic.191@singimail.rs, tuba@ieee.org,
{tbezdan,mzivkovic,nbacanin,istrumberger}@singidunum.ac.rs

Abstract. Artificial neural networks are very powerful machine learning techniques and they are capable to solve complex problems. In the artificial neural network, one of the most difficult challenges is to find the optimal values of the weights during the learning process. To address this issue, we propose a new hybridized metaheuristic method, called BAABC for weight connection optimization. The experiments are performed on two binary classification datasets. The obtained results are compared to other similar approaches where other metaheuristics are used. The obtained results show that the proposed algorithm can find the optimal weight connection values and achieve higher performance and the proposed BAABC outperformed the other methods.

Keywords: Artificail neural network · Metaheuristics · Algorithm hybridization · Bat algorithm · Artificial bee colony

1 Introduction

Artificial Neural Networks (ANNs) are inspired by biological human brain cells, and they are one of the most powerful techniques in machine learning. ANN is the component of Artificial Intelligence (AI) which has appeared to seem like it is in a rapid stage of development when put into practical use in real-world applications; tasks such as in speech, vision, and natural language processing the AI techniques and its maximum capabilities prove to be full of capped inadequacies and fragile systemic overload, thus forth, the crash in a real sense of the AI in need to perform many specific tasks at once. Much like in algorithmic methods for the purpose of problem-solving even each and every one of the AI techniques require specific input and systemic mapping of any given problem. The biggest issue here has to do with the "Rule-Based Approach" which requires necessary explicit state of rules. Biological neural networks here provide a possible solution in which it will present new method search approaches in order to solve and be able to input and output data from natural problems. This kind of problem is very complex and belongs to NP-hard optimization problems. Metaheuristic methods are proven to be efficient in NP-hard optimization problems, thus, in

D. Simian and L. F. Stoica (Eds.): MDIS 2020, CCIS 1341, pp. 52–66, 2021.
https://doi.org/10.1007/978-3-030-68527-0_4

this work, we propose a metaheuristic based approach for weight connection optimization in neural networks. Metaheuristic algorithm has many successful implementation in different real-life problems, such as [1–22].

The rest of the paper is organized as follows: Sect. 2 gives an overview of the artificial neural networks, Sect. 3 describes the original bat algorithm, Sect. 4 introduces the proposed method and describes the procedure of the optimization algorithm, Sect. 5 presents the simulation setup and the obtained experimental results, and Sect. 6 concludes the paper.

2   Artificial Neural Networks

Artificial neural network (ANN) is a constituent of artificial intelligence that is meant to mimic and emulate the functions and operations that of a human brain. In this section, we will further concentrate on the division strategy which focuses on the separation of the population and the generations so that the population and the generations can be split and categorized into divergent sections of the algorithm, Mantegna Lévy Flight, Particle Swarm Optimization, and Neighbourhood Search [23]. Through constant research and development, it has been verified that a good dose of exploration as well as exploitation is highly suggested in order to achieve any sort of evolutionary algorithms, swarm intelligence-based algorithms for example. During the exploration process, the algorithm performs a global search, and, on the other hand, the exploitation process is very important to look for the optimal solution by preforming a local search around the current best solution. That being said, in order to achieve good exploitation, automation looks diligently for a path consisted of high learning satisfying a level of efficiency to look for an immediate outcome. The BAABC algorithm used in this paper uses the same notion of a good balance between good exploration and good exploitation. The link between both exploration and exploitation also defines as well as are factors in the sheer creativity of artificial creativity [24].

Some different types of Artificial Neural Networks (ANNs) we will be discussing are Deep NNs, Convolutional NNs, and Feedforward NNs. Deep learning ANNs consist of several layers between the input and output layers. The application of deep learning has been recognized as a viable approach for big data analysis with suitable application to computer vision, several types of recognitions-such as speech recognition and pattern recognition, and natural language processing. Here we simply will be accentuating and quickly mentioning that certain applications of Deep learning techniques are found in several areas such as computer vision, speech recognition, and pattern recognition [25].

The next type of Neural Network we will touch upon will be Convolutional Neural Network (CNN). The most recent reports showcase high accuracy in computer vision and pattern recognition problem sets [26]. CNNs also are applied to the analysis of visual imagery. Likewise, they are highly proficient in the identification of faces, traffic signs, and everyday objects. This is due to the fact that the deep model applied in image super-resolution (SR) has demonstrated extraordinary performance, as opposed to, the older generational method of hand-crafted models when it comes to its efficiency and restoration assurance. This application of the Convolutional Neural Network is also known as the Super-Resolution Convolutional Neural Network (SRCNN).

Moreover, the next Neural Network we will discuss is the Feedforward Neural Networks. Due to the fact that the Feedforward NN was the first constructed NN and the most elementary type of Artificial Neural Network (ANN) devised it only constructs connections between its nodes rather than forming a cycle. This type of NN and its algorithms have brought data and statistical science and engineering communities a truly effective modus operandi for the assembly of nonlinear systems, accepting large sets of input data thus achieving successful outputs in application problems amongst the statistical and engineering problems sets revolved around classification and regression. To conclude the discussion of Feedforward Neural Networks we wanted to bring to light a class of Feedforward NN, Multilayer Perceptron (MLP). An MLP is consisted of at least three layers of nodes: an input layer, a hidden layer, and an output layer. All except for the input layer nodes, each node is a neuron that makes use of a nonlinear activation function. Multilayer Perceptrons (MLPs) are capable of differentiating data that is not linearly separable. MLPs use such supervised learning techniques called Back Propagation for training algorithms. The Back Propagation (BP) algorithm contributes a very useful method for the implementation and design of multilayer NNs [27].

3   Original Bat Algorithm

Bat Algorithm (BA) is a newly constructed and highly useful metaheuristic algorithm that works out Multi-objective RFID problems that prove to be well known hard optimization problems. Bats use sonar (bio-sonar) which is a certain method of echolocation that allows them to use for prey detection and echolocate their prey (track and locate). Bats rely on their sense of hearing and through this, they are able to discern the orientation of the surrounding space by emitting sound-wave pulses that ricochet back-and-forth within the spatial proximity. The pulse duration lasts approximately 8 to 10 m/s, with a constant frequency efficiency anywhere from 25 kHz to 150 kHz. The wavelengths correspondingly range from 2–14 mm [28]. Also, note that ray tracing was not used in this BA implementation as it is computationally expensive [29].

From Yang's analysis based on the echolocation behavior of bats, he constructed a metaheuristic algorithm; the bat algorithm based on three idealized rules [29]:

– Bats use their sense of distance from using echolocation, all while they can detect and analyze to a certain approximation of the distance between pray and any surrounding background barriers.
– In search for pray, bats fly in random order with a velocity v_i at position x_i with frequency f_i varying wavelength of λ and with a loudness of A_0.

Likewise, bats can automatically adjust the frequency of their emitted pulse rates at $r \in [0, 1]$.

- The loudness of their emitted calls vary from anywhere of a maximum initial value of A_0 to a minimum constant value of A_{min}

Our hybrid bat algorithm (BA) also uses a simplification method with:

- frequency of f,
- in such a range $[f_{min}, f_{max}]$ that together come to a;
- range of wavelength $[\lambda_{min}, \lambda_{max}]$.

For every bat the velocity is described as v_i^t and its position x_i^t and its iteration t in d(dimensional) search capacity. Here is the following bat motion exhibited by emendating its velocity and its position as follows [28, 29]:

$$f_i = f_{min} + (f_{max} - f_{min})\beta, \tag{1}$$

$$v_i^t = v_i^{t-1} + (x_i^{t-1} - x_{best})f_i, \tag{2}$$

$$x_i^t = x_i^{t-1} + v_i^t, \tag{3}$$

f_i is a frequency of bat i and $\beta \in [0, 1]$ is a variable/random number drawn from its distribution. x_{best} represents the current global best solution which is also put into account when being recalculated in each of its iterations. $\lambda_i f_i$ (the product) is the velocity increment; f_i as well as λ_i are used for the adjustment of the subjected velocity change while the other factor is fixed.

The local search procedure initiates a new solution for each bat in the algorithm for the current best position managing a random input implementation [29].

$$x_{new} = x_{old} + \epsilon A^t, \tag{4}$$

ϵ represents a random vector. The intervals $[-1, 1]$ and $A^t = < A_t^i >$ is the average loudness of all bats represented in the equation at the step of the time. Loudness A_i and the rate of pulse emission r_i, likewise, are affected during the cycles of the algorithm. This rate of change is represented through the mathematical equation [29]:

$$A_i^t = \alpha A_i^{t-1}, \ r_i^t = r_i^0[1 - exp(-\gamma t)], \tag{5}$$

$$A_i^t \rightarrow 0, \ r_i^t \rightarrow r_i^0, \ \text{while } t \rightarrow \infty \tag{6}$$

α and γ are constants. Only while for any α in between internals $(0, 1)$ and $\gamma > 0$ [28].

4 Hybridized Bat Algorithm

In this work, we propose a hybridized version of the original BA algorithm. By utilizing the artificial bee colony (ABC) algorithm [30] onlooker bee search mechanism, we are enhancing the exploration capability of the algorithm. The hybridized algorithm results in a better trade-off between exploration and exploitation and the possibility of getting stuck in the local optima is reduced.

At the beginning of the algorithm procedure, the initial solutions (individuals) in the population are generated randomly within a defined range of lower and upper bounds $[lb, ub]$:

$$x_{i,j} = lb_j + rand(ub_j - lb_j), \tag{7}$$

where $x_{i,j}$ denotes the ith solution and jth element. The lower and upper bound are denoted by lb_j and ub_j, respectively. The random number between 0 and 1 is denoted by $rand$.

In the next step, the solutions' fitness value is calculated by the fitness function and it is defined as:

$$F_{x_i} = \begin{cases} \frac{1}{f_{x_i}} & \text{if } f_{x_i} \geq 0 \\ 1 + |f_{x_i}| & \text{otherwise,} \end{cases} \tag{8}$$

where F_{x_i} denotes the fitness function of the solution x_i, and f_{x_i} refers to the objective function.

We are utilizing two ways for the solutions' position update, the original BA position update is used (Eq. (3)) if the iteration counter t is even, otherwise, if t is odd, the position of the solutions is being updated by the onlooker bee search mechanism. The location update by the onlooker bee search mechanism is calculated as follows:

$$x_{i,j}^t = x_{i,j}^{t-1} + rand \cdot (x_{i,j}^{t-1} - x_{k,j}^{t-1}), \tag{9}$$

where $x_{i,j}^t$ denotes the updated solution, $x_{i,j}^{t-1}$ indicates to the current solution's position, the neighbor solution's location is denoted by $x_{k,j}^{t-1}$, and $rand$ is a random number drawn from the uniform distribution.

In the succeeding step, the algorithm exploits the search space around the best location by using the random walk search process with small steps.

The probability selection method is responsible to make a decision if the newly generated solution will be accepted or not and it is calculated as follows:

$$p_i = \frac{F_{avg}}{\sum_{i=1}^n F_{x_i}} \tag{10}$$

where F_{x_i} represents the fitness value of the ith solution, F_{avg} denotes the average fitness value of all solutions in the population, and the selection probability is denoted by p_i.

The pseudocode of the hybridized bat algorithm with the ABC's onlooker search procedure is described in Algorithm 1.

5   Experiments and Results

This section describes the design of the experiment with the proposed BAABC approach, the simulation setup, and presents the obtained results.

The population of the algorithm represents a possible network structure, each solution in the population is represented by a one-dimensional vector, which contains the connection weights and biases. In this experiment, the neural network consists of the input layer, one hidden layer and the output layer, which is the classification layer. The size of the solution vector is calculated according to the following formula:

$$n_w = (n_x \times n_h + n_h) + (n_h \times n_o + n_o) \tag{11}$$

where the vector length is denoted by n_w, n_x is the size of input feature, n_h is the hidden unit number in the hidden layer, and n_o is the number of hidden units in the output layer.

The MSE (mean square error) is utilized for the loss function which is used to evaluate the solutions' fitness value.

Algorithm 1. Hybridized bat algorithm pseudocode

Randomly initialize population of N solutions
Initialize the velocity (v) of each solution, the pulse rate (r), loudness (A) and frequency (f)
Set the iteration counter to zero and define the number of the maximum iteration parameter ($MaxIter$)
while $t < MaxIter$ **do**
   for $i = 1$ to N (each solution in the population) **do**
      if t is even **then**
         Calculate the value of frequency by by utilizing Eq. (1)
         Calculate current velocity by using Eq. (2)
         Update the position by using the original BA position update Eq. (3)
      else
         Update the position by the onlooker search mechanism by employing Eq. (9)
      end if
      if $rand > r_i$ **then**
         Select the best solution in the population
         Generate a local solution around the best sultion by the random walk process Eq. (4)
      end if
      if $(p_i < A_i$ and $f(x_i) < f(x_{best}))$ **then**
         The new solution replaces the old solution
         Reduce the loudness A_i and increase the pulse rate r_i by utilizing Eq. (5)
      end if
   end for
   Evaluate each solution in the population according to the solutions' new position
   Sort the solutions in the population and save the current best solution x_{best}
end while
Return the best solution

$$MSE = \frac{1}{n} \sum_{i=1}^{n} (y_i - \hat{y}_i)^2 \qquad (12)$$

where n indicates to the number of, y and $\hat{y}$ denotes the actual value, and the predicted value, respectively.

The datasets are split into training (66%) and testing (33%) datasets. In both datasets, the features are normalized by using the following formula:

$$X_{norm} = \frac{X_i - X_{min}}{X_{max} - X_{min}} \qquad (13)$$

where X_{norm} denotes the normalized value, the ith input feature us denoted by X_i, X_{min} and X_{max} denotes the minimum and maximum value of the corresponding feature.

The elements in the solution vector are initialized between -1 and 1. In the experiment, we used 50 solutions in the population and the algorithm procedure is repeated 250 times. To obtain fair results, we repeated the experiment in 30 runs.

The parameters of the proposed BAABC method are summarized in Table 1.

Table 1. Hybridized bat algorithm control parameters

Parameter	Notation	Value
Population size	N	50
Maximum iteration	MaxIter	250
Minimum frequency	f_{min}	0
Maximum frequency	f_{max}	1
Constant minimum loudness	A_{min}	1
Maximum initial loudness	A_0	100
Constant parameter	α	0.9
Constant parameter	γ	0.9

For testing the performance of the proposed algorithm, we used two well-known test dataset, the Parkinson dataset [31] and the breast cancer dataset [32,33]. The first dataset has 22 features and 195 instances, while the second dataset has 9 features and 699 instances. Both datasets have two classes which indicate whether the patient is healthy or diagnosed with Parkinson's diseases or breast cancer.

For performance measurement, five different metrics are used:

Area under the curve (AUC): $AUC = \dfrac{1}{(TP+FP)(TN+FN)} \int_0^1 TP \, d \, FP$

Accuracy: $accuracy = \dfrac{TP+TN}{TP+FP+TN+FN}$

Specificity: $specificity = \dfrac{TN}{TN+FP}$

Sensitivity: $sensitivity = \dfrac{TN}{FN+TP}$

Geometric mean (g-mean): $g-mean = \sqrt{specificity \times sensitivity}$

In the metrics equations, TP and TN denotes the true positive value and the true negative value, while FP and FN indicates to the false positive and false-positive results.

The obtained experimental results are compared to other metaheuristic approaches (GOA, GA, PSO, ABC, FPA, BAT, FF, MBO, and BBO), and the results of these algorithms are taken from [34]. To make a fair comparison, the simulation configuration is done in a similar way as it is described in the [34].

The results of both datasets are presented in Table 2 and Table 3.

The best metrics of each algorithm are visualized in Fig. 1 and the convergence graph of the datasets' test accuracy performance is depicted in Fig. 2.

As observed, the proposed hybridized BA optimization algorithm outperforms other well-known swarm intelligence based algorithms. On the Parkinson dataset, the BAABC achieves the best results on all five metrics, and the second-best performed algorithm is GOA. In the case of specificity, three algorithms, BAABC, GOA, and FF achieved the highest result, and on the specificity metric, all algorithms resulted in value 1 and only FF has a result of 0.625. In the breast cancer dataset test results, BAABC has the best result in the AUC metric, while the second-best algorithm is ABC. Four algorithms, ABC, FPA, BAT, BBO resulted in the highest accuracy of 0.98319, BAABC, GOA, PSO, FF, and MBO are the second-best with 0.97899 best accuracies, and GA has the worst accuracy result 0.97479. On the sensitivity test, BAABC has the best performance of 0.99525, followed by the PSO, BAT, MBO with 0.99363. The specificity metric result of BAABC and ABC is equal, and it has the highest value of 1. And the fifth metric, the G-Mean is 0.98718 in case of the best algorithm on this metric, which ABC and BAABC is the second-best on the G-Mean metric.

Table 2. Evaluation results of training MLP networks for Parkinsons dataset

Algorithm	Metric	AUC	Accuracy	Specificity	Sensitivity	G-Mean
BAABC	AVG	0.94932	0.91761	0.80524	0.97311	0.86985
	STD	0.01230	0.02752	0.06251	0.06125	0.05478
	BEST	0.98972	0.97851	1.00000	1.00000	0.99158
	WORST	0.93591	0.75267	0.61528	0.91534	0.78211
GOA	AVG	0.91891	0.90746	0.72500	0.96471	0.83084
	STD	0.08401	0.03601	0.15536	0.01885	0.09294
	BEST	0.99387	0.97015	1.00000	1.00000	0.98020
	WORST	0.68873	0.82090	0.43750	0.92157	0.64169
GA	AVG	0.92823	0.88109	0.60000	0.96928	0.75794
	STD	0.02968	0.02055	0.12128	0.03364	0.06631
	BEST	0.96201	0.92537	0.87500	1.00000	0.87867
	WORST	0.84069	0.83582	0.43750	0.88235	0.66144
PSO	AVG	0.85118	0.84478	0.46042	0.96536	0.66365
	STD	0.03778	0.03026	0.08605	0.03203	0.06245
	BEST	0.90319	0.89552	0.62500	1.00000	0.75000
	WORST	0.70833	0.76119	0.31250	0.88235	0.53665
ABC	AVG	0.80098	0.83831	0.43750	0.96405	0.64645
	STD	0.09376	0.01842	0.08041	0.02962	0.05034
	BEST	0.92525	0.86567	0.68750	1.00000	0.77015
	WORST	0.56985	0.79104	0.31250	0.86275	0.55351
FPA	AVG	0.89011	0.86070	0.50417	0.97255	0.69764
	STD	0.04068	0.02458	0.08675	0.02280	0.05833
	BEST	0.95343	0.91045	0.68750	1.00000	0.81274
	WORST	0.80025	0.80597	0.37500	0.92157	0.60634
BAT	AVG	0.84641	0.86269	0.55625	0.95882	0.72112
	STD	0.11313	0.04156	0.16926	0.03865	0.10801
	BEST	0.98284	0.95522	0.93750	1.00000	0.90749
	WORST	0.62500	0.76119	0.25000	0.82353	0.50000
FF	AVG	0.88039	0.87214	0.98693	0.50625	0.70564
	STD	0.01810	0.02025	0.01296	0.06218	0.04496
	BEST	0.91176	0.91045	1.00000	0.62500	0.79057
	WORST	0.83701	0.83582	0.96078	0.43750	0.64834
MBO	AVG	0.85633	0.85075	0.47083	0.96993	0.67160
	STD	0.04241	0.03208	0.10346	0.02168	0.07894
	BEST	0.91544	0.89552	0.62500	1.00000	0.77491
	WORST	0.73775	0.76119	0.25000	0.90196	0.49507
BBO	AVG	0.93656	0.88159	0.55208	0.98497	0.73480
	STD	0.02311	0.02446	0.09297	0.01517	0.06249
	BEST	0.98039	0.92537	0.75000	1.00000	0.85749
	WORST	0.89951	0.82090	0.37500	0.94118	0.61237

Table 3. Evaluation results of training MLP networks for breast cancer dataset

Algorithms	Metric	AUC	Accuracy	Specificity	Sensitivity	G-Mean
BAABC	AVG	0.99781	0.97458	0.98152	0.98115	0.98812
	STD	0.01852	0.00285	0.01375	0.0423	0.06172
	BEST	0.99876	0.97899	1.00000	0.99525	0.98710
	WORST	0.98834	0.95825	0.97582	0.94784	0.95987
GOA	AVG	0.99536	0.97115	0.95309	0.98047	0.96665
	STD	0.00105	0.00514	0.01352	0.00553	0.00675
	BEST	0.99693	0.97899	0.97531	0.98726	0.97810
	WORST	0.99229	0.95798	0.92593	0.96178	0.94991
GA	AVG	0.99549	0.96751	0.94362	0.97983	0.96151
	STD	0.00072	0.00583	0.01796	0.00294	0.00884
	BEST	0.99709	0.97479	0.97531	0.98726	0.97492
	WORST	0.99418	0.95378	0.90123	0.97452	0.94022
PSO	AVG	0.99558	0.97045	0.94774	0.98217	0.96471
	STD	0.00100	0.00752	0.02331	0.00565	0.01126
	BEST	0.99756	0.97899	0.98765	0.99363	0.98107
	WORST	0.99402	0.95378	0.90123	0.96815	0.94022
ABC	AVG	0.98544	0.96891	0.96214	0.97240	0.96713
	STD	0.01499	0.00793	0.02592	0.00890	0.01161
	BEST	0.99772	0.98319	1.00000	0.98726	0.98718
	WORST	0.92593	0.94958	0.91358	0.94904	0.94047
FPA	AVG	0.99511	0.97241	0.95432	0.98174	0.96789
	STD	0.00090	0.00621	0.01838	0.00277	0.00922
	BEST	0.99670	0.98319	0.98765	0.98726	0.98427
	WORST	0.99292	0.95378	0.90123	0.97452	0.94022
BAT	AVG	0.97079	0.96218	0.93457	0.97643	0.95494
	STD	0.02875	0.01422	0.04541	0.00805	0.02250
	BEST	0.99568	0.98319	0.98765	0.99363	0.98127
	WORST	0.89188	0.92437	0.79012	0.96178	0.88605
FF	AVG	0.99619	0.97311	0.98089	0.95802	0.96938
	STD	0.00024	0.00324	0.00000	0.00951	0.00481
	BEST	0.99678	0.97899	0.98089	0.97531	0.97810
	WORST	0.99575	0.96639	0.98089	0.93827	0.95935
MBO	AVG	0.99453	0.96695	0.94074	0.98047	0.96031
	STD	0.00245	0.00770	0.02302	0.00708	0.01115
	BEST	0.99693	0.97899	0.98765	0.99363	0.98107
	WORST	0.98616	0.94958	0.87654	0.95541	0.93026
BBO	AVG	0.99550	0.97255	0.95514	0.98153	0.96822
	STD	0.00082	0.00478	0.01316	0.00256	0.00678
	BEST	0.99670	0.98319	0.97531	0.98726	0.98127
	WORST	0.99355	0.96639	0.93827	0.97452	0.95935

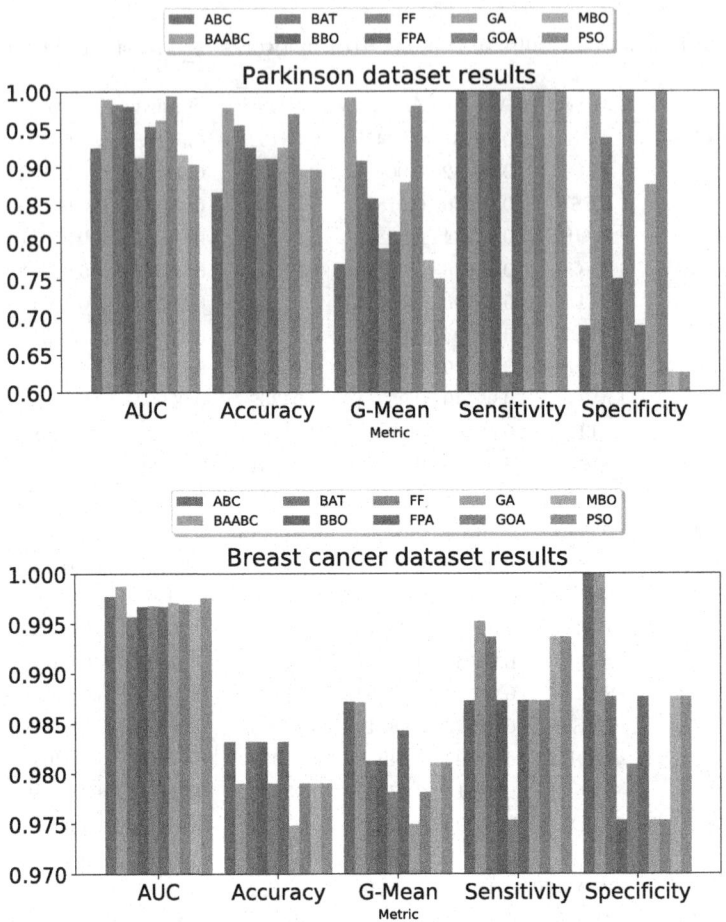

Fig. 1. Algorithm comparison of best result on five metrics

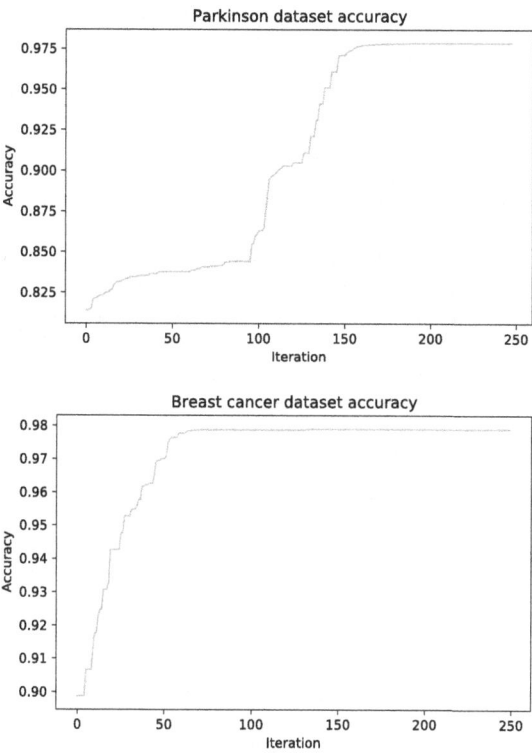

Fig. 2. Convergence graph

6 Conclusion

In the MLP training, during the gradient-based training, the algorithm can get stuck in the local optima and make the training difficult, to avoid this issue, in this work, we propose a stochastic approximation approach for weight connection optimization in MLP by a hybridized bat algorithm. The objective in the problem formulation is to find the right values of weights and biases to minimize the test error rate on the test datasets. The experiments are concluded on two medical datasets, breast cancer, and the Parkinson datasets. The obtained results are compared to other similar approaches where other metaheuristics are used, namely, GOA, GA, PSO, ABC, FPA, BAT, FF, MBO, and BBO. We used different metrics for the performance measurement, such as accuracy, specificity, sensitivity, geometric mean (g-mean), and area under the curve (AUC).

The obtained results show that the proposed algorithm can achieve higher performance than other comparable methods and the utilization of the onlooker search mechanism allows to avoid the local optima.

In future work, we plan to include more datasets with multi-classification problems as well as to optimize the number of hidden units in the hidden layer.

Acknowledgement. The paper is supported by the Ministry of Education, Science and Technological Development of Republic of Serbia, Grant No. III-44006.

References

1. Bacanin, N., Tuba, M.: Artificial bee colony (abc) algorithm for constrained optimization improved with genetic operators. Stud. Inf. Control **21**(2), 137–146 (2012)
2. Bacanin, N., Tuba, E., Bezdan, T., Strumberger, I., Tuba, M.: Artificial flora optimization algorithm for task scheduling in cloud computing environment. In: Yin, H., Camacho, D., Tino, P., Tallón-Ballesteros, A.J., Menezes, R., Allmendinger, R. (eds.) IDEAL 2019, Part I. LNCS, vol. 11871, pp. 437–445. Springer, Cham (2019). https://doi.org/10.1007/978-3-030-33607-3_47
3. Bacanin, N., Bezdan, T., Tuba, E., Strumberger, I., Tuba, M., Zivkovic, M.: Task scheduling in cloud computing environment by grey wolf optimizer. In: 2019 27th Telecommunications Forum (TELFOR), pp. 1–4. IEEE (2019)
4. Tuba, E., Strumberger, I., Bezdan, T., Bacanin, N., Tuba, M.: Classification and feature selection method for medical datasets by brain storm optimization algorithm and support vector machine. Procedia Comput. Sci. **162**, 307–315 (2019). 7th International Conference on Information Technology and Quantitative Management (ITQM 2019): Information technology and quantitative management based on Artificial Intelligence
5. Bezdan, T., Tuba, E., Strumberger, I., Bacanin, N., Tuba, M.: Automatically designing convolutional neural network architecture with artificial flora algorithm. In: Tuba, M., Akashe, S., Joshi, A. (eds.) ICT Systems and Sustainability. AISC, vol. 1077, pp. 371–378. Springer, Singapore (2020). https://doi.org/10.1007/978-981-15-0936-0_39
6. Bezdan, T., Zivkovic, M., Tuba, E., Strumberger, I., Bacanin, N., Tuba, M.: Glioma brain tumor grade classification from MRI using convolutional neural networks designed by modified FA. In: Kahraman, C., Cevik Onar, S., Oztaysi, B., Sari, I.U., Cebi, S., Tolga, A.C. (eds.) INFUS 2020. AISC, vol. 1197, pp. 955–963. Springer, Cham (2021). https://doi.org/10.1007/978-3-030-51156-2_111
7. Zivkovic, M., Bacanin, N., Tuba, E., Strumberger, I., Bezdan, T., Tuba, M.: Wireless sensor networks life time optimization based on the improved firefly algorithm. In: 2020 International Wireless Communications and Mobile Computing (IWCMC), pp. 1176–1181 (2020)
8. Bezdan, T., Zivkovic, M., Tuba, E., Strumberger, I., Bacanin, N., Tuba, M.: Multi-objective task scheduling in cloud computing environment by hybridized bat algorithm. In: Kahraman, C., Cevik Onar, S., Oztaysi, B., Sari, I.U., Cebi, S., Tolga, A.C. (eds.) INFUS 2020. AISC, vol. 1197, pp. 718–725. Springer, Cham (2021). https://doi.org/10.1007/978-3-030-51156-2_83
9. Bacanin, N., Tuba, M.: Firefly algorithm for cardinality constrained mean-variance portfolio optimization problem with entropy diversity constraint. Sci. World J. **2014** (2014)
10. Strumberger, I., Tuba, E., Bacanin, N., Jovanovic, R., Tuba, M.: Convolutional neural network architecture design by the tree growth algorithm framework. In: 2019 International Joint Conference on Neural Networks (IJCNN), pp. 1–8. IEEE (2019)
11. Magud, O., Tuba, E., Bacanin, N.: Medical ultrasound image speckle noise reduction by adaptive median filter. Wseas Trans. Biol. Biomed. **14** (2017)

12. Strumberger, I., Bacanin, N., Tuba, M., Tuba, E.: Resource scheduling in cloud computing based on a hybridized whale optimization algorithm. Appl. Sci. **9**(22), 4893 (2019)
13. Strumberger, I., Bacanin, N., Tuba, M.: Hybridized elephant herding optimization algorithm for constrained optimization. In: Abraham, A., Muhuri, P.K., Muda, A.K., Gandhi, N. (eds.) HIS 2017. AISC, vol. 734, pp. 158–166. Springer, Cham (2018). https://doi.org/10.1007/978-3-319-76351-4_16
14. Strumberger, I., Tuba, E., Bacanin, N., Beko, M., Tuba, M.: Modified and hybridized monarch butterfly algorithms for multi-objective optimization. In: Madureira, A.M., Abraham, A., Gandhi, N., Varela, M.L. (eds.) HIS 2018. AISC, vol. 923, pp. 449–458. Springer, Cham (2020). https://doi.org/10.1007/978-3-030-14347-3_44
15. Tuba, E., Strumberger, I., Zivkovic, D., Bacanin, N., Tuba, M.: Mobile robot path planning by improved brain storm optimization algorithm. In: 2018 IEEE Congress on Evolutionary Computation (CEC), pp. 1–8. IEEE (2018)
16. Strumberger, I., Sarac, M., Markovic, D., Bacanin, N.: Hybridized monarch butterfly algorithm for global optimization problems (2018)
17. Strumberger, I., Tuba, E., Bacanin, N., Zivkovic, M., Beko, M., Tuba, M.: Designing convolutional neural network architecture by the firefly algorithm. In: Proceedings of the 2019 International Young Engineers Forum (YEF-ECE), Costa da Caparica, Portugal, pp. 59–65 (2019)
18. Strumberger, I., Tuba, E., Bacanin, N., Zivkovic, M., Beko, M., Tuba, M.: Designing convolutional neural network architecture by the firefly algorithm. In: 2019 International Young Engineers Forum (YEF-ECE), pp. 59–65, May 2019
19. Strumberger, I., Bacanin, N., Tuba, M.: Enhanced firefly algorithm for constrained numerical optimization. In: 2017 IEEE Congress on Evolutionary Computation (CEC), pp. 2120–2127. IEEE (2017)
20. Zivkovic, M., Bacanin, N., Zivkovic, T., Strumberger, I., Tuba, E., Tuba, M.: Enhanced grey wolf algorithm for energy efficient wireless sensor networks. In: 2020 Zooming Innovation in Consumer Technologies Conference (ZINC), pp. 87–92. IEEE (2020)
21. Bacanin, N., Tuba, E., Zivkovic, M., Strumberger, I., Tuba, M.: Whale optimization algorithm with exploratory move for wireless sensor networks localization. In: Abraham, A., Shandilya, S.K., Garcia-Hernandez, L., Varela, M.L. (eds.) HIS 2019. AISC, vol. 1179, pp. 328–338. Springer, Cham (2021). https://doi.org/10.1007/978-3-030-49336-3_33
22. Strumberger, I., Tuba, E., Bacanin, N., Beko, M., Tuba, M.: Bare bones fireworks algorithm for the rfid network planning problem. In: 2018 IEEE Congress on Evolutionary Computation (CEC), pp. 1–8, July 2018
23. Tarkhaneh, O., Shen, H.: Training of feedforward neural networks for data classification using hybrid particle swarm optimization, mantegna lévy flight and neighborhood search. Heliyon **5**(4), e01275 (2019)
24. al Rifaie, M.M., Bishop, M.: Swarm intelligence and weak artificial creativity. AAAI (2013)
25. Liu, W., Wang, Z., Liu, X., Zeng, N., Liu, Y., Alsaadi, F.E.: A survey of deep neural network architectures and their applications. Neurocomputing **234**, 11–26 (2017)
26. Rahman, M.M., Islam, M.S., Sassi, R., Aktaruzzaman, M.: Convolutional neural networks performance comparison for handwritten bengali numerals recognition. SN Appl. Sci. **1**(12), 1660 (2019)

27. Leung, H., Haykin, S.: The complex backpropagation algorithm. IEEE Trans. Signal Process. **39**(9), 2101–2104 (1991)
28. Tuba, M., Bacanin, N.: Hybridized bat algorithm for multi-objective radio frequency identification (RFID) network planning. In: 2015 IEEE CONGRESS on Evolutionary Computation (CEC). pp. 499–506, IEEE (2015)
29. Yang, X.S.: A new metaheuristic bat-inspired algorithm. In: González, J.R., Pelta, D.A., Cruz, C., Terrazas, G., Krasnogor, N., et al. (eds.) NICSO 2010. SCI, vol. 284, pp. 65–74. Springer, Cham (2010). https://doi.org/10.1007/978-3-642-12538-6_6
30. Karaboga, D., Akay, B.: A modified artificial bee colony (ABC) algorithm for constrained optimization problems. Appl. Soft Comput. **11**(3), 3021–3031 (2011)
31. Little, M.A., McSharry, P.E., Roberts, S.J., Costello, D.A., Moroz, I.M.: Exploiting nonlinear recurrence and fractal scaling properties for voice disorder detection. Biomed. Eng. Online **6**(1), 23 (2007)
32. Mangasarian, O.L., Wolberg, W.H.: Cancer diagnosis via linear programming. University of Wisconsin-Madison Department of Computer Sciences, Technical report (1990)
33. Wolberg, W.H., Mangasarian, O.L.: Multisurface method of pattern separation for medical diagnosis applied to breast cytology. Proc Nat. Acad. Sci. **87**(23), 9193–9196 (1990)
34. Heidari, A.A., Faris, H., Aljarah, I., Mirjalili, S.: An efficient hybrid multilayer perceptron neural network with grasshopper optimization. Soft Comput. **23**(17), 7941–7958 (2019)

Architectural Design Optimization: Not an Usual Optimization Process

Elena Simona Nicoară$^{(\boxtimes)}$

Petroleum-Gas University in Ploiesti, 100680 Ploiesti, PH, Romania
snicoara@upg-ploiesti.ro

Abstract. Whereas optimization dowry is vastly and efficiently applied to many areas, both on academic and practice level, it had a poor influence on the architectural design practice. The complexity of such problems goes beyond the large number of constraints, decision variables and objective functions and beyond the difficulty of an accurate quantification of the customer's intentions. Not all the objectives are known in advance, the design progresses by incorporating constraints and objectives in stages, there is a continuous co-evolution between the problem formulation and the solution space and so on. All these characteristics made up of architectural design an interesting and challenging field of study through optimization glasses. In search of a better design (achieved by an architect), the constant seems to be a continuous travelling on different type problem-formulation spaces, which are vast, complex and significantly interdependent one with each other. The perspectives of classical optimization and of architectural design prove to be very different; therefore special approaches, such as machine learning and other artificial intelligence techniques are more appropriate to tackle architectural design, even if the results obtained so far are still limited in performance. The paper integrates the course of action made by researchers and practitioners in finding adequate approaches for modeling and solving architectural design optimization. By that, it constitutes an interesting learning experience with unusual optimization contexts.

Keywords: Architectural design · Optimization · Machine learning

1 Introduction

The vast constellation of optimization algorithms and models (with or without a solid mathematical background) and general or dedicated software tools allowed many different fertile directions for optimization in many real contexts, both for the theoreticians and for the practitioners. Also benchmark problems and test datasets are available for various fields: scheduling, bin packing, vehicle routing, network flow, quadratic assignment, travelling salesman etc.

On the other hand, there are still areas where computer-based optimization had a shallow practical influence. Such an area is architectural design (AD) [1], where a significantly smaller number of research studies were conducted, in contrast with engineering design for example. In this paper the reasons for that are investigated. We examine the differences between architectural design context and other fields' contexts

© Springer Nature Switzerland AG 2021
D. Simian and L. F. Stoica (Eds.): MDIS 2020, CCIS 1341, pp. 67–80, 2021.
https://doi.org/10.1007/978-3-030-68527-0_5

in order to understand why architectural design optimization (ADO) does not equally beneficiate from the optimization assets. We primarily have in mind the problem formulation (variables to be optimized, objectives plus constraints) and multi-objective optimization paradigms. The goal of this research consists in analyzing the course of action made by researchers and professionals in finding proper modeling and solving approaches for an unusual optimization context. Hence, the students and practitioners in optimization (and architecture) are the primary beneficiaries of such research, by following particular optimization challenges and their corresponding solutions.

In the '70s, architectural design problems were seen as strange or wicked problems firstly because their ill-structuring hallmark [2]. To note that this perspective belongs to researchers or scientists or even practitioners who aim to structure in certain common ways a context in order to tackle it. The vast majority of optimization models and optimization solving methods were designed having these "common ways" in mind. The human mind, generally, was used to follow the same paths, the same thinking direction, regarding also structuring. Such a perspective proves to be very useful in other areas, but the "wicked" architectural design may reveal novel approaches of optimization.

All researchers in AD point to the specificity of the area, which lead to the necessity to differently approach the optimization. Among the *characteristics of AD*, we mention: the existence of a large number of constraints of many types (geometric, dimensional, topology and performance constraints), large number of variables and vast number of objectives, the difficulty of translating the qualitative intentions of the customer into quantitative objectives, the difficulty to accurately merge objectives, the fact that many usual objectives are contained in other objectives, the need to incorporate the constraints and objectives step by step (because they are not known in advance but they emerge during optimization), the need to simulate solutions during optimization process, multimodality, irregularity of many objective functions. Moreover, a clear distinction between constraints and objectives on one hand and between variables and objectives on the other is determinant for the solution quality, and there is a continuous co-evolution between problem formulation and solution space [3–6].

The complexity of AD problems surpasses the large number of constraints (of many types: geometric, dimensional, topological and performance constraints), variables and objectives and the difficulty of accurate quantification of the client's requirements [2, 7]. Moreover, a clear distinction between constraints and objectives on one hand and between variables and objectives on the other is determinant for the solution quality. In architectural design often such kind of misinterpretations of problem statement are reported. In some cases, for example, the building structure have to be the variable to be optimized, in other cases the same building structure have to be seen as objective or even constraint. In addition, a continuous co-evolution between the problem formulation and the solution space occurs in AD.

To restrain the field of action for design search, some few major initial objectives are often considered (both by the customer and the architect) and a small set of variables to optimize these objectives. Present challenges in the architectural design focus on energy saving, cost-effectiveness, improving safety and occupancy comfort, and health. Therefore, the most used objectives are related to building energy performance, useful daylight illuminance, daylight glare probability, safety (including carbon emissions), geometry rationalization, design sustainability, structure, layout, ventilation, wind loads, life cycle cost and indoor environmental quality [2, 3, 7–12]. Nevertheless, complete formulations for design objectives are near impossible to be made, as [13] states.

Among the variables to be optimized we mention: structure, daylight availability, general shape for building code/regulations, circulation, layouts, views, dimensions, aspect ratio, relations between spaces and orientation [1, 3, 7].

As for constraints, area, aspect ratio, orientation, relations between spaces, dimensions, views, energy performance, fire safety regulations, budget, acoustics, wind turbulence, accessibility for the disabled, building use, comfort, sunlight penetration, accessibility, privacy, and structure could be used [2, 3, 7, 10]. In architectural design the constraints are limitative, but while the design process advances and the design becomes clearer, they became guiding tools for the designer [2].

Particular design problems expressed by architects are [1]:

- layout optimization considering the user-defined site adjacencies;
- design for shading devices with minimum material used and constraints regarding solar radiation, building structure and manufacturing aspects;
- planarization of paneling on complex surfaces with minimal cost, considering various types of fabrication constraints;
- structural form-finding;
- thermal and electrical capacity management, related to the building shape and building occupancy.

Because the designers' perspective is not centered on the structure of variables, constraints and objectives, we will not find a taxonomy for design problems as it is in other areas. Nevertheless, in the research studies and the architecture courses we find somewhat frequently architectural design problems such as:

- architectural layout design problem: finding the best adjacencies between functional spaces under given constraints;
- building energy consumption problem;
- structural design problem (with reducing structural weight);
- sustainable building design problem, which include energy consumption and structural weight;
- floor plan design problem (Fig. 1 shows a solution of such problem);
- shape optimization and so on.

From the common perspective of an optimization algorithm designer/user, one will attempt to find correspondence with known optimization problems, such as: the square packing, stock cutting, bin packing, convex hull, Voronoi diagrams, facility location, general assignment and various graph algorithms. Optimization (as developed so far) is a tool for AD rather than a tool analogue with architectural design [13].

As [7] notice, from the architect's point of view, the design concept is regularly fixed and optimization is required only for improving a certain performance on this concept. The demand addressed to computer science community is for modeling tools necessary for the initial design stage.

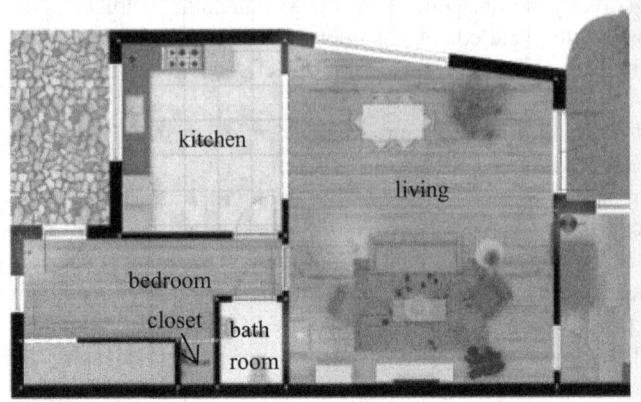

Fig. 1. Example of solution for a floor plan design problem [14]

A problematic integration of optimization with architectural design is obvious. Wortmann [3] motivates this by the limited application of parametric design and performance simulations, the lack of knowledge and incentives for black-box optimization and inefficient optimization methods in the literature.

The paper presents in the following the multi-objectivity handling in AD problem. Existing modeling and solving techniques for AD problems are synthetized next, in correlation with software tools preponderantly used by practitioners and researchers. For this, results of two surveys [1, 10] conducted on architectural designers and computer scientists acting in AD area were used.

2   Multi-objectivity in ADO

Though every architectural design process is multi-objective, the classical approach of multi-objective optimization has a low popularity in ADO. [15] is considered the first paper bringing together multi-objectivity and architectural design.

The survey [1] in 2015–2016 was conducted over 165 architectural trainees and architects, from 34 countries, having 1–46 years work experience. Besides the need for multi-objective optimization tools comparing to single-optimization tools (78%) revealed by the survey, 91% of respondents expressed their need to choose promising solutions during the optimization process. Another survey [10] was held in 2016–2018 with only 18 respondents (from academia, students and professionals) who already use regularly ADO software tools. 50% of the respondents typically optimize only a single objective. Those who attempt to optimize with multiple objectives use preponderantly for this aim penalty functions and weighted sums; only a few of them apply multi-objective optimization algorithms. Respondents additionally unveiled a need for efficiency for software tools (to rapidly find good solutions), for clustering methods regarding the final different solutions, for an overview of all possible designs. Additionally, architects prefer Pareto-based optimization because it provides many alternative good solutions.

Regarding the methods to handle multi-objectivity and to evaluate candidate solutions, most published researches focused on the aggregation and on the Pareto-based methods. We consider that alternation of objectives worth to be tested, being more appropriate to ADO than the others.

Recently, [13] examines eight consensual assumptions on the multi-objective optimization appropriateness for ADO in contrast with single-objective optimization. As a conclusion of this critical analysis is that, despite MOO seems to a first view to be the most appropriate model for ADO, it is impractical to simultaneously optimize all the objectives in an architectural design problem, because they are not known in advance. Moreover, the relation between objectives could not be completely known and understood in the initial stage. Most of the objectives occur during the design, as co-evolution model of ADP shows, also from communication with the client in various stages of the design process. Other results of this analysis, related to multi-objectivity, show that:

- the needed quantification for problem formulation is difficult because the decision maker does not see the big picture, which is the Pareto front;
- the initial objectives are often included one in the other, requiring a strong analysis before design begins and during the design process develops;
- visual simulation of candidate solutions is important in AD, so more than three objectives brings a new problem in respect with this;
- inadequate approximations of Pareto fronts can lead to error for the visual designer;
- if MOO algorithms are efficient and robust is not yet provable, primarily because very few benchmark problems exist in ADO and, secondary, because small instances were employed so far.

Based on all these conclusions, using common MOO is adequate only when it is important to understand tradeoffs between conflicting objectives and tens of thousands of optimization steps are possible [13]. Also, optimization proved so far it may inform, but not replace the design made by human.

The survey [1] offers also interesting results on optimization software tools. 93% of the 165 respondents want to understand the optimization methods used in the optimization tools, 91% want interactively choose (partial) solutions during the optimization process and 54% prefer full control over the optimization procedure. 82.2% prefer to get as outcome a few high-quality solutions, not a single solution, fact that recommends population-based heuristic methods to be incorporated in such tools.

3 First and Second Generation Modeling and Solving Approaches

Due to the specificity and complexity of architectural design problems, modeling and solving procedures followed in time distinctive directions.

The common habit of architectural designers to reduce the set of objectives and constraints associated to a rigorous initial design problem in order to simplify the problem and to search the solution in a small set of manageable solutions (by generating and testing them) led to the generate-and-test model.

In 1990, the cognitive mechanisms in solving architectural design problem were investigated. Figure 2 depicts an example of knowledge base representation. The control strategy here is detected by the production rules system and the design solutions are obtained through the activation of constraints and the corresponding rules stored in memory [16].

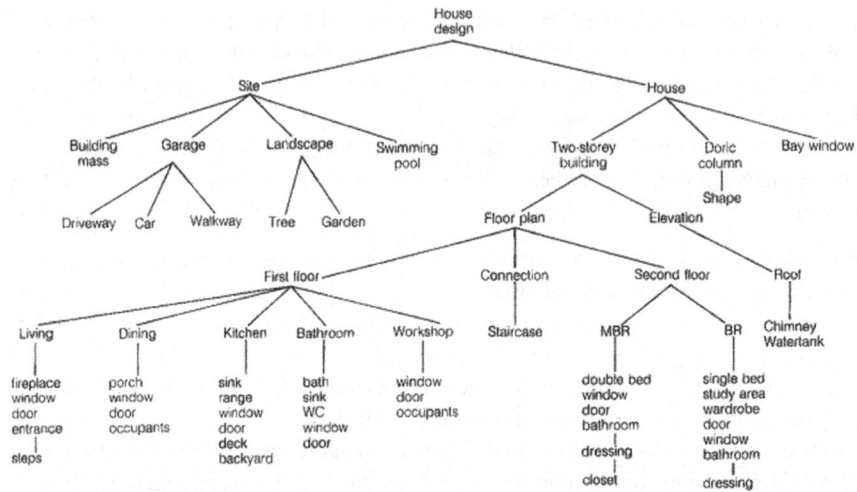

Fig. 2. Example of knowledge base representation for the cognitive model in AD [16]

As the author remarks, in the model the efficient design solutions are strongly dependent on the user's ability "to select rules in the schema and the ability to develop new schema to test newly generated design units".

A later modeling approach, co-evolution, is based on the paradigm in which the design process usually develops by an iterative co-evolution of the problem space and the solution space [8, 9], pretty much like a research project is done. Starting ideas generally significantly transforms themselves during the design process, unveiling additional objectives, which enchain other types of ideas in order to be attained. This model explains in an accurate manner the design cycles, but the complexity of implementing it led to the conclusion that it only allows a deeper understanding of the process itself [3].

In synthesis, the main conventional modeling techniques in the first stages of computational AD are: *analysis-synthesis* (associated with the cognitive model), *generate-and-test* and *co-evolution*. A hybrid model of the last two, proposed in 2018 is the performance-informed design space exploration [3].

Once metaheuristic search started to obtain significant results in many fields, also for AD this became a challenge. The first intuition of researchers produced tests with *evolutionary methods* used for single-objective and multi-objective problems. Heuristic methods focused on solving constraints, where constraints were employed as design drivers. In 1996 a *genetic algorithm* based design model for problem design exploration was proposed [8], rooting in the idea that the requirements are not known in advance. Here, the requirements are determined by the design process, making use of the genotype.

The main goal of genetic algorithms in AD is to influence the spatial structure without an impact on the visual form [17].

EvoArch, an optimization tool based on a genetic algorithm, was proposed and applied in architectural layout design problems [12]. Figure 3 depicts a solution generated using EvoArch.

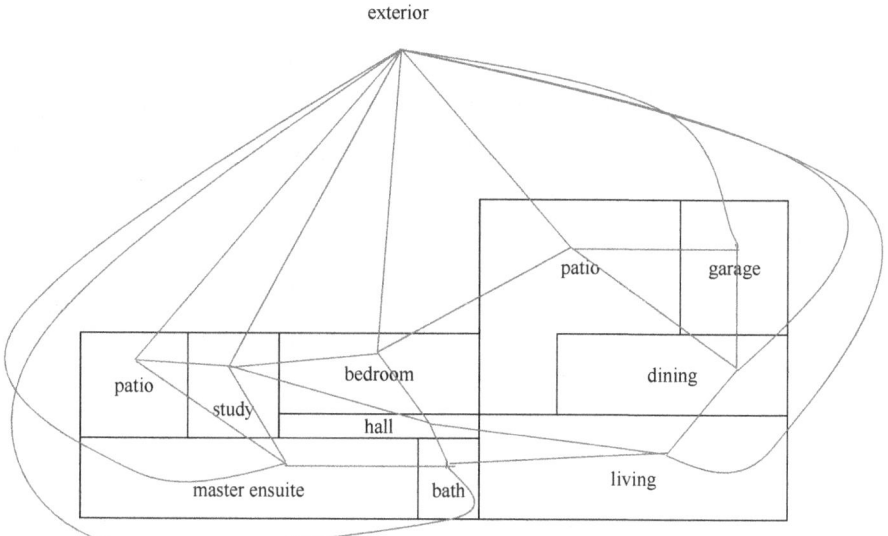

Fig. 3. Architectural space topology of a floor plan generated using EvoArch (adapted from [12])

Another genetic algorithm successfully applied on two relative small design problems generated the solutions presented in Fig. 4 [7].

Later, other metaheuristics were tested: swarm intelligence (especially Particle Swarm Optimization), simulated annealing, variable neighborhood search, covariance matrix adaptation evolution strategy (CMA-ES) [1, 4–7, 18, 19].

An example of difficult problem solved with swarm intelligence was reported in [17], where a spatial structure (an interior staircase) had to be generated to best fit between two perpendicular predefined areas so as to avoid interception with any building elements in neighborhood (see Fig. 5).

Metaheuristics are recommended only as a "last choice method" [20]. For simulation-based optimization problems, as ADO, black-box optimization methods are adequate. These methods attempt to do the best exploration of the design space, while locally exploit promising areas in that space.

74   E. S. Nicoară

Global black-box techniques are: direct search, model-based methods and meta-heuristics. Model-based methods build models of mathematical formulations of the problem, which are faster (but less accurately) generated than simulations. Through this mechanism, an approximation of the design space is done. In each iteration, the method searches a good design, simulates it and accordingly updates the model. By this, an improvement of the model during optimization is obtained. Comparing with other types of optimization techniques, these methods need additional computing steps to build and search the model at every iteration.

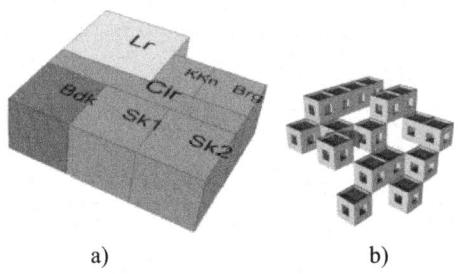

a)   b)

Fig. 4. Examples of design optimal solutions generated by a genetic algorithm: a) solution is a floor plan and constraints are area, aspect ratio, orientation and relations between spaces, b) solution is a configuration of housing units [7]

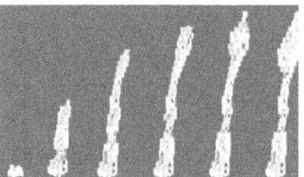

Fig. 5. Stages of form creation with swarm intelligence [17]

However, model-based optimization methods are rarely used in ADO, even if they have a good potential for optimizing building designs "based on time-intensive, structural and environmental simulations" [20]. RBFOpt, a model-based optimization algorithm, proved to be the most stable nondeterministic algorithm on two problems: one with 15 discrete variables and a single objective and one with 40 continuous variables and two conflicting objectives [20].

A comparative analysis on many black-box optimization methods showed that global methods such as RBFOpt and DIRECT (a direct search algorithm), and simulated annealing are likely to yield the best results. If constraints and/or objective functions need computationally expensive simulations, the model-based methods generally return the largest improvement in a small number of evaluations, but may be overridden by other techniques in the long run. The direct search and model-based search are the best choices if the objective function is almost convex. Metaheuristics remains a good solution if function evaluations are computationally cheap [5].

Nevertheless, simulation-based methods are more appropriate to solve ADOs that population-based metaheuristics, where tens of thousands of solutions have to be evaluated by simulation [20].

Results on applying classical optimization on ADPs show that generally *optimization may inform, but not replace the design made by human.*

In the AD process, in search of a good or better design, the constant seems to be a continuous movement on different type problem-formulation spaces, which are vast, complex and significantly interdependent one with each other. An architect passes the design process through this vast network of spaces simplifying the problem. He/she reduces the searched space by selecting in the beginning a small set of possible designs in a given solution space. Then, during the design process, some movements on other spaces are made to improve the result, based on new objectives and constraints. Experience and prior knowledge of the designer is determinant for an (initial) design solution [21]; it is known that architects always make analogy of the current situation with their past knowledge, following a certain design thinking.

Rather than try to accord the architects' perspective with the classical optimization perspective, a different encompassing of optimization had to be searched for.

4   Third Generation Modeling and Solving Approaches

The limitations of the first and second generation approaches, both in modeling and solving ADO, have various facets. [22] clearly caught them for the floor plan generation, the main and most frequent AD problem, which has an exponential growth in complexity: 1) The local search optimization methods need an expert to manually design the initial layout; 2) The shape grammars (procedural architecture), on the other hand, use static rules and they are to be manually introduced in the system; 3) The probabilistic models successfully capture abstract patterns in data-driven approaches, but the synthesizing real-world-ready scenes remains an open problem. These techniques are therefore restricted to simple and small instances.

Because structure can't be clearly understood in ADPs, *learning approaches* (where analogue data input and output are used) were proposed later, with strong focus on computational intelligence and machine learning.

Machine learning (ML) was tested for architectural design since 1990, but with shallow results and being limited in AD tools. In 2004 a learning model in design was proposed [23], where the model was based on analytical, formal representation of cognitive activities in the design. The computational tool, LinD, revealed that the design implies hidden learning and often complex patterns in the used data.

Only lately, around 2012, more favorable results were reported in the research community. An example is capitals' design generation with ML [17], see Fig. 6. The AD problem is considered *a data processing problem.* The algorithm becomes a co-designer by generating possible spatial variations of a certain form, which can be used as final solutions or as an input for a future design. The goal is to learn decision making on basis of the architects' expertize.

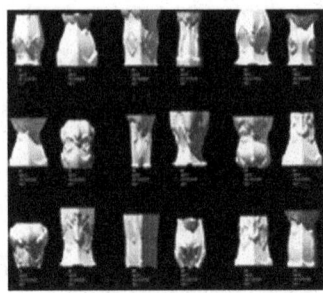

Fig. 6. Capitals automatic design with machine learning [17]

Generally, ML and *artificial neural networks* (ANNs) approaches the design as a *learning problem*. The algorithm generates possible spatial variations of the form, inputs are coordinates, surface vector, volume, outdoor illuminance etc., datasets can be samples which allow analysis of local deformations, and solution can be for example 3D variations of forms.

A 3-layer ANN framework which combines context-aware design-specific data and building performance models to improve building performance predictions during design is presented in [24]. It includes a technique for feature ranking, which allows the architect to assess the relative impact of factors on prediction and to analyze the contextual factors on human-building interaction.

ANNs and *deep learning* bring forth conceptual design [25], allowing the client's requirements and contextual information to be encoded into textual or visual descriptors. The goal is here that ANN learns mapping the features to conceptual ideas, using text-to-image or text-to-3D models synthesis.

Also computer vision, recommender systems, clustering, dimensionality reduction [25] and hybrid models are investigated and tested. *Computer vision* may employ in AD neural style transfer, where encoding the design is made either by simple sequences of images or by geometric properties. The design style may be transferred from other buildings or other architects, obtaining style variations, which could be used in datasets. *Recommender systems* may propose adequate design templates and avoid poor performance design templates, by using extrapolation, learning the frequently used operators by the architect and so on.

Regarding *hybridization,* [26] reports:

- emergent parameterization, obtained from parametric models and generative processes, which generate high volumes of design variants; and
- performance based design methodology, from the conjuction of ML and simulation, where adaptation of the ongoing behaviour is present.

The last one requires a good understanding of the interactions within the underlying structural, mechanical, thermodynamic etc. data on the behavior of the design items and, additionally, efficient computational tools for the complex models. This methodology allows designers to decipher meaning in the unknown, to find patterns or even anomalies and to take better decisions.

Even creation of *domain specific languages* from building regulations, previous pozitive and negative designs and other items is a germinal idea for ADPs.

All these ML-flavoured techniques are still incipient and limited for ADO. Reasons for that are [25]: a need of balanced training sets (which is hard to find in AD), a need for vast mature datasets and a reduced decision explanation ability.

In contrast with other professional areas, in architecture, ML expertise can not be separate from professionals [27]. Here, the interaction between the designer and the design significantly facilitates better outcomes. As a consequence, a main solution lies in architectural education, where ML is adopted by the architects and architecture students. Hence, the architects' creativity is mixed with employing neural network models. There are reported developed ML prediction models to analyze datasets and predict optimized spatial configurations for buildings [27].

While practitioners with vast experience ground in solving problems on their intuition, ML can act similarly and predict out of previous simulation results how new systems would behave. In the near future we can expect also artificial intelligence personal assistants in AD [17], which learn the designer's behavior and patterns of thinking, augmenting the solving ability and the human creativity.

5 Software Tools for ADO

Parametric design software packages frequently employed by practitioners are [1]: Grasshopper3D for Rhinoceros3D (89.9%), Dynamo for Autodesk Revit (6.6%) and Digital Project (0.7%). The preferred programming and scripting languages are, according to the same survey, Python in Grasshopper3D (28.6%), Processing (26.2%), RhinoScript in Rhinoceros3D (13.5%), C# in Grasshopper3D (12.7%), Visual Basic in Grasshopper3D (11.1%), MEL in Autodesk Maya (3.2%) and Python Script in Blender (0.7%).

Optimization tools adequate to handle a design exploration process are classified by [11] into four types: standalone, CAD-based, simulation-based and Building Information Modelling (BIM)-based tools.

The standalone class consists is generic optimization tools and comprise algorithms configurable for design problems. They may be supplemented with CAD tools using high programming knowledge. Examples are GenOpt, ModelCenter and MATLAB. As for *CAD-based tools*, the optimization is integrated into the CAD application; with basic visual programming coding, an architect may explore the design space. Galapagos in Grasshopper Rhinoceros3D is such a tool. Optimo in Dynamo is another one. *Simulation-based and BIM-based optimization tools* try to optimize a clear set of design variables while satisfy clear objectives and high-resolution building information is needed for the solution execution. No programming skills are required, the algorithm is integrated within the simulation or the BIM application. jEPlus + EA and ThermalOpt proved adequate for optimization of building system, which do not significantly modify the building form. H.D.S Beagle application, in contrast, shows a good flexibility regarding "encoding complex building forms in the massing design stage without need for programming". The drawbacks refer to a limited set of objectives provided and the need to encode the design concept as a parametric model.

Grasshopper holds various sub-tools as Galapagos, Opossum, Silvereye, Octopus and Goat. Another platforms useful to ADO are Design Space Exploration, Autodesk Refinery, DesignBuilder (including Pareto-based NSGA_II algorithm), Optimo (including NSGA_II), modeFRONTIER [5, 10], Revit and FloorPlanner [14].

As for ML tools, [26] presents:

- Libraries: TensorFlow, Keras, Caffe, CNTK, Accord.NET and
- Services: Microsoft Azure, Amazon Machine Learning and Google Cloud Platform.

6 Conclusions and Future Work

Common optimization for AD may not be successfully adopted due to the multiple structural differences of ADPs against the usual optimization areas. First of all, ADPs are multi-disciplinary; they incorporate many different solution spaces. On the other hand, large number of constraints, variables and objectives are present in ADPs, it is difficult to translate the clients' requirements into objectives, simulating solutions is necessary during optimization process and many more issues to consider. This specificity induces to ADPs a special type complexity. In this context, the challenge consists in finding optimization paradigms or algorithms which prove to be useful tools.

We motivate the problematic integration of the usual optimization perspective in AD by an incompatibility between the common perspective in the optimization community and the designer's perspective. In searching of a good or better design, the constant is a continuous moving on different type problem-formulation spaces, which are huge, complex and significantly interdependent one with each other. Rather than try to match these perspectives, which is probably unfruitful, a different encompassing of optimization will be beneficial. This should start from the idea of a perfect design process which creates and updates a dynamic structure of variables and values of them, in strong correspondence with the objectives and eventually the constraints.

The modeling techniques evolved since 1990 from the analysis-synthesis, generate-and-test and co-evolution approaches to metaheuristic paradigms (evolutionary computation, swarm intelligence, simulated annealing) and lately to machine learning (ANNs, computer vision and deep learning), recommender systems, clustering, dimensionality reduction and hybrid models. Solving problems intimately followed the modeling approaches and results were obtained, using every approach, with different efficiency.

Machine learning approaches, where the design is considered a learning problem and the goal is that the system learn decision making on basis of architects' expertize, seems at the moment to be the most appropriate course of action to be developed. The obtained results are only relative satisfactory, because AD projects are increasingly built upon vast heterogeneous datasets and models. The conjunction ML-ADO is still an incipient one, because we need balanced training sets (which is hard available in AD), mature datasets and adequate decision explanation function.

Architectural design optimization, remaining an open area both for the research community and for the practitioners, we see further research and tests in three directions. One of them is similarity-based machine learning, where task-specific similarity/dissimilarity functions and non-metric representations and data are used. The second pursues to exploit the resemblance of ADPs with computer architecture design problems, where ML showed good results and research increased fifty times in 2016–2019 comparing to 2004 [28]. Another idea is to test on AD surrogate-based optimization, where approximations for the expensive evaluations functions are to be created and used with optimization algorithms.

References

1. Cichocka, J.M., Browne, W.N., Rodriguez, E.: Optimization in the architectural practice. An international survey. In: Janssen, P., Loh, P., Raonic, A., Schnabel, M..A. (eds.) Proceedings 22nd CAADRIA Conference 2017, Hong Kong, CN, pp. 387–397 (2017)
2. Pauwels, P., Strobbe, T., Derboven, J., De Meyer, R.: Analysing the impact of constraints on decision-making by architectural designers. In: Proceedings 14th EuropIA 2014 conference on the Advances in Design Sciences and Technology, Nice, France, Architecture, City & Information Design (2014)
3. Wortmann, T.: Efficient, visual, and interactive architectural design optimization with model-based methods. Ph.D. thesis, Singapore University of Technology and Design (2018)
4. Cichocka, J.M., Browne, W.N., Rodriguez, E.: Evolutionary optimization processes as design tools. In: Proceedings 31th International PLEA Conference Architecture In (R) Evolution, Bologna, Italy (2015)
5. Wortmann, T., Nannicini, G.: Black-box optimization methods for architectural design. In: Chien, S., Choo, M.A., Schnabel, W., Nakapan, M.J., Kim, S.R. (eds.) Proceedings 21st CAADRIA Conference 2016, Hong Kong, Living Systems and Micro-Utopias: Towards Continuous Designing, pp. 177–186 (2016)
6. Vierlinger, R.: Multi objective design interface. Ph.D. thesis, Technischen Universitat Wien (2013)
7. Strobbe, T., Pauwels, P., Verstraeten, R., De Meyer, R.: Metaheuristics in architecture: using genetic algorithms for constraint solving and evaluation. In: Proceedings 14th CAADFutures Conference, Liège, Belgium (2011)
8. Maher, M.L., Poon, J.: Modeling design exploration as co-evolution. Comput.-Aided Civ. Infrastruct. Eng. 11(3), 195–209 (1996)
9. Dorst, K., Cross, N.: Creativity in the design process: co-evolution of problem– solution. Des. Stud. 22(5), 425–437 (2001)
10. Wortmann, T.: Architectural design optimization - results from a user survey. In: Architecture Across Boundaries 2019, Suzhou, China, vol. 1 (2019)
11. Chen, K.W., Choo, T.S., Norford, L.K.: Enabling algorithm-assisted architectural design exploration for computational design novices. Comput.-Aided Des. Appl. 16(2), 269–288 (2019)
12. Wong, S., Chan, K.: EvoArch: an evolutionary algorithm for architectural layout design. Comput. Aided Des. 41, 649–667 (2009)
13. Wortmann, T., Fischer, T.: Does architectural design optimization require multiple objectives? A critical analysis. In: Proceedings 25th CAADRIA Conference, Hong Kong, vol. I, pp. 365–374 (2020)
14. Floorplanner. https://floorplanner.com/. Accessed 04 Aug 2020
15. Radford, A., Gero, J.: On optimization in computer aided architectural design. Build. Environ. 15(2), 73–80 (1980)

16. Chan, C.S.: Cognitive processes in architectural design problem solving. Des. Stud. **11**(2), 60–80 (1990)
17. Cudzik, J., Radziszewski, K.: Artificial intelligence aided architectural design. In: Proceedings of the 36th eCAADe, Lodz, vol. 1 (2018)
18. Emmerich, M.T.M., Deutz, A.H.: A tutorial on multiobjective optimization: fundamentals and evolutionary methods. Nat. Comput. **17**(3), 585–609 (2018). https://doi.org/10.1007/s11047-018-9685-y
19. Rutten, D.: Galapagos: on the logic and limitations of generic solvers. Architect. Des. **83**(2), 132–135 (2013)
20. Wortmann, T.: Model-based optimization for architectural design - optimizing daylight and glare in grasshopper. Technol. Archit. + Des. **1**(2), 176–185 (2017)
21. Schön, D.: The Reflective Practitioner: How Professionals Think in Action. Basic Books, New York (1983)
22. Racec, E., Budulan, S., Vellido, A.: Computational intelligence in architectural and interior design: a state-of-the-art and outlook on the field. In: Proceedings of the 19th Catalan Conference on Artificial Intelligence (2016)
23. Sim, S.K., Duffy, A.H.B.: Evolving a model of learning in design. Res. Eng. Design **15**(1), 40–61 (2004)
24. Chokwitthaya, C., Zhu, Y., Dibiano, R., Mukhopadhyay, S.: A machine learning algorithm to improve building performance modeling during design. MethodsX **7**, 1–15 (2020)
25. Belém, C.G., Santos, L., Leitão, A.M.: On the impact of machine learning architecture without architects? In: Proceedings of CAAD Futures 2019, Daejon, South Korea (2019)
26. Tamke, M., Nicholas, P., Zwierzycki, M.: Machine learning for architectural design: practices and infrastructure. Int. J. Archit. Comput. **16**(2), 123–143 (2018)
27. Khean, N., Fabbri, A., Haeusler, M.H.: Learning machine learning as an architect, how to? In: Proceedings of the 36th eCAADe, Lodz, vol. 1 (2018)
28. Penney, D.D., Chen, L.: A survey of machine learning applied to computer architecture design. Cornell University archive. https://arxiv.org/abs/1909.12373 (2019)

Intelligent Systems for Decision Support

The Virtual Doctor: The Online Tool to Organise Unscheduled Visits for Cystic Fibrosis (CF) Patients Using Machine Learning

Aine Curran$^{(\boxtimes)}$ (ID), Tamara Vagg (ID), and Sabin Tabirca (ID)

School of Computer Science and Information Technology, University College
Cork, Western Road, Co., Cork, Ireland
118224339@umail.ucc.ie, {tv3,tabirca}@cs.ucc.ie

Abstract. It has been predicted that there will be an influx in the number of adult Cystic Fibrosis (CF) patients over the next five years. This increase will add further strain to the currently overworked system. Patients with CF are required to meet with their Multidisciplinary Team (MDT) approximately every four months. If a patient has any health-related concerns in-between these appointments, they need to undergo a phone evaluation. These phone calls can be sporadic and require a member of the MDT to be on hand causing a strain on resources and time. The Virtual Doctor project aims to reduce this effect by creating an online evaluation service, where patients interact with the Virtual Doctor using speech recognition and speech synthesis technologies within their own web browser. The Virtual Doctor will work by generating a report from the patient's visit which is then evaluated and sent to the MDT to action. The patient will also be given feedback from the visit which outlines when the team will be in contact. This paper will discuss the design requirements that need to be met for both the CF patient and the MDT, as well as the implementation of the Virtual Doctor tool. The tool has also been placed under a pilot study with 12 participants, where their experience using the tool was rated. The overall results were positive which shows that this tool may benefit both patients and medical staff allowing patients to learn about their own condition and health status while improving time and resource management.

Keywords: Virtual Doctor · 3D avatar · Machine learning · Text-To-Speech · Speech recognition · Interactive

1 Introduction

Cystic fibrosis (CF) is a hereditary disease, caused by a gene deficiency. A major symptom of CF is the production of thick viscid mucus. This mucus can cause damage to the patients organs and can give rise to intestinal obstruction, failure to thrive, and recurrent infections of the lungs [1]. Life expectancy for CF patients is lower than the average person, which can range from their teenage years to their 50s, depending on how well managed their disease is. CF Ireland is an organisation set up in the 1960s by parents of CF patients to improve the treatment and facilities for people with CF in Ireland. According to the organisation, there is approximately 1,300 children and adults

© Springer Nature Switzerland AG 2021
D. Simian and L. F. Stoica (Eds.): MDIS 2020, CCIS 1341, pp. 83–97, 2021.
https://doi.org/10.1007/978-3-030-68527-0_6

living with CF in Ireland, making Ireland the highest rated in the world. For CF cases per capita [2]. It has also been predicted that there will be an influx in the number of CF patients by 2025, with an increase of 20% for paediatrics and by 75% for adults [3].

In Ireland, CF patients are required to visit specialised CF centres every three-four months. These centres contain their own Multidisciplinary Team (MDT), consisting of clinical nurse specialists, dieticians, respiratory consultants, respiratory technicians, medical social worker, clinical psychologist and physiotherapists [4]. Patients are required to contact the MDT via phone if they have any issues regarding their health in between the scheduled visits. They are required to take an evaluation via the phone before being referred for an unscheduled appointment. These calls can be sporadic in nature and may overwhelm the MDT staff and their resources. With the expected increase in CF patients, there has never been a more crucial time to come up with a new evaluation system that patients can access from their own home.

This research aims to create a web-based 'Virtual Doctor' for patients suffering from CF to interact with via speech. The Virtual Doctor will generate an evaluation of the patient's current condition which will then be transferred to the MDT for further evaluation. The main of the purpose system is to get CF patients to be able to receive treatment more quickly by improving the efficiency of the unscheduled appointments. The research hopes to achieve this by developing and designing a Virtual Doctor to the specifications of CF patients and their MDT. This project was done in collaboration with the Adult CF unit at Cork University Hospital. Throughout the project, several meetings were arranged with the MDT, ensuring all medical information was reviewed and validated by the MDT.

Within this paper, there will be a literature review that will examine other projects and tools that are currently using similar technologies within the medical field as well as critique existing works. It will also outline a study on the acceptance of avatars to the user and what are the factors needed to create successful Virtual Doctor tool. The next section will discuss the design of the tool, including the requirements needed for the patient and the user, as well as the technical requirements. Followed by the implementation of the Virtual Doctor and discussion of the technologies used to develop the tool. Finally, an evaluation of the tool will be shown in detail from the beginning stages of development to the final product evaluation.

1.1   Literature Review: Remote Healthcare

Inhealthcare [5]. Due to the ageing population, lack of funding and higher standards in the healthcare system, there is enormous pressure in developed countries' healthcare systems [6]. There has never been a more important time for innovative solutions to improve their efficiency guaranteeing that patients receive treatment faster without reducing the quality of care. While remote healthcare is not yet wide-spread, some organisations are turning to digital health solutions, as it is low cost and efficiencies which can be delivered at scale. In recent years the National Health Service (NHS) has implemented the platform Inhealthcare [7], a digital health platform that specialises in providing care to patients from their own home.

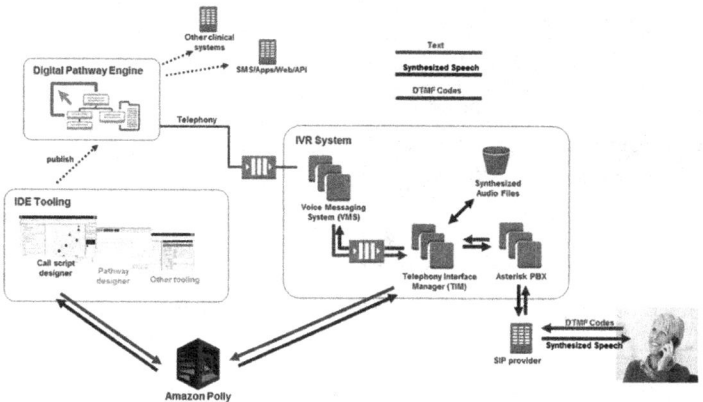

Fig. 1. Inhealthcare's architecture [7]

The Inhealthcare platform is an automated telephony system that works alongside Amazon Polly, which generates the synthesized speech that is then streamed down the telephone line. It is built using established healthcare software systems and enables clinical protocols/pathways to be modelled, developed, tested and executed. Inhealthcare's architecture can be seen in Fig. 1. The platform works by using a digital pathway engine to automatically generate remote communications. The integrated development environment (IDE) is used to build clinical pathways and protocols, which can be then passed to the digital pathway engine. The automated telephone calls are constructed from an element within the IDE known as the call script designer. To manage these calls a Voice Messaging System (VMS) is used to decipher the call script, manage the data of a phone call and return the data to the digital pathway engine. Information is then received as synthesized speech generated from Amazon Polly, and the caller responds via key pressing on their telephone keypad [7].

1.2 Literature Review: Avatars in Healthcare

While Inhealthcare is a huge advancement in healthcare it lacks some considerations, for example, patients hearing may be impaired due to illness so relying purely on audio may come with downfalls. The downfalls can be eliminated by providing an interface where text can be displayed along with the audio. An interface also provides the ability to have someone or something to guide patients throughout the visitation process. As avatars have become more popular over the last decade, it is not surprising that organisations are beginning to adopt the recent developments of avatars and use them for medical applications with little to no cost for the patient. Two projects that use avatars as a major design in their interfaces are SPARX [8] and Sense.ly [9] (see Fig. 2).

SPARX [8]. SPARX is an application that helps young people manage their mental health, seen on the left in Fig. 2. SPARX engages with young people through an interactive fantasy game. Patients are able to create their own avatar therapists [10].

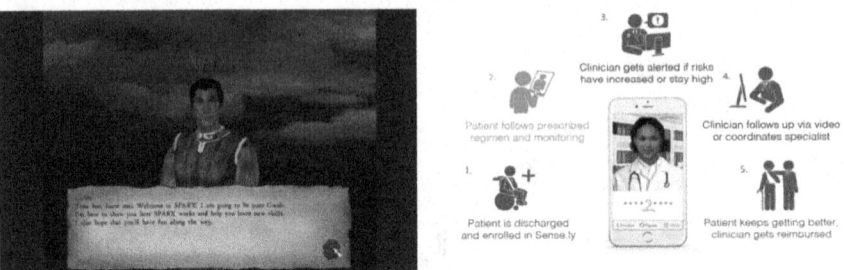

Fig. 2. Images of SPARX [8], and Sense.ly [9].

SPARX aims to deliver cognitive behavioral therapy as a treatment to manage clinical depression. The results from the studies on this project showed a reduction in depression, anxiety, and an improvement in their quality of life. If a patient's mental health does not improve by the end of the game, they will be contacted by a medical professional. The results further shown that young patients were more likely to open up about their mental health issues to a virtual therapist rather than a person present in front of them. SPARX is a self-help resource, meaning patients learn how to man-age their own mental health without any contact from the clinician throughout the treat-ment. The game provides a cost-effective, solution for young people who would not normally be able to afford or even have access to one-on-one therapy sessions [10].

Sense.ly [9]. Sense.ly is an application to help patients with chronic diseases manage and monitor their illness, seen on the right in Fig. 2. The application provides an en-gaging visitation for the patient via an assistant avatar. The virtual assistant helps patients to track their health symptoms providing both text and speech input options for the patient. Sense.ly applies clinical protocols and adopts their triage algorithms, these algorithms are used in medicine as a process to determine the urgency of the patient's needs and treatment by the severity of their condition. If the patient's condition is critical, the patient is directed to a clinic. During a visitation to the Sense.ly application a report is generated containing the patients' symptoms. The symptoms are then matched with data from the clinic's algorithm, an analysis developed, and diagnosis is made. Sense.ly offers over 20 conditions and protocols. Sense.ly's success results shows patients willingness to adopt to new technologies, lower cost and a reduction in readmissions [11].

1.3   Literature Review: Users and Avatars

It is clear medicine is willing to adapt to new technologies and use avatars to provide a more efficient level of care, however it is important to understand what makes the user accept or reject an avatar application. In recent years there has been significant research into the user's experience with avatars.

There are various considerations to be made when implementing an avatar into an application such as rendering styles, gender and realistic behaviour. McDonnell *et al.*, held a study where ten rendering styles were applied to an avatar and presented to users for evaluation [12]. These styles ranged from abstract to realistic. The results showed the users favoured the more realistic avatar and the more abstract avatar. While the avatars that were within the middle ranges were rejected by the user. It is thought that middle range characters can take users too much cognitive energy to categorize due to their unfamiliar appearance [12].

Garau *et al.,* held a study where users were to rate their acceptance of avatars that were basic and non-gender specific versus more realistic and gender specific [13]. It was found that the user's accepted the realistic avatars over basic non-gender characters. The avatars were then given the functionality of random eye gaze and informed eye gaze. The results of this study showed that the basic avatar with the random eye gaze was the least preferred by users and the higher realism avatar with informed eye gaze was favoured overall. This indicates that the user's acceptance of an avatar is not just based on app appearance but also behaviour. These factors needed to be considered when developing the Virtual Doctor as it is important to get the patients acceptance [13].

McDonnell et al. evaluation, carried out a second round of experiments [12]. The subjects were asked to try to detect whether their avatars were telling the truth or lying. They presented avatars with pre-recorded truths and lies from two human actors to the subjects as well as videos of humans telling truths and lies. It was found that there was no difference in the scoring between the virtual avatars or with real video interactions. This shows that people interact with virtual avatars in the same way as real humans, people rely on certain cues in the human voice, or in the audio track. Meaning speech of the Virtual Doctor may be one of the most important factors when developing this tool [12].

1.4   Literature Review: Critical Points

From the literature review Sect. 1.2 and 1.3, it is clear that the current healthcare system is under pressure and innovative solutions are required. The NHS are currently looking at new innovations within healthcare system as seen in the digital health platform, Inhealthcare. While this platform may have its issues, it is currently guaranteeing that patients receive treatment faster without reducing the quality of care, and shows patients are seeking alternative solutions that offer remote healthcare services. The SPARX and Sense.ly applications show that patients can engage with virtual avatars, with regards to their health. It was found that patients were more willing to share personal information with these avatars, as seen from the SPARX application. In addition to this, virtual-medical-avatars can be used as a cost-effective solution for both patients and the healthcare system.

Furthermore, in Sect. 1.3, it is determined which rendering style is best for user acceptance and that 'realism' isn't enough for users to fully invest in virtual avatars. Some key notes which are identified to support the design of the proposed Virtual Doctor are: firstly, gender specific avatars improve patient acceptance when compared to non-gender characters. Secondly, having elements of human behaviour can improve the acceptability of the avatar. Finally, the speech synthesis is an important aspect to the avatar as users may rely on cues in the speech more so than the visual representation.

2 Design

2.1 Application Design and Requirements

Virtual Doctor Interface. As the Virtual Doctor will be dealing with personal information, a login system is required. Before accessing the Virtual Doctor, the patient will be greeted with a login page. Upon logging in the patient will enter the Virtual Doctor homepage, where the doctor avatar awaits within a 3D scene of a CF clinic room. The doctor avatar will be programmed to talk to the patient. It will also exhibit human behaviours like blinking and eye movement. A panel will contain text with questions of the avatar, having the text with the audio will aid the user in understanding what the doctor is asking.

There will also be several other panels present on the homepage each serving a different purpose. The first panel will be a text box with a submit button, this text box will serve as a text input option for the patients to answer the doctor's questions. The next panel that will be presented to the client on the homepage is the information panel. This panel helps guide the patient through the questions while giving suggestions to the type of answers the doctor expecting. For example, when the Virtual Doctor asks the patient 'how long have you been having this issue?', a list of suggestions will appear on the screen such as 24 h, less than a week and over a week. Giving the patient some direction on how to answer the questions. At the start of the visit, this panel will have a short guild on how to use the web application as well as the start button so patients can start the questionnaire in their own time. The last panel will be a 'You say' panel. It is important to consider speech recognition failures and to understand what went wrong. This panel displays the patient's input answers so they can correct themselves if the answers were misinterpreted by the Virtual Doctor.

Speech Synthesis. A major consideration for this project is the communication between the patient and Virtual Doctor. This requires a voice for the Virtual Doctor to be generated, this was achieved by using Text-To-Speech (TTS) or Speech Synthesis. Speech Synthesis artificially simulates human speech from text through a computer. The text goes through various stages during the speech synthesis process. The first stage is pre-processing, which eliminates the ambiguity around how words need to be read. It also handles homographs, which are words that are spelt the same but not necessarily pronounced the same and have different meanings. The second stage is text normalisation, this is where the text is converted into expressions suitable for speech. An example of this is the conversion of '10%' to 'ten percent' or '½' would convert into 'half'. These expressions are then converted into units of sound called phonemes, which are the smallest unit of speech distinguishing one word from another. Once the first two stages are complete the next stage involves the computer using the phonemes to convert text into a sequence of sounds. The final stage is waveform production, which comprises sound generation techniques and human recordings being used to mimic the human voice and read out the complete text [14]. The out-put speech from the Speech Synthesis will then need to be lip-synced with the mouth animations of the 3D Virtual Doctor avatar.

Speech Recognition. Following this, the Virtual Doctor will 'listen' to the patient response and be able to document the patient's answers using Speech Recognition. Speech Recognition works by converting physical sound to an electrical signal using a microphone. It then gets converted to digital data with an analogue-to-digital converter. Several models can be applied to the digital data to transcribe the audio to text. Most modern speech recognition systems rely on what is known as a Hidden Markov Model (HMM). The HMM works by breaking down the speech signal into roughly ten-millisecond fragments. Each fragment is then mapped to a vector of real numbers. A sequence of these vectors is used for the final output of the HMM and are matched to correlating phonemes to decode the speech into text. Finally, algorithms are used to predict the word that produces the given sequence of phonemes [15]. The text generated from the speech recognition will be documented and stored in the database. The design of the flow of this project can be seen below in Fig. 3.

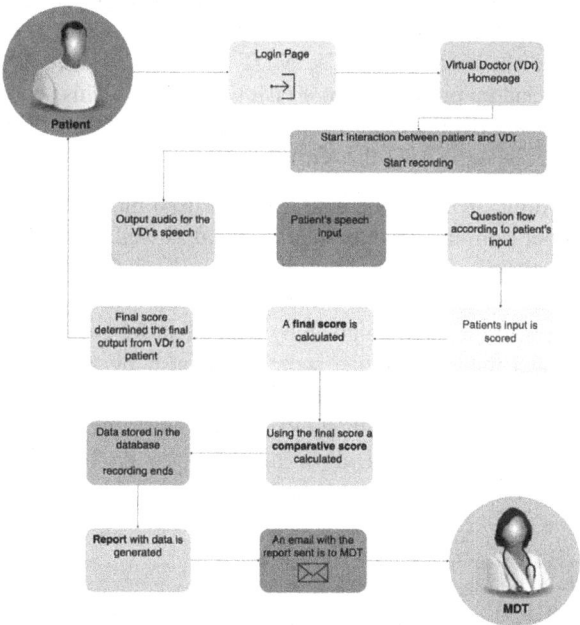

Fig. 3. Design flow of the proposed Virtual Doctor system.

2.2 MDT's Requirements

The Questionnaire. The MDT required a particular question to be asked in a set order. These questions were validated by CF clinicians and the MDT and needed to be categorised into groups, so the MDT can evaluate the patient's condition. The categories are chest, sinus, gastrointestinal and other. Once the patient identifies the issue, the questions branch out into another subsection of categories gathering more details on the said issue for the MDT to evaluate, for example, the first chosen issue may be chest, within the chest section are subsections such as cough, shortness of breath and chest pain. Once the issue is identified fully a series of questions will be output to the patient and the patient can answer generating the input. Patients will have the option to go back and select another subsection if needed. If they have no further issues within the selected category, then they will be required to answer the questions from the additional symptoms branch. Once the questionnaire is finished the Virtual Doctor needs to indicate to the patient how long they can expect to wait until they hear from the MDT which is generated from the scoring system.

Scoring. The MDT also required a system to evaluate the rate of negative change in a patient's condition for this a scoring system will be implemented. The severity score generated from the responses given by the patient to each of the questions will be used as a comparison against 'baseline score', a score based on how the patient would normally feel, resulting in the 'comparison score'. This new score represents the rate of change in the patient's condition which will be forwarded to the MDT along with the severity score and patient's answers. For example, if the patient receives a severity score and a higher comparison score. The MDT will be alerted that this patient may need attention soon and be prioritised for a visit to the MDT. This scoring system was developed and validated over several meetings with the CF MDT. The severity score determines the final response from the Virtual Doctor.

Report and Recording. To store the data gathered from the patients visit, a database is required. The input from the patient's visit is stored as well as baseline data that will be used as a comparison to the data from the visits. From the data gathered a report needs to be generated for the MDT to evaluate via email. This report will contain the user's details along with the answers given by the patient for each question along with the severity score and the comparative score. Each visit by a patient to the Virtual Doctor will also be recorded from the start of the questionnaire to the end. The MDT will be able to sign into their own Virtual Doctors account where they can access these recordings via Webpage. This will allow the MDT to go back and listen to the recording if ever an issue may arise, such as the Virtual Doctor misinterpreting the patient.

3 Implementation

Before entering the Virtual Doctor's, homepage patients are greeted by a login screen which requires patients to enter their username and password. This was developed using Node.js [16], Express [17] and MySQL [18]. The information entered by the patients gets checked against data stored in a MySQL database. If the details entered are correct the patient gets redirected to the Virtual Doctor's homepage or if they are entered wrong the patient is presented with an error message (Fig. 4).

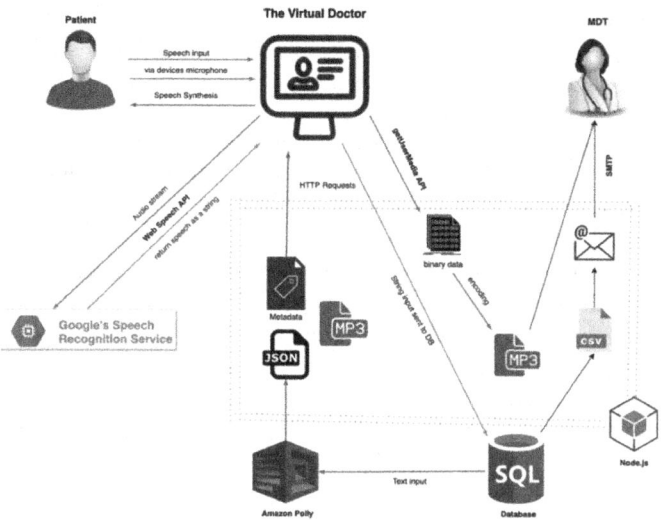

Fig. 4. The Virtual Doctor's architecture.

3.1 Avatar Doctor

The doctor avatar was created using a model in the format of SEA3D [19] and implemented into the 3D rendering environment Three.js as the foundation [20]. The Virtual Doctor was developed from a project by Paul Hodgson [21], where he used a female model that contained several animations that represent human visemes, which are the mouth shapes and the positions of the face that humans have when speaking. These animations were then used with a synchronized speech audio stream created using the Amazon Polly TTS API [22].

Amazon Polly is a web service that allows developers to create a human-like speech from text input. The text can either be written in a plain text format. Amazon Polly uses deep learning and neural technology to synthesize speech to create life-like speech. It creates the neural voices using a sequence-to-sequence model which aims to map a fixed-length input with a fixed-length output where the lengths of the input and output may be different. The synthesized speech output is available in several formats, for this project the text to be converted into the MP3 format. The major benefit of using Amazon Polly for this project is that it can break down the text input into metadata using speech marks. Speech marks are used to describe the speech that is being synthesized, they contain the visemes that make up the sentence, the start and end of words in a sentence. Using the speech marks along with the synthesized speech audio stream provides the ability to synchronize speech with facial animation such as lip-syncing, which was used to create the doctor avatar for this project.

3.2 Speech Recognition

For the patient's input a Speech Recognition API was required, for this, the Web Speech API was used [23]. The Web Speech API is event-based and handles all communication to Google's web-based speech recognition service via a HTTPS POST request meaning developers do not directly interact with the web service but rather communicate through events, therefore eliminating the need for any Natural Language Processing (NLP) packages.

The Speech Recognition API works by accessing the device's microphone which receives the speech input of the patient. The API allows JavaScript to have access to the browser's audio stream and convert it to text. Once the speech recognition service is successful in receiving a result the transcript property is set to get a string containing the individual recognised results. These string results are then used to control the flow of the conversation between the Virtual Doctor and the patient. To achieve this the string is searched for keywords using JavaScript methods and regular expression. If the keywords are found, they determine the flow of the conversation along with other methods that will be discussed further on in this paper.

As the API is an event-based architecture allows programs to asynchronously process speech. However, for this project the speech recognition needs to start after each question is asked by the Virtual Doctor and progress to the next question once the recognition is complete. Therefore, this project needed to have a way to synchronize the operations. To do this a promise is applied to the speech recognition API's code. If no error occurs when the promise is resolved. However, if the error event is triggered then it is rejected. When the final results are available or rejected the promise created will be resolved.

3.3 Control Question Flow

The questions need to be presented to the patient in the flow that was required by the MDT. Firstly, an event listener is added to listen out for when the audio stops playing. Once the audio ends the speech recognition API starts. The flow of conversation is controlled using two different methods. The first method calls the audio file along with the visemes. It does this by passing the ID of the question which corresponds to the ID of these files. This method is deployed once a keyword has been matched to the patient's input. For example, if the patient says 'sinuses' the ID input for this method will be sinuses which can correspond to elements in an array. This calls the audio and visemes files with the ID sinuses and will output the synthesized speech and animations that create the interpretation of the Virtual Doctor saying 'You are here because of your sinuses'. This example will move to the next question once the sinuses sentence is finished. It does this by identifying the last sentence said where the ID is sinuses once that condition is met the next question/sentence will be presented to the patient.

The second method used to control the flow of the conversation relies on the order of the question in the database, as a lot of the questions are one after the other. A switch statement was used along with the index of the questions from the database. A 'key', which is assigned to a questions ID, is used as the switch expression which is evaluated once. The 'key' is then compared with the values of each case which are the question IDs from the database. If there is a match, the associated block of code is executed. Each function has its own condition that needs to be met. To move to the next question in the database the questions index gets incremented. This loops through the questions and only stops once the condition has been met. Once true the loop breaks and the results from the visit are sent to the database.

3.4   Scoring and Report

To generate the scoring system a variable named 'score' is used. The variable may get incremented depending on the patient's input. The final score is used to represent the patient's current health condition as well as determine the final output of the Virtual Doctor to the patient. It is also used to compare against the baseline score which generates a comparative score. The patient's input and scores from the visitation are stored to a MySQL database, once the series of questions is complete the data gets stored in the database. A report is then created by retrieving the data from the database and saving the file as a Comma-Separated Values (CSV) file using the Node module File System [24]. The CSV file/report is then sent to the MDT using the Node module Nodemailer [25].

To obtain the recording from the visitation the JavaScript Web API MediaDevices [26] as used which provides access to media devices such as the microphone. The getUserMedia API handles streaming from users' microphones, however, the streamed audio is stored as a binary format. This project requires the audio to be saved as an external MP3 file to do this; the binary data needs to be sent to the server and then encoded using the Busboy NPM [27] and FFmpeg NPM [28]. Busboy NPM is used to save all incoming files to disk. Once the file is saved node's FFmpeg library was used to convert the binary file to mp3 format. The fnExtractSoundToMP3() method is a built-in class in the FFmpeg package that converts audio files to mp3. The recording for this project is set to start when the start button is pressed and at the end when the final statement is made by the Virtual Doctor. The complete Virtual Doctor tool can be seen in Fig. 5.

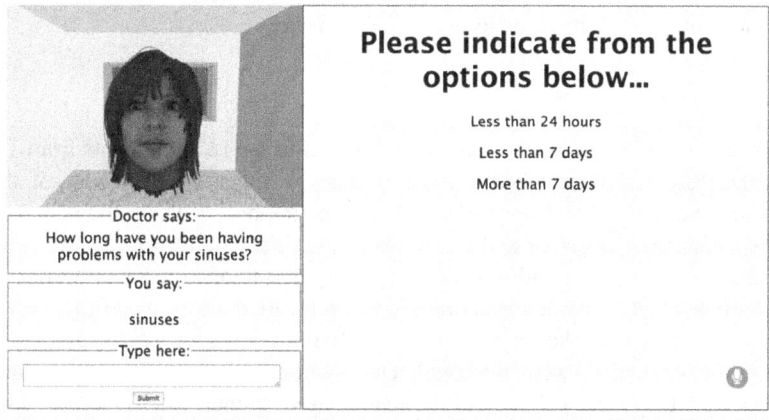

Fig. 5. The Virtual Doctor asking the patient a question from the sinus category.

4   Evaluation

It was important for this system to be tested in the early stages of development as it mainly relies on speech recognition. The speech recognition API needed to be determined as suitable for this project in the early stages or identify if a different speech recognition API may be required. This testing of a project is known as formative evaluation. The Virtual Doctor will be tested in the final stages of development using summative evaluation to check the quality of the project or if any further issues may arise. To assess the quality of the final prototype a user study will be conducted with 12 participants. The participant's satisfaction will be scored, the scores will be generated from a questionnaire once the participants visit to the Virtual Doctor is completed.

4.1   Formative Evaluation

For the formative evaluation, two females and two males, in the age range of 27–47 and originating from Ireland and Italy, were asked to run the program and were given instructions to select 'sinus branch' as their reason for visiting the Virtual Doctor. They were then instructed to continue answering the questions asked by the Virtual Doctor while being monitored. There were no issues from speech recognition API retrieving the data of the different accents, however, other issues arose from this initial testing. One issue identified was that the participants did not know when the speech recognition was occurring, meaning the users may speak too soon or too late leading to mistranslations of the users. To eliminate this problem, an indicator was added to the homepage to show the user when the speech recognition begins and ends. Given the user a better gauge to know when the appropriate time was to answer. Another issue with the prototype was that if the speech recognition was misheard the Virtual Doctor system would have to restart and the participants had to start answering questions from the beginning which they found frustrating to deal with. To solve this problem a function was added that allowed the Virtual Doctor to indicate that there was a miscommunication and ask the user to repeat their answer.

4.2   Summative Evaluation

For the summative evaluation, the project was tested using 12 participants from Trinity College Dublin, which varied in a much wider age range (17–50). Out of the 12 participants five females and seven males and demographics of the participants ranged from Irish, Japanese, Bolivian and Portuguese. Each participant was asked to run the project from the beginning and select any category and to answer the questions asked by the Virtual Doctor. Participants were given little guidance outside of the Virtual Doctor as this program should be easy to understand without the need for outside guidance, however, they were observed. Once participants had finished trialing the Virtual Doctor, they were given a questionnaire based on their experience of using the Virtual Doctor. A Likert Scale was applied to the participant's inputs to generate a scoring system. The questions included:

1. How would you rate the quality of the service?

    a. Likert Scale from 1–5 (Very high quality – very low quality)
2. I found the process of using the Virtual Doctor too Complex?

    a. Likert Scale from 1–5 (Strongly agree – strongly disagree)
3. I found the characters on the screen easy to read

    a. Likert Scale from 1–5 (Strongly agree – strongly disagree)
4. I would learn how to use this service fast

    a. Likert Scale from 1–5 (Strongly agree – strongly disagree)
5. I felt comfortable using the service

    a. Likert Scale from 1–5 (Strongly agree – strongly disagree)
6. I would prefer to use this service than talk over the phone

    a. Likert Scale from 1–5 (Strongly agree – strongly disagree)
7. I found the audio quality to be:

    a. Likert Scale from 1–5 (Very high quality – very low quality)
8. I would use the service again

    a. Likert Scale from 1–5 (Strongly agree – strongly disagree)
9. I found this service to be fast and efficient

    a. Likert Scale from 1–5 (Strongly agree – strongly disagree)
10. Do you have any other comments, questions, or concerns?

    a. Free text answer

The feedback from the participants was positive, the results show the participants were mostly satisfied with the interface, the quality of the audio and how easy the program is to use. The highest possible score from the usability survey is 45. Seven participants rated the Virtual Doctor over 38, while only one gave an overall satisfactory rating under 25 meaning the participants were satisfied using the application.

Figure 6 shows the results of each individual question. The question that gave the lowest rating were question 4, 'I would learn how to use this service fast'. While this can be a major concern for the Virtual Doctor, none of the ratings received was below a 3.5, while still scoring high this may need to be a consideration in the future development of the Virtual Doctor. Other considerations were identified by the comment second of the questionnaire. Two comments were made about the speech recognition failing to identify what was being said. However, it was found that nine out of 12 said they would prefer to use the Virtual Doctor than talk to a person over the phone, proving the Virtual Doctor reached the goals it set out in terms of usability. A factor that may need to be taken into consideration is that all these subjects were in good health and not affected by any illness. The next stage for the Virtual Doctor will be to see how well it will be received by the CF patients.

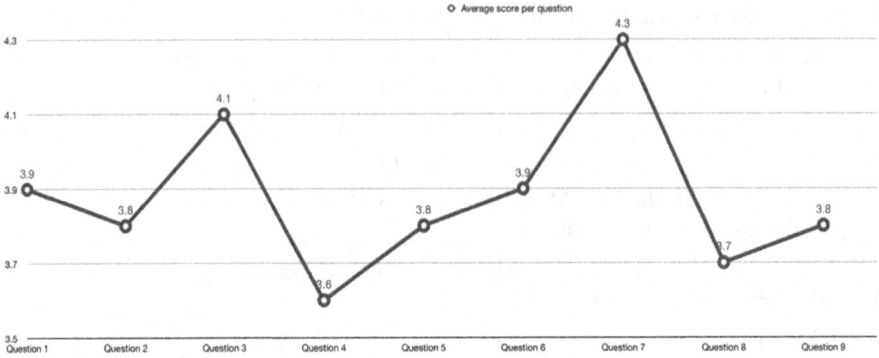

Fig. 6. Average rating per question.

5 Conclusion

Cystic Fibrosis patients need to attend clinical appointments every three-four months to help manage their condition. During this time if a patient falls ill, they must contact their MDT for a phone evaluation before organising an unscheduled visit. These visits are sporadic and can put enormous pressure on CF unit's staff and resources. There is a need to develop intelligent solutions which can assist the MDT with these unscheduled visits while also catering to the needs of the patients and the healthcare system. The Virtual Doctor aims to optimise these unscheduled appointments and give patients the opportunity to receive treatment frequently and effectively without lowering the quality of care. This research investigates the design, implementation, and preliminary evaluation of a Virtual Doctor. The Virtual Doctor is designed to engage in conversation with the patient by asking a series of questions, the patient's responses are used to evaluate the patient's current condition before reporting back to the MDT for further analyze. It is a web-based solution that allows the patient to communicate via speech or text to a virtual avatar using machine learning technology. A scoring system has been implemented which generates final output of the Virtual Doctor and determines the current health status of the patient. The Virtual Doctor has been preliminary evaluated using Usability Testing techniques. The results from these tests show the potential for the Virtual Doctor while further clinical testing is due to take place in the near future.

References

1. Cystic fibrosis. https://www.irishhealth.com/article.html?con=526. Accessed 16 Aug 2020
2. What is Cystic Fibrosis | Cystic Fibrosis. https://www.cfireland.ie/about-cf/what-is-cystic-fibrosis#. Accessed 02 Feb 2020
3. Burgel, P.-R., et al.: Future trends in cystic fibrosis demography in 34 European countries. Eur. Respir. J. **46**, 133–141 (2015)
4. Goździk, J., Majka-Sumner, L., Cofta, S., Nowicka, A., Piorunek, T., Batura-Gabryel, H.: Challenges in care of adult CF patients–the specialist cystic fibrosis team. Rocz Akad Med. Bialymst. **50**, 42–45 (2005)

5. Inhealthcare | Digital health and remote patient monitoring technology. https://www.inhealthcare.co.uk/. Accessed 16 Aug 2020
6. Peter, S.: Reforming Markets in Health Care: An Economic Perspective. McGraw-Hill Education, London (2000)
7. Using Amazon Polly to Deliver Health Care for People with Long-Term Conditions, Amazon Web Services. https://aws.amazon.com/blogs/machine-learning/using-amazon-polly-to-deliver-health-care-for-people-with-long-term-conditions/. Accessed 16 Aug 2020
8. SPARX. https://www.sparx.org.nz/. Accessed 16 Aug 2020
9. Sensely: Character-based Enterprise Virtual Assistant Platform. https://www.sensely.com/. Accessed 16 Aug 2020
10. Merry, S.N., et al.: The effectiveness of SPARX, a computerised self help intervention for adolescents seeking help for depression: randomised controlled non-inferiority trial. BMJ **344**, e2598 (2012)
11. Sense.ly Integrates Mayo Clinic's Triage Algorithms to Enhance Their Virtual Nurse App. https://hitconsultant.net/2017/05/10/sensely-integrates-mayo-clinics-triage-algorithms/. Accessed 02 Feb 2020
12. McDonnell, R., Breidt, M., Bülthoff, H.H.: Render me real?: investigating the effect of render style on the perception of animated virtual humans. ACM Trans. Graph. **31**, 1–11 (2012)
13. Garau, M., Slater, M., Vinayagamoorthy, V., Brogni, A., Steed, A., Sasse, M.A.: The impact of avatar realism and eye gaze control on perceived quality of communication in a shared immersive virtual environment. In: Proceedings of the SIGCHI Conference on Human Factors in Computing Systems, Ft. Lauderdale, Florida, USA, pp. 529–536 (2003)
14. What is Speech Synthesis? - Definition from Techopedia. https://www.techopedia.com/definition/3647/speech-synthesis. Accessed 02 Feb 2020
15. The Ultimate Guide To Speech Recognition With Python – Real Python. https://realpython.com/python-speech-recognition/. Accessed 16 Aug 2020
16. Node.js. https://nodejs.org/en/. Accessed 09 Feb 2020
17. Express - Node.js web application framework. https://expressjs.com/. Accessed 09 Feb 2020
18. MySQL. https://www.mysql.com/. Accessed 02 Feb 2020
19. SEA3D. https://sunag.github.io/sea3d/. Accessed 09 Feb 2020
20. three.js – JavaScript 3D library. https://threejs.org/. Accessed 09 Feb 2020
21. Hodgson, G.P.: https://github.com/prh1/avatar. Accessed 09 Feb 2020
22. Amazon Polly, Amazon Web Services. https://aws.amazon.com/polly/. Accessed 09 Feb 2020
23. Web Speech API. https://developer.mozilla.org/en-US/docs/Web/API/Web_Speech_API. Accessed 02 Feb 2020
24. File system | Node.js v14.8.0 Documentation. https://nodejs.org/api/fs.html. Accessed 23 Aug 2020
25. Nodemailer. https://nodemailer.com/about/. Accessed 09 Feb 2020
26. MediaDevices.getUserMedia().   https://developer.mozilla.org/en-US/docs/Web/API/MediaDevices/getUserMedia. Accessed 16 Aug 2020
27. busboy, *npm*. https://www.npmjs.com/package/busboy. Accessed 16 Aug 2020
28. ffmpeg, *npm*. https://www.npmjs.com/package/ffmpeg. Accessed 16 Aug 2020

Canonical Decomposition of Basic Belief Assignment for Decision-Making Support

Jean Dezert[1] and Florentin Smarandache[2(✉)]

[1] The French Aerospace Lab, Palaiseau, France
`jean.dezert@onera.fr`
[2] Department of Mathematics, University of New Mexico, Gallup, NM, USA
`smarand@unm.edu`

Abstract. We present a new methodology for decision-making support based on belief functions thanks to a new theoretical canonical decomposition of dichotomous basic belief assignments (BBAs) that has been developed recently. This decomposition based on proportional conflict redistribution rule no 5 (PCR5) always exists and is unique. This new PCR5-based decomposition method circumvents the exponential complexity of the direct fusion of BBAs with PCR5 rule and it allows to fuse quickly many sources of evidences. The method we propose in this paper provides both a decision and an estimation of the quality of the decision made, which is appealing for decision-making support systems.

Keywords: Decision-making · Belief functions · PCR5

1 Introduction

This paper deals with the decision-making support problem from many sources of evidence characterized by belief functions (BF) defined over a same frame of discernment. Belief functions introduced by Shafer [1] are appealing to model epistemic uncertainty. They are well-known and used in the artificial intelligence community to fuse uncertain information and to make a decision. However, many debates in scientific community started with Zadeh's criticism [2,3] - see additional references in [4] - have bloomed on the validity of Dempster's rule of combination and its counter-intuitive behavior (not only in high conflicting situations, but also in low conflicting situations as well). That is why many rules of combination have been developed by different researchers [5] (Vol. 2) over the last decades. In this work we consider only the rule based on the proportional conflict redistribution principle no 5 (PCR5 rule) to combine basic belief assignments (BBAs). This choice is motived not only by its conflict redistribution principle, but also by its ability to generate a unique canonical decomposition of any dichotomous BBA that will be convenient for decision-making from many sources of evidence.

This paper is organized as follows. After a brief recall of basics of belief functions in Sect. 2, we present succinctly the canonical decomposition of a (dichotomous) BBA in Sect. 3 based on [6]. Then we propose a new decision-making

© Springer Nature Switzerland AG 2021
D. Simian and L. F. Stoica (Eds.): MDIS 2020, CCIS 1341, pp. 98–112, 2021.
https://doi.org/10.1007/978-3-030-68527-0_7

support methodology that exploits this canonical decomposition in Sect. 4 for working in a general framework with many (non dichotomous) sources of evidences, with basic illustrative examples. Conclusions are given in Sect. 5.

2   Basics of Belief Functions

2.1   Definitions

The answer[1] of the problem under concern is supposed to belong to a given finite discrete frame of discernment (FoD) $\Theta = \{\theta_1, \theta_2, \ldots, \theta_n\}$, with $n > 1$. All elements of Θ are mutually exclusive[2]. The set of all subsets of Θ (including empty set $\emptyset$ and Θ) is the power-set of Θ denoted by 2^Θ. A Basic Belief Assignment (BBA) given by a source of evidence is defined [1] as $m(\cdot) : 2^\Theta \to [0, 1]$ satisfying $m(\emptyset) = 0$ and $\sum_{A \in 2^\Theta} m(A) = 1$. The quantity $m(A)$ is the mass of belief of A. Belief and plausibility functions are respectively defined from $m(\cdot)$ by

$$Bel(A) = \sum_{B \in 2^\Theta | B \subseteq A} m(B) \qquad (1)$$

and

$$Pl(A) = \sum_{B \in 2^\Theta | A \cap B \neq \emptyset} m(B) = 1 - Bel(\bar{A}). \qquad (2)$$

where $\bar{A}$ is the complement of A in Θ.

$Bel(A)$ and $Pl(A)$ are usually interpreted respectively as lower and upper bounds of an unknown (subjective) probability measure $P(A)$. A is called a Focal Element (FE) of $m(\cdot)$ if $m(A) > 0$. When all focal elements are singletons then $m(\cdot)$ is called a *Bayesian BBA* [1] and its corresponding $Bel(\cdot)$ function is equal to $Pl(\cdot)$ and they are homogeneous to a (subjective) probability measure $P(\cdot)$. The vacuous BBA (VBBA for short) representing a totally ignorant source is defined as[3] $m_v(\Theta) = 1$. A dogmatic BBA is a BBA such that $m(\Theta) = 0$. If $m(\Theta) > 0$ the BBA $m(\cdot)$ is nondogmatic. A simple BBA is a BBA that has at most two focal sets and one of them is Θ. A FoD is a dichotomous FoD if it has only two elements, say $\Theta = \{A, \bar{A}\}$ with $A \neq \emptyset$ and $A \neq \Theta$. A dichotomous BBA is a BBA defined over a dichotomous FoD.

2.2   PCR5 Rule of Combination

The combination of distinct sources of evidence characterized by their BBAs is done by Dempster's rule of combination in Shafer's mathematical theory of evidence [1]. The justification and behavior of Dempster's rule (corresponding to the normalized conjunctive rule) have been disputed from many counter-examples involving high and low conflicting sources (from both theoretical and

---

[1] I.e. the solution, or the decision to take.
[2] This is so-called Shafer's model of FoD [5].
[3] The complete ignorance is denoted Θ in Shafer's book [1].

practical standpoints) as reported in [4]. Many alternatives to Dempster's rule are now available [5], Vol. 2. Among them, we consider in the sequel the PCR5 rule which transfers the conflicting mass only to the elements involved in the conflict and proportionally to their individual masses, so that a more sophisticate and precise distribution is done with the PCR5 fusion process. The PCR5 rule is presented in details (with justification and examples) in [5], Vol. 2 and Vol. 3. We only briefly recall for convenience its formula for the fusion of two BBAs, which is symbolically noted as $m_{PCR5} = PCR5(m_1, m_2)$, where $PCR5(\cdot, \cdot)$ represents the PCR5 fusion rule for two BBAs. With this PCR5 rule, one has $m_{PCR5}(\emptyset) = 0$, and $\forall X \in 2^\Theta \setminus \{\emptyset\}$

$$m_{PCR5}(X) = m_{Conj}(X) + \sum_{\substack{X_2 \in 2^\Theta \\ X_2 \cap X = \emptyset}} [\frac{m_1(X)^2 m_2(X_2)}{m_1(X) + m_2(X_2)} + \frac{m_2(X)^2 m_1(X_2)}{m_2(X) + m_1(X_2)}] \quad (3)$$

where $m_{Conj}(X) = \sum_{\substack{X_1, X_2 \in 2^\Theta \\ X_1 \cap X_2 = X}} m_1(X_1) m_2(X_2)$ is the conjunctive rule, and where all denominators in (3) are different from zero. If a denominator is zero, that fraction is discarded. Extension of PCR5 for combining qualitative BBA's can be found in [5], Vols. 2 & 3. All propositions/sets are in a canonical form. A variant of PCR5, called PCR6 has been proposed by Martin and Osswald in [5], Vol. 2, for combining $s > 2$ sources. The general formulas for PCR5 and PCR6 rules are also given in [5], Vol. 2. PCR6 coincides with PCR5 when one combines two sources. The difference between PCR5 and PCR6 lies in the way the proportional conflict redistribution is done as soon as three (or more) sources are involved in the fusion.

3   Canonical Decomposition of a Dichotomous BBA

Because the canonical decomposition of a dichotomous BBA has been presented in details in [6], we only make a succinct presentation here. A FoD is a dichoto-mous FoD if it is made of only two elements, say $\Theta = \{A, \bar{A}\}$ with $A \cup \bar{A} = \Theta$ and $A \cap \bar{A} = \emptyset$. A is different from Θ and from Empty-Set because we want to work with informative FoD. A dichotomous BBA $m(\cdot) : 2^\Theta \to [0, 1]$ has the general form

$$m(A) = a, \quad m(\bar{A}) = b, \quad m(A \cup \bar{A}) = 1 - a - b \quad (4)$$

with $a, b \in [0, 1]$ and $a + b \leq 1$.

The canonical decomposition problem consists in finding the two following simpler BBAs m_p and m_c of the form

$$m_p(A) = x, \quad m_p(A \cup \bar{A}) = 1 - x \quad (5)$$

$$m_c(\bar{A}) = y, \quad m_c(A \cup \bar{A}) = 1 - y \quad (6)$$

with $(x, y) \in [0, 1] \times [0, 1]$, such that $m = Fusion(m_p, m_c)$, for a chosen rule of combination denoted by $Fusion(\cdot, \cdot)$. The simple BBA $m_p(\cdot)$ is called the *pro-BBA* (or pro-evidence) of A, and the simple BBA $m_c(\cdot)$ the *contra-BBA* (or contra-evidence) of A. The BBA $m_p(\cdot)$ is interpreted as a source of evidence providing an uncertain evidence in favor of A, whereas $m_c(\cdot)$ is interpreted as a source of evidence providing an uncertain contrary evidence about A. In [6], we proved that this decomposition always exists and is unique if we use the PCR5 fusion rule. In the vacuous BBA case when $a = 0$ and $b = 0$, the BBA $m(\cdot)$ can be interpreted as the PCR5 fusion of two degenerate pro- and contra-evidences BBAs $m_p(\cdot)$ and $m_c(\cdot)$ which coincide with the vacuous BBA with $x = 0$ and $y = 0$. Hence any (Bayesian, or non Bayesian) dichotomous BBA $m(\cdot)$ can be always interpreted as the result of the PCR5 fusion of these two (pros and cons) aspects of evidence about A. It is worth noting that this type of canonical decomposition is different of Smets' canonical decomposition problem [7] which needs to work with generalized simple BBA which are not *stricto sensu* valid BBAs as defined by Shafer [1].

For the case of dichotomous dogmatic BBA, the expression of solutions x and y of canonical decomposition are as follows [6]:

- if $a = b$ and $a + b = 1$ then $a = b = 0.5$ and $x = y = 1$;
- if $a < b$ then $x < y$, and we have $y = 1$ and $x = \frac{a+\sqrt{a^2+4a}}{2}$;
- if $a > b$ then $x > y$, and we have $x = 1$ and $y = \frac{b+\sqrt{b^2+4b}}{2}$.

For the case of dichotomous non-dogmatic BBA, the expression of solutions x and y of the canonical decomposition do not have simple analytical expression because one has to find x and y solutions of the system

$$a = x(1-y) + \frac{x^2 y}{x+y} = \frac{x^2 + xy - xy^2}{x+y} \tag{7}$$

$$b = (1-x)y + \frac{xy^2}{x+y} = \frac{y^2 + xy - x^2 y}{x+y} \tag{8}$$

under the constraints $(a,b) \in [0,1]^2$, and $0 < a+b < 1$. In fact, we have proved in [6] that $x \in [a, a+b] \subset [0,1]$ and $y \in [b, a+b] \subset [0,1]$, but the explicit expression of x and y are very complicated to obtain analytically (even with modern symbolic computing systems like Mathematica™, or Maple™) because after algebraic calculation, and for $x \neq 1$, one has to solve the following quartic equation which has at most four real solutions with only a valid one in $[a, a+b]$

$$x^4 + (-a-2)x^3 + (2a+b)x^2 + (a+b-ab-b^2)x + (-a^2-ab) = 0 \tag{9}$$

and then compute y by $y = (a+b-x)(1-x)$.

Once the numerical values are committed to a and to b the numerical (approximate) solutions x and then y can be easily obtained by a standard numerical solver. For instance, with Matlab™ we can use the `fsolve` command, and this is what we use to make the canonical decomposition of dichotomous non-dogmatic BBA.

3.1 Canonical Decompositions from Other Well-Known Rules

In [6] we did prove that this type of canonical decomposition cannot be obtained by the conjunctive rule only, because if m_p and m_c exist and if $x > 0$ and $y > 0$ then $m_{Conj}(\emptyset) = x \cdot y > 0$ which means that $m = Conj(m_p, m_c)$ is not a proper BBA as defined by Shafer's. If we use the disjunctive rule of combination we will always obtain the vacuous BBA as the result[4] of $Disj(m_p, m_c)$ because $m_p(A)m_c(\bar{A})$, $m_p(A)m_c(A \cup \bar{A})$, $m_p(A \cup \bar{A})m_c(\bar{A})$ and $m_p(A \cup \bar{A})m_c(A \cup \bar{A})$ will all be committed to the uncertainty $A \cup \bar{A}$. So for any choice of m_p and m_c we always get same result (the vacuous BBA) when using the disjunctive rule making the canonical decomposition of non vacuous dichotomous BBA m just impossible. Due to the particular simple form of BBAs $m_p(\cdot)$ and $m_c(\cdot)$, Yager's rule [8] and Dubois-Prade rule [9] coincide, and we have to search x and y in $[0, 1]$ such that $m(A) = a = x(1 - y)$ and $m(\bar{A}) = b = (1 - x)y$. Assuming[5] $y < 1$, one gets from the first equation $x = a/(1 - y)$. By replacing x by its expression in the second equation $y - xy = b$ we have to find y in $[0, 1)$ such that (after basic algebraic simplifications) $y^2 + (a - b - 1)y + b = 0$. This 2nd order equation admits one or two real solutions y_1 and y_2 if and only if the discriminant is null or positive respectively, that is if $(a-b-1)^2 - 4b \geq 0$. However this discriminant can become negative depending on the values of a and b. For instance, for $a = 0.3$ and $b = 0.6$, we have $(a - b - 1)^2 - 4b = -0.71$ which means that there is no real solution for the equation $y^2 - 1.3 \cdot y + 0.6 = 0$. Therefore, in general (that is for all possible values a and b of the BBA m), the canonical decomposition of the BBA $m(\cdot)$ cannot be obtained from Yager's and Dubois & Prade rules of combination. If we use the averaging rule, we are searching x and y in $[0, 1]$ such that $m(A) = a = (x + 0)/2$ and $m(\bar{A}) = b = (0 + y)/2$, which means that $x = 2a$ and $y = 2b$ with x and y in $[0, 1]$. So, if $a > 0.5$ or $b > 0.5$ the canonical decomposition is impossible to make with the averaging rule of combination. Therefore, in general, the averaging rule is not able to provide a canonical decomposition of the BBA $m(\cdot)$.

If we consider the canonical decomposition of a dichotomous non-dogmatic BBA $(a + b < 1)$ using Dempster's rule of combination [1], denoted $DS(m_p, m_c)$, we have to obtain x and y in $[0, 1]$ such that[6] $xy \neq 1$ and

$$m(A) = a = \frac{x(1 - y)}{1 - xy} \tag{10}$$

$$m(\bar{A}) = b = \frac{y(1 - x)}{1 - xy} \tag{11}$$

---

[4] $Disj(m_p, m_c)$ denotes symbolically the disjunctive fusion of m_p with m_c.

[5] Taking $y = 1$ would mean that $x(1 - y) = 0$ but $m(A) = a$ with $a \neq 0$ in general, so the choice of $y = 1$ is not possible.

[6] The third equality $m(A \cup \bar{A}) = 1 - a - b = \frac{(1-x)(1-y)}{1-xy}$ being redundant with (10) and (11) is useless.

with the constraints $0 < x < 1$ and $0 < y < 1$. Therefore,

$$x = \frac{a}{1-y+ay}, \quad y \neq \frac{1}{1-a} \tag{12}$$

and we solve the equation $y - xy + bxy = b$ with x expressed as function of y as above. We get the equation for $a \neq 1$

$$(a-1)y^2 + (1+b-a)y - b = 0 \tag{13}$$

whose two solutions are $y_1 = b/(1-a)$ and $y_2 = 1$ - see [6] for details.

For the case $a \neq 1$, the second "solution" $y_2 = 1$ implies $x = \frac{a}{1-y_2+ay_2} = \frac{a}{a} = 1$ which is not an acceptable solution[7] because one must have $xy \neq 1$. The solution (x, y) of the decomposition problem for $a \neq 1$ is actually given by the first solution y_1, that is

$$y = y_1 = \frac{b}{1-a} \in [0, 1) \tag{14}$$

$$x = \frac{a}{1-y+ay} = \frac{a}{1-b} \in [0, 1) \tag{15}$$

The analysis of the case $a = 1$ corresponding to the dogmatic BBA given by $m(A) = a = 1$, $m(\bar{A}) = b = 0$, $m(A \cup \bar{A}) = 1 - a - b = 0$ shows that this BBA is not canonically decomposable by Dempster's rule. Why? Because one has to solve with $0 \leq x, y \leq 1$ and $1 - xy \neq 0$ the system of equations $(x - xy)/(1 - xy) = 1$ and $(y - xy)(1 - xy) = 0$ which is satisfied for $x = 1$ and $y \in [0, 1)$, that is any value in $[0, 1)$ can be chosen for y. Similarly, for the case $(a, b) = (0, 1)$ one has to solve with $0 \leq x, y \leq 1$ and $1 - xy \neq 0$ the system of equations $(x - xy)/(1 - xy) = 0$ and $(y - xy)/(1 - xy) = 1$ which is satisfied for $y = 1$ and $x \in [0, 1)$, that is any x value in $[0, 1)$ can be chosen. Therefore one sees that for the case $(a, b) = (1, 0)$ and the case $(a, b) = (0, 1)$ there is no unique decomposition of these dogmatic BBAs from Dempster's rule of combination. More generally, any dogmatic BBA $m(A) = a$, $m(\bar{A}) = b$ with $a + b = 1$ is not decomposable from Dempster's rule of combination for the case when $(a, b) \neq (1, 0)$ and $(a, b) \neq (0, 1)$ - See Theorem 4 with its proof in [6].

In summary, the canonical decomposition based on Dempster's rule of combination is possible only for nondogmatic BBA with $0 < a < 1$, $0 < b < 1$ and $a + b < 1$ and we have $x = \frac{a}{1-b}$ and $y = \frac{b}{1-a}$. Dempster's rule does not allow to obtain a canonical decomposition if the BBA is a Bayesian (dogmatic) dichotomous BBA.

Example where Dempster's canonical Decomposition Is Possible

Consider $m(A) = a = 0.6$, $m(\bar{A}) = b = 0.2$ and $m(A \cup \bar{A}) = 1 - a - b = 0.2$. The solution (x, y) of the decomposition of $m(\cdot)$ based on Dempster's rule is

$$x' = \frac{a}{1-b} = \frac{0.6}{1-0.2} = 0.75 \quad \text{and} \quad y' = \frac{b}{1-a} = \frac{0.2}{1-0.6} = 0.50$$

---

[7] Otherwise the denominator of (10) and (11) will equal zero.

Therefore, the pro- and contra- evidential BBAs m_p and m_c are given by

$$m_p(A) = x = 0.75, \quad m_p(A \cup \bar{A}) = 1 - x = 0.25$$
$$m_c(\bar{A}) = y = 0.50, \quad m_c(A \cup \bar{A}) = 1 - y = 0.50$$

It can be verified that $DS(m_p, m_c) = m$.

If we make the PCR5-based canonical decomposition, we will obtain in this example $x \approx 0.6861$ and $y \approx 0.3628$. Therefore, the pro- and contra- evidential BBAs m_p and m_c based on the PCR5-based canonical decomposition are

$$m_p(A) = x = 0.6861, \quad m_p(A \cup \bar{A}) = 1 - x = 0.3139$$
$$m_c(\bar{A}) = y = 0.3628, \quad m_c(A \cup \bar{A}) = 1 - y = 0.6372$$

It can be verified that $PCR5(m_p, m_c) = m$.

In the case where Dempster's rule can be applied for making the canonical decomposition (that is when $a + b < 1$) we see that the canonical values (parameters) x and y can be very different from those obtained with PCR5 rule as shown in the previous example. This is normal because the principles of conflicting information redistribution of Dempster's rule and PCR5 rule are very different, and there is no link between parameters x and y obtained with Dempster's rule versus those obtained from PCR5. In PCR5 rule the conflict is a refined conflict, i.e. the conflict is split into partial conflicts, so in PCR5 the total conflict is more accurately redistributed than in Dempster's rule because each partial conflict is redistributed only to the elements involved into it, while in Dempster's rule the total conflict is redistributed to all focal elements, therefore even the elements that were not involved in the conflict receive conflicting mass, which is inaccurate.

It is worth noting that the internal conflict of m based on Dempster's rule will be in this example $xy = 0.75 \cdot 0.5 = 0.375$, whereas the internal conflict of m based on PCR5 rule will be only $xy \approx 0.6861 \cdot 0.3628 \approx 0.2489$. In fact we can attest that the internal conflict obtained from PCR5-based canonical decomposition is always lesser (or equal) to the internal conflict obtained from Dempster-based canonical decomposition. Although such claim cannot be proved algebraically[8], we can always make a fine sampling of (a, b) values in $[0, 1)$ satisfying $a + b < 1$ to evaluate numerically x and y and compare the internal conflict xy to the internal conflict, denoted $x'y' = \frac{a}{1-b} \cdot \frac{b}{1-a}$, obtained with Dempster-based canonical decomposition. In doing this we see that the difference $\Delta = x'y' - xy$ is always greater (or equal) to zero as clearly shown in Fig. 1. This means that the PCR5-based canonical decomposition is more efficient than Dempster-based canonical decomposition because it always yield pro- and contra-evidences which are less conflicting when using PCR5 rule than when using Dempster's rule, which is normal.

---

[8] Because there is no simple analytical expressions for solutions x and y of PCR5-based canonical decomposition.

Difference between internal conflicts based on Dempster's and PCR5 decompositions

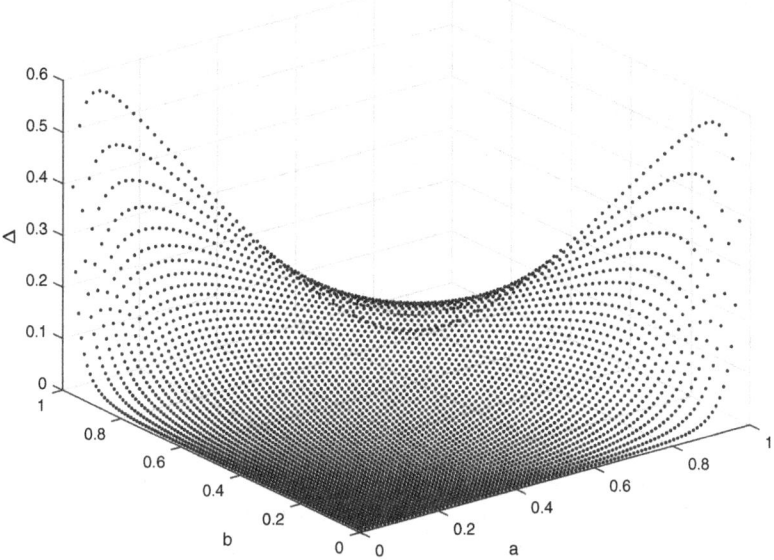

Fig. 1. Plot of $\Delta = x'y' - xy$ as function of a and b.

It is important to keep in mind that Dempster-based canonical decomposition is only possible for non-dogmatic BBAs (when $a + b < 1$) but cannot be obtained with dogmatic BBAs, whereas PCR5-based canonical decomposition works for all types of dichotomous BBAs (dogmatic and non-dogmatic ones).

3.2 Simple Example of PCR5-Based Canonical Decomposition

Let consider $m(A) = 0.3$, $m(\bar{A}) = 0.4$ and $m(A \cup \bar{A}) = 1 - m(A) - m(\bar{A}) = 0.3$, therefore $a = 0.3$ and $b = 0.4$. The quartic Eq. (9) becomes

$$x^4 - 2.3x^3 + x^2 + 0.42x - 0.21 = 0 \qquad (16)$$

The four solutions of this quartic equation are approximately[9]

$$x_1 \approx 1.5203, \qquad x_2 \approx -0.4243, \qquad x_3 \approx 0.7942, \qquad x_4 \approx 0.4099$$

One sees that x_1 and x_2 are not acceptable solutions because they do not belong to $[0, 1]$. If we take $x_3 \approx 0.7942$ then will get $y_3 = (a + b - x_3)/(1 - x_3) = (0.7 - x_3)/(1 - x_3) \approx -0.4576$. We see that $y_3 \notin [0, 1]$ and therefore the pair (x_3, y_3) cannot be a solution of the PCR5-based canonical decomposition problem for the BBA $m(\cdot)$ of this example. If we take $x_4 \approx 0.4099$ then will get

---

[9] The solutions can be easily obtained with the *roots* command of Matlab[TM].

$y_4 = (a+b-x_4)/(1-x_4) = (0.7-x_4)/(1-x_4) \approx 0.4916$ which belongs to $[0,1]$. So the pair $(x_4, y_4) \in [0,1]^2$ is the unique solution of the canonical decomposition problem. Therefore the canonical masses $m_p(\cdot)$ and $m_c(\cdot)$ are given by

$$m_p(A) \approx 0.4099, \quad m_p(A \cup \bar{A}) \approx 0.5901 \quad \text{and} \quad m_c(\bar{A}) \approx 0.4916, \quad m_c(A \cup \bar{A}) \approx 0.5084$$

It can be verified that $PCR5(m_p, m_c) = m$.

3.3 Advantages and Limitation of PCR5-Based Decomposition

The PCR5-based canonical decomposition offers the following advantages:

1. It is well justified theoretically.
2. It gives us access to the simpler pro- and contra-evidences $m_p(\cdot)$ and $m_c(\cdot)$ which are unique and always exist for any possible (dogmatic, or non-dogmatic) dichotomous BBA $m(\cdot)$.
3. It allows to define clearly the notion of internal conflict of a dichotomous source of evidence simply as $K_{int}(m) \triangleq m_p(A)m_c(\bar{A})$.
4. It always provides less conflicting pro- and contra-evidences than what we would obtain with Dempster's rule when considering non-dogmatic dichotomous BBA $m(\cdot)$. This proves the superiority of PCR5-based canonical decomposition over Dempster's-based canonical decomposition in general.
5. It allows also to adjust or revise[10] quite easily a dichotomous source of evidence (if needed) according to the knowledge one has on it by reinforcing or discounting its pro- or contra-evidential BBA.
6. It can be easily achieved with classical numerical solvers on the shelf.
7. The decomposition can be done off-line for many sampled (a, b) values at any precision we want, and stored in computer memory for working directly with $m_p(\cdot)$ and $m_c(\cdot)$ instead of making the decomposition on the fly. This is of prime importance for real-time applications where this method could be used.
8. It allows to establish efficient fast[11] suboptimal PCR5 fusion scheme, see [10] for details, examples and evaluations.

The only important limitation of this PCR5-based canonical decomposition is that it applies only to dichotomous BBAs, and it seems very difficult (maybe impossible) to use or to extend it for making directly some new canonical decomposition of non dichotomous BBAs. Because of this limitation the use of PCR5-based canonical decomposition appears, at first glance, quite restrictive for being really useful in applications involving non dichotomous BBAs. Of course in applications working with dichotomous BBAs (like those in robotics or for autonomous vehicle navigation using belief-based perception based on grid occupancy) this PCR5-based canonical decomposition may have a great interest. In fact we have already used it for belief-based inter-criteria analysis in [11] and that is why we do not present our results in this work. Nevertheless we will show in the next section how this PCR5-based canonical decomposition could be used for the decision-making support in a more general context involving many non-dichotomous BBAs. This is a problem which has not been addressed in [6].

---

[10] This point is not detailed here because is out of the scope of this paper.
[11] Where the complexity is linear with the number of dichotomous BBAs to fuse.

4   Decision-Making Using PCR5-Based Decomposition

In this section we propose a new simple general decision-making scheme based on PCR5-based canonical decomposition of dichotomous BBA. We consider $S > 2$ distinct sources of evidence characterized by their BBAs[12] $m_s^\Theta(\cdot)$ defined over the same (possibly non dichotomous) FoD $\Theta = \{\theta_1, \ldots, \theta_n\}$, with $n > 1$.

Can we exploit the PCR5-based canonical decomposition in this context to make a decision? How? We answer positively to the first question and explain in details how we can proceed. For this, we need to express the problem in the framework of dichotomous BBAs that has been presented in the previous section. More precisely, suppose one has a BBA $m^\Theta(.)$ defined on 2^Θ with $|\Theta| \geq 2$, then based on Bel and Pl formulas (1)–(2), it is always possible to calculate $Bel^\Theta(X)$ and $Pl^\Theta(X)$ for any $X \in 2^\Theta$. From $Bel^\Theta(X)$ and $Pl^\Theta(X)$ one can always build a simpler coarsened dichotomous BBA on the dichotomous (coarsened) FoD $\Theta_X \triangleq \{X, \bar{X}\}$ if $X \neq \emptyset$ and $X \neq \Theta_X$ as follows

$$m^{\Theta_X}(X) = Bel^\Theta(X) \tag{17}$$

$$m^{\Theta_X}(\bar{X}) = 1 - Pl^\Theta(X) \tag{18}$$

$$m^{\Theta_X}(X \cup \bar{X}) = Pl^\Theta(X) - Bel^\Theta(X) \tag{19}$$

Hence, $Bel^{\Theta_X}(X) = m^{\Theta_X}(X) = Bel(X)$ and $Pl^{\Theta_X}(X) = m^{\Theta_X}(X) + m^{\Theta_X}(X \cup \bar{X}) = Bel^\Theta(X) + Pl^\Theta(X) - Bel^\Theta(X) = Pl^\Theta(X)$. This dichotomous BBA $m^{\Theta_X}(\cdot)$ can always be decomposed canonically into its pro- and contra-evidences $m_p^{\Theta_X}(.)$ and $m_c^{\Theta_X}(.)$.

Therefore, instead of combining $S > 1$ non dichotomous BBAs $m_s^\Theta(.)$ for $s = 1, 2, \ldots, S$ altogether from which a decision is classically drawn, we propose to make the decision from the set of all combined coarsened BBAs relatively to each possible dichotomous frame of discernment Θ_X. Of course this decision-scheme is only suboptimal because the whole information is not processed (combined) altogether, but separately using only the coarsened (less informative) BBAs $m_s^{\Theta_X}(X)$. However, this method allows to use fast suboptimal PCR5 fusion of $m_s^{\Theta_X}(X)$ thanks to PCR5-based canonical decomposition as presented in [10] which can be applied with many (hundreds or even thousands) sources of dichotomous BBAs. With this simple suboptimal decision-scheme we can easily restrict the domain $\mathcal{D}$ on which the decisions can be made, for instance $\mathcal{D}$ can be chosen as the set of singletons of 2^Θ, or any other subset of 2^Θ depending on the application under concern as it will be shown in the next section. The generic steps of the method we propose are as follows:

- **Inputs:** BBAs $m_s^\Theta(\cdot)$, $s = 1, \ldots, S$, and the decision domain $\mathcal{D} \subset 2^\Theta$.
- **Step 1:** For $s = 1, \ldots, S$, coarsening of $m_s^\Theta(\cdot)$ into dichotomous BBA $m_s^{\Theta_X}(\cdot)$, for each $X \in \mathcal{D}$ based on (17)–(19).
- **Step 2:** For $s = 1, \ldots, S$, PCR5-based canonical decomposition of $m_s^{\Theta_X}(\cdot)$ to get pro- and contra-evidences $m_{p,s}^{\Theta_X}(\cdot)$ and $m_{c,s}^{\Theta_X}(\cdot)$.

---

[12] For clarity, we need to introduce in the notations a superscript to indicate the FoD we are working on.

- **Step 3:** Conjunctive fusion of all the pro-evidences $m_{p,s}^{\Theta_X}(\cdot)$ to get $m_p^{\Theta_X}(\cdot)$.
- **Step 4:** Conjunctive fusion of all the contra-evidences $m_{c,s}^{\Theta_X}(\cdot)$ to get $m_c^{\Theta_X}(\cdot)$.
- **Step 5:** PCR5 fusion of $m_p^{\Theta_X}(\cdot)$ with $m_c^{\Theta_X}(\cdot)$ to get $m_{PCR5}^{\Theta_X}(\cdot)$ for $X \in \mathcal{D}$.
- **Step 6:** Decision-making from the set of the combined coarsened dichotomous BBAs $\{m_{PCR5}^{\Theta_X}(\cdot), X \in \mathcal{D}\}$ to get the final decision $\hat{X} \in \mathcal{D}$.
- **Output:** the final decision $\hat{X} \in \mathcal{D}$

In steps 3 and 4 we use the conjunctive fusion because there is no conflict between all pro-evidences $m_{p,s}^{\Theta_X}(\cdot)$, and there is also no conflict between all contra-evidences $m_{c,s}^{\Theta_X}(\cdot)$, $s = 1, \ldots, S$. The steps 1 to 5 do not require high computational burden and they can be done very quickly, specially if PCR5-based decompositions have been done off-line (as they should be) [10].

We must detail a bit more the principle of the decision-making for the step 6. Actually, the decision-making for step 6 can be interpreted as a decision-making problem from a set or coarsened BBAs $m_{PCR5}^{\Theta_X}(\cdot)$ defined over different dichotomous FoD Θ_X which are all the different coarsenings of the whole (refined original) FoD Θ. In this paper we propose two methods to make the decision from the set of coarsened BBAs $\{m_{PCR5}^{\Theta_X}(\cdot), X \in \mathcal{D}\}$.

4.1   Method 1 for Step 6

This method is very simple. We take the decision $\hat{X}$ corresponding to the largest value of $m_{PCR5}^{\Theta_X}(X)$, that is

$$\hat{X} = \arg \max_{X \in \mathcal{D}}(m_{PCR5}^{\Theta_X}(X)) \tag{20}$$

If there exist several arguments having the largest value (i.e. there is a tie), we select the one whose $m_{PCR5}^{\Theta_X}(\bar{X})$ is smaller.

Example 1 (without tie): Suppose $\Theta = \{A, B, C, D, E\}$ and we want to make a decision/choice only among the elements of $\mathcal{D} = \{A, B, C\}$. Suppose after applying steps 1–5 we get the following 3 BBAs

$$m_{PCR5}^{\Theta_A}(A) = 0.3, \quad m_{PCR5}^{\Theta_A}(\bar{A}) = 0.2, \quad m_{PCR5}^{\Theta_A}(A \cup \bar{A}) = 0.5$$
$$m_{PCR5}^{\Theta_B}(B) = 0.1, \quad m_{PCR5}^{\Theta_B}(\bar{B}) = 0.5, \quad m_{PCR5}^{\Theta_B}(B \cup \bar{B}) = 0.4$$
$$m_{PCR5}^{\Theta_C}(C) = 0.4, \quad m_{PCR5}^{\Theta_C}(\bar{C}) = 0.3, \quad m_{PCR5}^{\Theta_C}(C \cup \bar{C}) = 0.3$$

The decision will be $\hat{X} = C$ because $m_{PCR5}^{\Theta_C}(C) > m_{PCR5}^{\Theta_A}(A) > m_{PCR5}^{\Theta_B}(B)$.

Example 2 (with tie). We consider same $m_{PCR5}^{\Theta_B}(.)$ and $m_{PCR5}^{\Theta_C}(.)$ as in Example 1 but $m_{PCR5}^{\Theta_A}(.)$ is given by $m_{PCR5}^{\Theta_A}(A) = 0.4$, $m_{PCR5}^{\Theta_A}(\bar{A}) = 0.2$, and $m_{PCR5}^{\Theta_A}(A \cup \bar{A}) = 0.4$. In this case, there is a tie between A and C because $m_{PCR5}^{\Theta_A}(A) = m_{PCR5}^{\Theta_C}(C) = 0.4$. But because $m_{PCR5}^{\Theta_A}(\bar{A}) < m_{PCR5}^{\Theta_C}(\bar{C})$ we will take $\hat{X} = A$ as the final decision.

The interest of this method is above all its simplicity, but it does not allow to quantify the quality (trustfulness) of the decision which is often useful and required in decision-making support systems, and that is why we propose a second method for the decision-making of step 6.

4.2   Method 2 for Step 6

This second method is a bit more sophisticate but it circumvents the exponential complexity of the direct PCR6 fusion of $S \geq 2$ BBAs defined on non dichotomous FoD Θ. Once the step 5 is accomplished we propose to fuse altogether the (coarsened) dichotomous $m_{PCR5}^{\Theta_X}(\cdot)$ and to apply the decision-making method based on the distance between the belief intervals [12]. Because the fusion must operate on the same common frame, we need just to express each BBA $m_{PCR5}^{\Theta_X}(\cdot)$ as a dichotomous BBA on Θ which is denoted $m_{PCR5}^{\Theta_X \uparrow \Theta}(\cdot)$. This is done very easily by just expressing each $\bar{X}$ as the disjunction of all elements of Θ included in $\bar{X}$. The fusion of BBAs $m_{PCR5}^{\Theta_X \uparrow \Theta}(\cdot)$ is done by the weighted averaging rule of combination, where each weighting factor depends on the decisioning-making easiness of the BBA $m_{PCR5}^{\Theta_X}(\cdot)$ to fuse. The easier the decision-making, the higher the weighting factor. We summarize this method 2:

1) For each $X \in \mathcal{D}$, establish $m_{PCR5}^{\Theta_X \uparrow \Theta}(\cdot)$ from $m_{PCR5}^{\Theta_X}(\cdot)$
2) For each $X \in \mathcal{D}$, compute the weighting factor $w(X)$ of $m_{PCR5}^{\Theta_X \uparrow \Theta}(\cdot)$ by

$$w(X) = \frac{1}{C}(1 - h(m_{PCR5}^{\Theta_X \uparrow \Theta})) \tag{21}$$

where $C = \sum_{X \in \mathcal{D}}(1 - h(m_{PCR5}^{\Theta_X \uparrow \Theta}))$ is a normalization factor, and where $h(m_{PCR5}^{\Theta_X \uparrow \Theta}) = H(m_{PCR5}^{\Theta_X \uparrow \Theta})/H_{\max} \in [0,1]$ is the normalized pignistic entropy of the BBA $m_{PCR5}^{\Theta_X \uparrow \Theta}$ defined by $H(m_{PCR5}^{\Theta_X \uparrow \Theta}) = -\sum_{X \in 2^\Theta} BetP(X) \log_2(BetP(X))$ and $BetP(X)$ is the pignistic probability of X [13], and $H_{\max} = \log_2 |\Theta|$.
3) Make the weighting average of $m_{PCR5}^{\Theta_X \uparrow \Theta}(\cdot)$ for all $X \in \mathcal{D}$ to get the BBA

$$m^{\Theta}(\cdot) = \sum_{X \in \mathcal{D}} w(X) m_{PCR5}^{\Theta_X \uparrow \Theta}(\cdot) \tag{22}$$

4) From $m^{\Theta}(\cdot)$ make the decision based on minimum of belief-interval distance [12], that is
$$\hat{X} = \arg \min_{X \in \mathcal{D}} d_{BI}(m^{\Theta}, m_X^{\Theta}) \tag{23}$$

where m_X^{Θ} is the BBA focused on X that is $m_X^{\Theta}(X) = 1$ and $m_X^{\Theta}(Y) = 0$ if $Y \neq X$, and where $d_{BI}(.,.)$ is the belief-interval distance defined by (see [12] for details, justification and examples)

$$d_{BI}(m_1, m_2) \triangleq \sqrt{N_c \cdot \sum_{X \in 2^\Theta} d_W^2(BI_1(X), BI_2(X))} \tag{24}$$

where $N_c = 1/2^{|\Theta|-1}$ is a normalization factor to have $d_{BI}(m_1, m_2) \in [0, 1]$, and $d_W(BI_1(X), BI_2(X))$ is the Wassertein's distance between belief intervals $BI_1(X) \triangleq [Bel_1(X), Pl_1(X)] = [a_1, b_1]$ and $BI_2(X) \triangleq [Bel_2(X), Pl_2(X)] = [a_2, b_2]$ given by

$$d_W([a_1, b_1], [a_2, b_2]) \triangleq \sqrt{\left[\frac{a_1 + b_1}{2} - \frac{a_2 + b_2}{2}\right]^2 + \frac{1}{3}\left[\frac{b_1 - a_1}{2} - \frac{b_2 - a_2}{2}\right]^2}$$

5) The quality (or trustfulness) of the decision is given by

$$q(\hat{X}) \triangleq 1 - \frac{d_{BI}(m, m_{\hat{X}})}{\sum_{X \in \mathcal{D}} d_{BI}(m, m_X)} \quad (25)$$

$q(\hat{X}) \in [0, 1]$ becomes maximum (equal to one) when $d_{BI}(m^\Theta, m_{\hat{X}}^\Theta)$ is zero, which means that $m^\Theta(\cdot)$ is focused only on $\hat{X}$. The higher $q(\hat{X})$ is, the more confident in the decision $\hat{X}$ we are. When there exists a tie between multiple decisions $\{\hat{X}_j, j > 1\}$, then the prudent decision corresponding to their disjunction $\hat{X} = \cup_j \hat{X}_j$ should be preferred (if allowed), or we can apply the method 1 to resolve the tie, or in desperation select randomly $\hat{X}$ among the elements $\hat{X}_j$ involved in the tie.

Of course we could adopt a more complicate method where the averaging fusion could operate on all the possible dichotomous BBAs related with each element $X \in 2^{\Theta \backslash \{\emptyset, \Theta\}}$ instead of $X \in \mathcal{D}$, but this would substantially increase the computational burden. Because the decision $\hat{X}$ must be constrained to belong to $\mathcal{D}$, we restrict the fusion to be applied only for the dichotomous BBAs related to these elements only. By doing this we can reduce substantially the computational burden if $|\mathcal{D}|$ is much lesser than $2^{|\Theta|}$.

For convenience, we show how works the method 2 in the previous Example 1 using the same Θ and $\mathcal{D} = \{A, B, C\}$. We have to make the weighted average of the three following BBAs

$m_{PCR5}^{\Theta_A \uparrow \Theta}(A) = 0.3$, $m_{PCR5}^{\Theta_A \uparrow \Theta}(B \cup C \cup D \cup E) = 0.2$, $m_{PCR5}^{\Theta_A \uparrow \Theta}(A \cup \bar{A} = \Theta) = 0.5$

$m_{PCR5}^{\Theta_B \uparrow \Theta}(B) = 0.1$, $m_{PCR5}^{\Theta_B \uparrow \Theta}(A \cup C \cup D \cup E) = 0.5$, $m_{PCR5}^{\Theta_B \uparrow \Theta}(B \cup \bar{B} = \Theta) = 0.4$

$m_{PCR5}^{\Theta_C \uparrow \Theta}(C) = 0.4$, $m_{PCR5}^{\Theta_C \uparrow \Theta}(A \cup B \cup D \cup E) = 0.3$, $m_{PCR5}^{\Theta_C \uparrow \Theta}(C \cup \bar{C} = \Theta) = 0.3$

with $B \cup C \cup D \cup E = \bar{A}$, $A \cup C \cup D \cup E = \bar{B}$ and $A \cup B \cup D \cup E = \bar{C}$. The pignistic entropies are respectively equal to $H(m_{PCR5}^{\Theta_A \uparrow \Theta}) \approx 2.1710$, $H(m_{PCR5}^{\Theta_B \uparrow \Theta}) \approx 2.3201$ and $H(m_{PCR5}^{\Theta_C \uparrow \Theta}) \approx 2.0754$, and their normalized values are $h(A) \approx 2.1710/2.3219 = 0.9350$, $h(B) \approx 2.3201/2.3219 = 0.9992$ and $h(C) \approx 2.0754/2.3219 = 0.8938$. From Eq. (21) we get the weighting factors $w(A) \approx 0.37803$, $w(B) \approx 0.00463$ and $w(C) \approx 0.61734$, and the weighted average BBA is

$$m^{\Theta}(A) = w(A)m_{PCR5}^{\Theta_A\uparrow\Theta}(A) + w(B)\cdot 0 + w(C)\cdot 0 \approx 0.1134$$

$$m^{\Theta}(B) = w(A)\cdot 0 + w(B)m_{PCR5}^{\Theta_B\uparrow\Theta}(B) + w(C)\cdot 0 \approx 0.0005$$

$$m^{\Theta}(C) = w(A)\cdot 0 + w(B)\cdot 0 + w(C)m_{PCR5}^{\Theta_C\uparrow\Theta}(C) \approx 0.2469$$

$$m^{\Theta}(B\cup C\cup D\cup E) = w(A)m_{PCR5}^{\Theta_A\uparrow\Theta}(B\cup C\cup D\cup E) + w(B)\cdot 0 + w(C)\cdot 0 \approx 0.0756$$

$$m^{\Theta}(A\cup C\cup D\cup E) = w(A)\cdot 0 + w(B)m_{PCR5}^{\Theta_B\uparrow\Theta}(A\cup C\cup D\cup E) + w(C)\cdot 0 \approx 0.0023$$

$$m^{\Theta}(A\cup B\cup D\cup E) = w(A)\cdot 0 + w(B)\cdot 0 + w(C)m_{PCR5}^{\Theta_C\uparrow\Theta}(A\cup B\cup D\cup E) \approx 0.1852$$

$$m^{\Theta}(\Theta) = w(A)m_{PCR5}^{\Theta_A\uparrow\Theta}(\Theta) + w(B)m_{PCR5}^{\Theta_B\uparrow\Theta}(\Theta) + w(C)m_{PCR5}^{\Theta_C\uparrow\Theta}(\Theta) = 0.3761$$

From Eq. (24) we get $d_{BI}(m^{\Theta}, m_A^{\Theta}) \approx 0.6818$, $d_{BI}(m^{\Theta}, m_B^{\Theta}) \approx 0.7541$ and $d_{BI}(m^{\Theta}, m_C^{\Theta}) \approx 0.5874$ because $d_{BI}(m^{\Theta}, m_C^{\Theta}) < d_{BI}(m^{\Theta}, m_A^{\Theta}) < d_{BI}(m^{\Theta}, m_B^{\Theta})$. Thus the final decision must be $\hat{X} = C$ because it corresponds to the smallest d_{BI} distance value. This decision is the same as with method 1. Based on Eq. (25) one has $q(\hat{X} = C) \approx 0.7096$ indicating a pretty good trustful decision because it is much greater than 0.5. If one have preferred $\hat{X} = A$ (the second best choice) then $q(\hat{X} = A) \approx 0.6630$ which is a bit worse, and for $\hat{X} = B$ one gets the least trustful decision because $q(\hat{X} = B) \approx 0.6273$. Note that a more optimistic attitude (if preferred) could be obtained by replacing the BetP probability by the DSmP probability [5] (Chap. 3 of Vol. 3) in the entropy derivation.

5   Conclusions

In this work we have presented a very new methodology for decision-making under uncertainty in the framework of belief functions thanks to the unique PCR5-based canonical decomposition of any (dogmatic or non-dogmatic) dichotomous BBAs. We have shown that this new canonical decomposition provides less conflicting contra- and pro-evidences with respect to the decomposition based on Dempster's rule when the latter can be applied. Any BBAs defined on a general (non dichotomous) frame of discernment can be transformed into a set of coarsened dichotomous BBAs that can always be decomposed canonically and combined easily and quickly in one PCR5 fusion step to get a suboptimal fusion result for each element of the decision space under consideration. The final decision can be made in two ways: either by a simple comparative analysis of masses of elements of the decision space, or on the minimization of belief-interval distance which also offers the advantage of quantifying the quality of the decision. The evaluation of this new methodology for real applications is under progress and it will reported in forthcoming publications.

References

1. Shafer, G.: A Mathematical Theory of Evidence. Princeton University Press, Princeton (1976)
2. Zadeh, L.A.: On the validity of Dempster's rule of combination. ERL Memo M79/24, Department of EECS, University of California, Berkeley, U.S.A. (1979)

3. Zadeh, L.A.: A simple view of the Dempster-Shafer theory of evidence and its implication for the rule of combination. Al Mag. **7**(2), 85–90 (1986)

4. Dezert, J., Tchamova, A.: On the validity of Dempster's fusion rule and its interpretation as a generalization of Bayesian fusion rule. Int. J. Intell. Syst. **29**(3), 223–252 (2014)

5. Smarandache, F., Dezert J. (eds.): Advances and Applications of DSmT for Information Fusion, vols. 1–4. American Research Press, Rehoboth (2004–2015)

6. Dezert, J., Smarandache, F.: Canonical decomposition of dichotomous basic belief assignment. Int. J. Intell. Syst. **35**(7), 1105–1125 (2020)

7. Smets, P.: The canonical decomposition of a weighted belief. In: Proceedings of International Joint Conference on Artificial Intelligence, San Mateo, CA, USA, pp. 1896–1901 (1995)

8. Yager, R.: On the Dempster-Shafer framework and new combination rules. Inf. Sci. **41**, 93–138 (1987)

9. Dubois, D., Prade, H.: Representation and combination of uncertainty with belief functions and possibility measures. Comput. Intell. **4**, 244–264 (1988)

10. Dezert, J., Smarandache, F., Tchamova, A., Han, D.: Fast fusion of basic belief assignments defined on a dichotomous frame of discernment. In: Proceedings of Fusion 2020, Pretoria, South Africa (2020)

11. Dezert, J., Fidanova, S., Tchamova, A.: Fast BF-ICrA method for the evaluation of MO-ACO algorithm for WSN layout. In: Proceedings of FedCSIS International Conference, Sofia, Bulgaria (2020)

12. Han, D., Dezert, J., Yang, Y.: Belief interval based distances measures in the theory of belief functions. IEEE Trans. SMC **486**, 833–850 (2018)

13. Smets, P., Kennes, R.: The transferable belief model. Artif. Intell. **66**(2), 191–234 (1994)

KOI: An Architecture and Framework for Industrial and Academic Machine Learning Applications

Johannes Richter[1,2](✉) [iD], Johannes Nau[1] [iD], Michael Kirchhoff[1] [iD],
and Detlef Streitferdt[1] [iD]

[1] Technische Universität Ilmenau, 98693 Ilmenau, Germany
{johannes.richter,johannes.nau,michael.kirchhoff,
detlef.streitferdt}@tu-ilmenau.de
[2] Göpel electronics, 07745 Jena, Germany

Abstract. A novel framework is presented, which simplifies the integration of machine learning into systems for industrial inspection and testing. In contrast to most approaches utilizing a centralized setup, the proposed work follows an edge-computing paradigm. The scope is not limited to inspection tasks but includes all requirements connected to such tasks. The support for continual and distributed learning, as well as distributed accumulation of training data, is a crucial feature of the proposed system. An integrated user rights management allows for the collaboration of multiple people with different background of expertise and tasks on the same machine learning models. Through platform-independent design and the use of a progressive web app as a user-interface, this framework supports the deployment in heterogeneous systems. Separation of concerns and clean object-oriented design makes the framework highly extensible and adaptable to other domains.

Keywords: Machine learning · Deep learning · Software architecture · Distributed learning · Industrial Machine Learning · Edge computing.

1 Introduction

Over the last years, machine learning techniques such as deep learning have seen an increased interest from academic and industrial researchers. Throughout many fields and problems, research has shown outstanding results. Most notably, the classification performance in fields such as machine vision has risen to and above human performance in specific fields. This increase in performance makes machine learning, and especially deep learning approaches a promising tool for industrial automated classification tasks.

Several machine learning libraries emerged over the last years [1–6], which simplified the design and training process for machine learning models. However, there are still industrial applications where novel experts seek machine

---

Source code available: https://github.com/koi-learning.

© Springer Nature Switzerland AG 2021
D. Simian and L. F. Stoica (Eds.): MDIS 2020, CCIS 1341, pp. 113–128, 2021.
https://doi.org/10.1007/978-3-030-68527-0_8

learning solutions, but existing frameworks and workflows just cannot satisfy the use cases and requirements. Industrial inspections often pose situations where sophisticated machine learning software would be beneficial, but the collection and preparation of vast amounts of training data are just not feasible. For further references, the examples below illustrate the given situation. The discussed problem and the solution, therefore, are not limited to those examples.

The field of electronic testing equipment benefits a lot from tailored machine learning solutions. Existing algorithms are not flexible enough to perform well on ever so slightly changing features, so human experts must verify the decisions made by the test machine. These extra checks cost time and money and can be a tedious task to perform.

The example of such test equipment, for this paper, with human re-evaluation steps is automatic optical inspection, called AOI for short. Highly optimized inline systems use sophisticated illumination and optical setups to acquire multispectral two-dimensional and three-dimensional observations of electronic components. An inspection program checks for any defects based on the multitude of pictures acquired. Each program consists of a hierarchy of steps that ultimately use various algorithms for defect detection. If the system suspects a defect, the process routes the defective product out of the production line to a repair- or verification-station. Here, a human expert inspects the product, re-evaluates the machine decision, and possibly repairs the product. If the human expert verifies the defect, he can supply further information, like the error class, which describes the nature and maybe the cause of the defect. Due to constantly changing features of the inspected objects and deviations in the manufacturing process itself, debugging of inspection programs is time-consuming. Like many machine learning approaches, developing inspection programs, requires a multitude of examples to include all acceptable process deviations in the final program. Manufacturers cannot maintain prolonged programming and debugging cycles with high-mix low-volume processes. The result is a demand for the automatic creation of inspection programs.

Nevertheless, even with fully automatic programming, the burden of debugging remains. To avoid wasting any valuable production time, initial inspection programs have a high pseudo defect (steps incorrectly marked as defective) rate. Manual inspection must compensate for the high rate of pseudo defects. This tradeoff enables the manufacturer to start production with as little programming effort as possible while maximizing the chance to find all defects. The manual inspection step is also prone to errors. Human operators can confuse error-classes and mark real errors as pass and pseudo defects as real errors. This inspection process has the potential for machine learning to vastly improve cost and time effectiveness.

Any viable machine learning solution should allow continual learning. Continual learning allows for quick deployment, followed by successive improvements of the models [7]. The classification and training task needs to be divided between multiple people to minimize the impact of personal bias. Here, active learning [8] is the best solution. Multiple users can respond to requests from the model for additional insights. By randomly assigning experts to the response tasks, the risk of biased responses from a single person, and thus the adverse effects on the model, are reduced.

Up until now, there is no machine learning model known, which applies to general inspection tasks. Because the inspection tasks may vary from domain to domain and often even in the same domain, we suspect that such a model requires some form of general artificial intelligence. The fact that the necessary training data is kept secret by the inspection companies highly increases the problem of finding a model that satisfies the requirements of all domains. There are simply too many features and edge cases that a model needs to be aware of, so we believe that the application of artificial intelligence will produce domain- and often problem-specific machine learning models, at least for the foreseeable future.

So, a solution is needed that enables machine learning engineers to deploy their models in industrial and academic environments and manage the whole lifecycle of the mentioned models. The solution should avoid boilerplate code to keep the development steps simple. Furthermore, a software feature is needed, which allows continual learning to adapt models to an evolving focus and changing sets of features. The rest of the paper is organized as follows: Sect. 2 of this paper will discuss requirements, while Sect. 3 discusses the state-of-the-art regarding this research. Section 4 will show the proposed architecture, including architectural diagrams. In Sect. 5, an example scenario helps to illustrate the user experience of the system.

2   Requirement Engineering

This section assesses the given requirements to provide the needed knowledge for the proposed system. The first subsection describes the stakeholders of the system, while the second subsection lists the requirements.

2.1   Stakeholders and Use Cases

An analysis of a real-world inspection solution in the field of electronic manufacturing leads to the following stakeholders and use-cases that were generalized to match a wide range of industrial applications.

Machine Learning Engineer. The ML-Engineer defines models for an academic, research, development, or production environment. He/She wants the development process to have an uncluttered workflow and wants to be free in his/her choice of ML-Toolkits and ML-Frameworks. When working problem-oriented, he/she needs a set of simple processes to train, observe, and deploy new versions or iterations of his models.

Production Supervisor. The production supervisor aims towards short development cycles and as little debugging effort as possible to run an economical production process. Machine learning solutions should not prolong the introduction of new products, and the training needed for the system users to be able to develop inspection programs. Through assigned roles and user privileges, the supervisor wants to define who can make changes to machine learning solutions, or who can use an instance for inference.

Test System Operator/ML-User. The ML-User observes the inspection process while re-evaluating faults detected by the inspection system. The users assist the machine learning system with their abstraction abilities and domain knowledge by responding to label requests issued by instances.

2.2   Requirements

The requirements define the necessary features that must be implemented in the proposed system. Future development could go beyond these requirements, but for productive usage, the proposed system has to fulfill these requirements. The requirements were elicited in an industrial/university cooperative context.

Functional Requirements

- Deployment of different machine learning models in productive environments
  Rationale: Deploy new machine learning models on a variety of inspection systems with different applications and test strategies. The system should be able to manage and deploy multiple models on different systems at the same time.
- Iterative training of models using additional insights and data
  Rationale: As complete datasets are hard to acquire, the system should be able to train models on incomplete datasets and continue or re-run the training at a later time, while preserving the already achieved classification quality. This early training enables fast deployment of machine learning in productive environments, with as minimal initial effort as possible.
- Asynchronously collecting samples from different sources
  Rationale: Different processes and systems can become a valuable source of training data. The ability to asynchronously accumulate training samples frees the developer from searching and composing training sets in advance.
- Support for label-requests for continual and active learning
  Rationale: The models can, at any time, request additional insight into the data. As this feature is part of the deployment system itself, the programming overhead for the machine learning developer minimizes. In an environment where the validity of labels can not be guaranteed, the active request for additional data insight and the distribution of workload between users increases the value of the training samples over time.

- User- and role-management to control access to models and instances
  Rationale: Controlling what models and instances a user can see or interact with, enables many deployment scenarios, where different people work cooperatively using the same model. The role management prevents malicious changes and adversarial training input to productive machine learning models. The user management enables different persons to share the same computation hardware without sharing models, instances, or samples.
- Lifecycle management of models and instances
  Rationale: When using continual training for machine learning, users want to preserve already achieved classification quality. Especially in a productive environment, the ability to go back to previous states of training is crucial. When wrong inputs lead to stuck training progress, the lifecycle management enables to user to continue training from a previous much more viable state.

Non-functional Requirements

- Edge computing allowing for inference execution right on the target system
  Rationale: Using the typically performant computational hardware of the inspection systems instead of a centralized inference system lowers the overall cost of setup. It makes the classification available where it is needed, even when offline.
- Graphical user-interface for most common actions and observation
  Rationale: Being able to interact with the system and performing the most common task using a graphical interface increases the acceptance of the system and flattens the learning curve when first working with the proposed system.
- Adaptive deployment for different heterogeneous systems
  Rationale: The support for heterogeneous systems and the use of platform-independent software enables the user to integrate the proposed system into many existing inspection systems and environments, without being forced to use a specific operating system or library.
- Minimum boilerplate code for unhindered development
  Rationale: If the proposed system has as little as possible impact on the model development, the users are free to implement whatever they need. As there is no general approach to artificial intelligence, and there are many libraries and frameworks with different development paradigms available, the proposed system should not burden the user with unnecessary restrictions.
- Distributed computation to enable the concurrent training of multiple models
  Rationale: By controlling which hardware computes which model, the user can steer the performance of the overall system. Crucial or frequently updating models can run on dedicated high-performance hardware. Multiple users can share the same computation hardware for non-time-critical problems.

3   State of the Art

This section reviews state-of-the-art solutions, regarding deployment and life-cycle management for continual machine learning in academic and industrial environments.

There are runtime environments and inference frameworks, like *NVidia TensorRT* [9], available, alongside open standards for exchanging trained models, like *ONNX* [10]. As they focus on the performant deployment of pre-trained models, they are out of scope for this research as this paper aims towards the lifecycle management of iteratively training machine learning systems.

Cortex.dev [11] is an open-source stack for machine learning engineering and deployment. The critical difference between the proposed approach and said software is that *Cortex.dev* deploys machine learning models as web-based APIs, while the proposed architecture aims for local execution of inference tasks. Therefore, this paper provides an edge computing solution instead of a cloud computing solution.

Tensorflow Serving is the official software toolkit for serving machine learning models build and trained using the TensorFlow software system [1,2]. Like *Cortex.dev*, *TensorFlow Serving* aims for a web-based solution for running models for inference. The documentation suggests the use of Docker images containing the serving API for local inference [12].

Both systems are designed for deployment and miss the continual learning focus in their design, even though they do not prevent the user from doing so.

The open-source library *horovod* [13] enables the user to scale single GPU training software up towards multiple GPUs to speed up the training process. Distributing computation between multiple GPUs is not the aim of this project, and *horovod* does not provide deployment to the previously described edge systems. Furthermore, *horovod* could be used together with the proposed system.

Allegro.ai Trains [14] is a set of tools for machine learning, which aims for cooperative and concurrent development and training of machine learning models. It has a rich feature set available for academic and industrial developers to organize developing machine learning code and optimizing hyperparameters. As a development tool, it lacks the focus on integration into products, edge computing, and build-in mechanics for continual learning.

4   Proposed Architecture

This paper proposes a new software architecture and system called "KOI". The name is a German abbreviation for "Kognitive Objektorientierte Inspektion", which translates to object-oriented cognitive inspection. Key features of this architecture are its simple object-oriented design, separation of concerns through multiple components, and the vast amount of deployment scenarios it supports. The following section describes the architectural decisions made while developing this software. Layered C4-Diagrams [15] describe every main component of the system. Some deployment scenarios will conclude this section.

4.1   General Objects

A pure object-oriented approach simplifies the communication between stake-holders and describes the interfaces of the system's components. Consent about the objects passed around and computed in a complex system makes the entire system design more accessible. The proposed system has three fundamental objects as shown in Fig. 1. All stakeholders and components need to understand those objects. These objects are Models, Instances, and Samples.

Model. Models are a formal definition of machine learning systems. As such, they hold the executable code, parameter definitions, and meta information needed for training and inference. Experts in machine learning design models with one or multiple applications in mind. Thus, models are not executable on their own, as they lack problem-centered samples and state information.

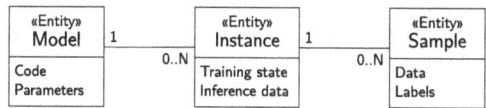

Fig. 1. Entity Relationship model of general Objects

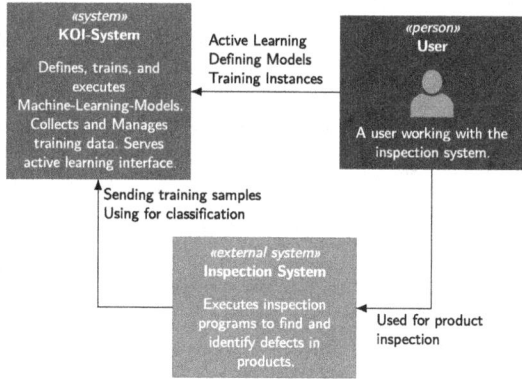

Fig. 2. System Context diagram for KOI-System

Instance. Instances are non-abstract versions of models. Application experts instantiate, parameterize, and assign a set of data samples to them. Instances have a training state, to continue training, and hold all information, like weights and biases, needed for inference.

Sample. A sample is a piece of atomic information consumed by the instance while training. It consists of multiple data entities of high-dimensional tensors identified by a unique key, as well as an arbitrary number of labels represented as tensors as well. Once finalized and marked for consumption, samples do not accept further data tensors but can accumulate additional labels.

4.2 Software Architecture Description

The KOI-System embeds into a system context as shown in Fig. 2, where an inspection system uses it for classification or regression tasks. The inspection system optionally sends samples for training proposes. For configuration and management tasks, the system also needs to interact with one or multiple users. In the case of continual and active learning applications, the users also needs to be able to supervise the training process and respond to requests of the system to provide labels for unknown or ambiguous data.

Figure 3 shows the components that are part of the high-level architecture. The central KOI-API component handles the persistence of the objects described in Sect. 4.1 on top of a role-based authorization schema. The usage of a relational database meets our requirement to express the object structure from Fig. 1. Since there is the need to save multiple megabytes or even gigabytes of data, there is also a file-based storage, which will be referenced by the relational database through unique identifiers. The API-component provides an interface

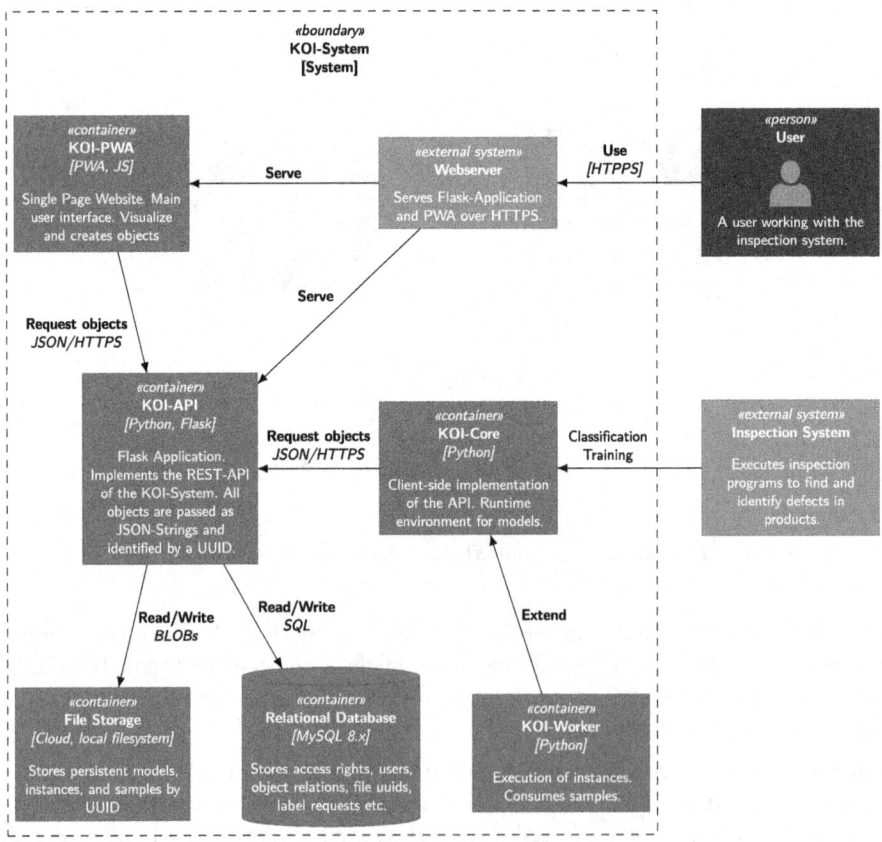

Fig. 3. Container diagram for KOI-System

to the system as a REST-API [16]. All other components solely communicate with this API. Two of those components are the KOI-PWA and the KOI-Core. The KOI-PWA component is a progressive web application [17] and will present the end-user with a GUI. At the same time, the KOI-Core is a Python-library [18] wrapping the REST-API and integrates additional logic for the execution of machine learning models for training or inference. The system's user has to embed the KOI-Core component in the actual inspection devices software. Furthermore, the component KOI-Worker, which uses this library to train models, is part of this architecture. The API does not compute or train any machine learning models. This is the sole task of the worker program.

The following subsections will describe the architecture of each component in detail.

KOI-API. The system component called KOI-API is the central logic of the whole software system, as it handles all objects and actions carried out on said objects. Figure 4 shows the structure of this component. An object-relationship-model lays the foundation of object management. This model defines the properties and relations of any object handled by the system. Another benefit of

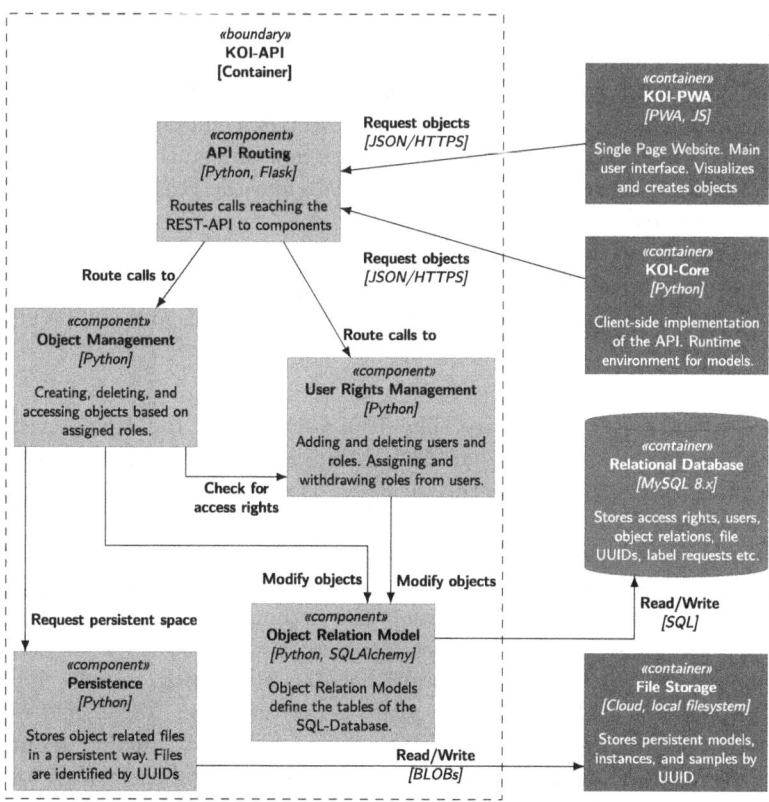

Fig. 4. Component diagram for KOI-API

this model is the little to no effort needed to serialize all objects into an SQL-database. Every object, which is accessible to other components, carries a UUID following RFC4122 [19] for easy identification.

The persistence sub-module handles the storage of large binary objects opaquely. Fields stored using this sub-module are still available through the entity-relationship-model, and thus are only referenced in the database using the unique file id. This opaque reference allows the cost and computation efficient separation of the high-volume data and the database. The persistence module can handle more than just the local filesystem. The module could also implement the usage of networked file systems, cloud services, or could use a database itself.

The KOI-API connects all other components by exposing a software interface through HTTPS following the REST-paradigm. Every object is serialized and sent as a JSON-encoded string. As this component is the central control logic of the system, it also manages any user access control and grants user rights. Access control is role-based, using three types of roles. Each type has different rights associated with the system or an object. There are general roles granting access to administrative actions, such as adding or deleting users and roles, or adding new models to the system. Roles associated with models can allow the user to make changes to a model or instantiate it for training and inference. Like model roles, roles concerning instances grant users rights to carry out actions on the associated instance. A set of bits with fixed length encode the roles. Each bit encodes whether a right is granted ("true") or not granted ("false"). Users can have any number of roles with any number of objects. The system determines if a user can act upon and inspect a specific object by combining all his assigned roles on the object in question. When creating a new object, the user gets the role 'owner', granting all rights on this object, including the right to assign further roles to other users.

The routing module selects the appropriate submodule for incoming requests and responds to malformed requests with adequate HTTP status codes. The KOI-API uses the FLASK-Framework [20] to implement web communication and the routing.

KOI-Core and KOI-Worker. The KOI-Core wraps the REST-API in a Python package and thus makes all API objects available for any other software written in Python. The architecture in Fig. 5 is a reference implementation, so we chose Python for convenience, openness, and readability of the code. Other implementations of the KOI-Core are possible and encouraged by this software design.

All API objects have a counterpart in the python implementation as opaque object proxies. These proxies lazily fetch the fields of any object when accessing the members of said objects. A simple caching mechanism reduces repeated requests to the API; this limits the load on the interface. All objects register into an object pool. The purpose of the object pool is to further reduce the load on the interface by holding already fetched objects and their fields for later reuse. The pool identifies the objects by their unique identifier assigned by the

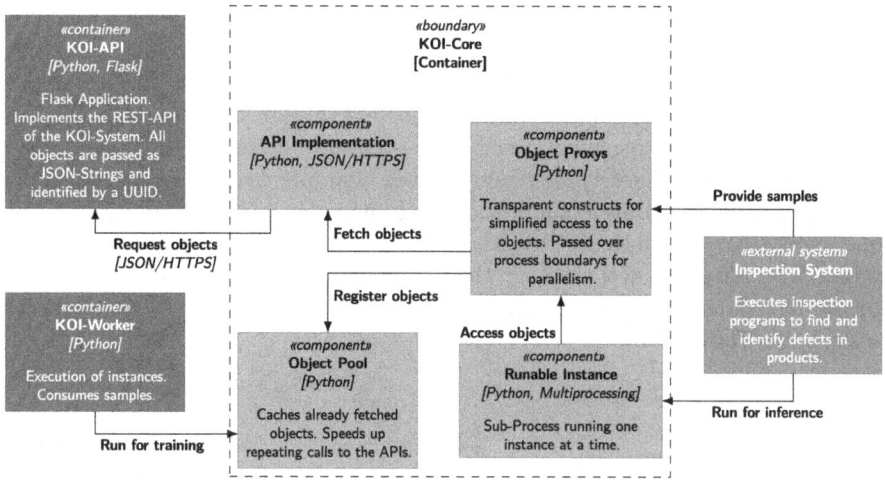

Fig. 5. Component diagram for KOI-Core

KOI-API. Additionally, the pool handles the creation of new objects through the KOI-Core. The pool requests the creation of new objects using the KOI-API and provides the object proxy.

A module handles any communication with the KOI-API. This module provides a set of functions describing actions carried out by the KOI-API. These functions take and return proxy objects and send HTTP-Requests with appropriate JSON-encoded objects to the KOI-API. This abstraction frees the remaining code components of any request- and response-based communication logic and encapsulates the JSON-encoding of all objects. Any other component uses the proxy objects and the object pool.

The KOI-Worker is a background worker based on the KOI-Core package with a command-line interface. After successfully authenticating with a user-defined API-Server, the worker starts polling available tasks from the API and executes them one by one. Multiple workers can communicate with the same API and work on different models concurrently. The user access roles and thus the privileges assigned to the worker's user account determine which models any worker executes.

KOI-PWA. The KOI-PWA is the central graphical user interface for general users and administrators of the KOI-system alike. It enables the users to add and configure new models for experimentation or deployment in a streamlined workflow. Through its responsive web-based interface, users can create and edit roles and grant access rights to other users. Another massive benefit of this user-interface is its ability to inspect samples assigned to a specific instance and respond to label-requests.

Because this software aims for heterogenous hardware-systems, we decided to implement the user interface as a progressive web. The web-technology enables the application to run with consistent user experience on many platforms and provides an easy and cross-platform mechanism for visualization plugins. Models can specify additional visualization and label-request plugins for the user to view or listen to samples. These plugins implement a simple JavaScript interface.

As the KOI-PWA only communicates through the API with all other components, it is easily replaceable by other user interfaces with a broader or a more problem-specific set of actions. Like the KOI-Core, it is a reference implementation, so replacing or complementing it with other components is encouraged by the design of the system.

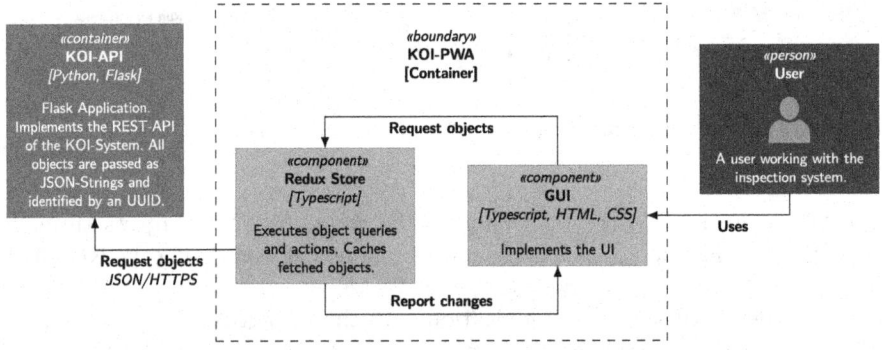

Fig. 6. Component diagram for KOI-PWA

As Fig. 6 shows, the KOI-PWA mainly consists of two components:

1. The Graphical User Interface, which is developed based on Lit-Elements from the Polymer-Project [21].
2. The redux store [22] together with its surrounding logic.

This architecture loosely resembles a Model-View-Controller architectural pattern, where the GUI is the view, Redux-Store-State resembles the model, and the action logic dispatched to the redux-store represents the controller. However, because the application is directly working on the API, the logic is tightly coupled with its state, as it is CRUD-Logic (Create, Retrieve, Update, and Delete). Therefore, we choose to split the application only into the two components shown in Fig. 6.

4.3   Deployment Options

A vital advantage of this software design is the vast amount of deployment options it enables. The following section describes three deployment scenarios to illustrate the versatility of the proposed system architecture. Other systems and services can serve the persistent data and database used by the KOI-API independently from all following deployments.

Single System Deployment. All described components run on the same target system. The system collects its dataset and trains an individual instance of a model using its computation hardware. Although the benefits of collaboration, shared datasets, and training distribution are lost using this approach, it enables the same model deployment mechanisms as the other examples. Furthermore, the user can easily upgrade the system to become part of one of the other deployment scenarios.

Multiple Systems/Personal Cloud. Multiple systems fulfilling the same inspection task, work cooperatively. By sharing data collected on a subset of systems, the overall dataset gives a broader view of the given problem. Transferring the training task to one of the systems or an external training server frees the target systems from the additional computational load. Through sharing a dataset and training, all connected systems have the same performance when executing for inference.

Machine Learning as a Service. A machine learning expert develops a model and serves individual instances to a set of customers. Each customer connects an arbitrary number of systems to his instance and collects his dataset while sharing only the model with the others. Personal workers perform the training procedure. These workers run in a cloud service or onsite, hosted by the machine learning expert or the consumer himself.

5   User Experience

This section illustrates the targeted user experience using a simple example. The example shows what actions a user has to take in order to train and deploy a new machine learning model using the proposed architecture. Regarding the system setup, the following is assumed: The user has control over a high-performance computer running the KOI-API and hosting the KOI-PWA, as well as running a KOI-Worker process. The system also hosts the database and file storage. This setup is the same as the first deployment example from the previous section.

5.1   Model Development and Preparation

The user develops a machine learning model contained in the file __model__.py. The development can happen on his personal computer and does not have to involve the system hosting the KOI-System. The .py-file must contain the implementation of essential entry points. The KOI-System uses these entry points, to work with the model. More sophisticated models consisting of multiple files of code and auxiliary data are possible as well. A single zip-file accumulates all needed files.

Using the KOI-PWA, the user creates a new empty model and assigns it a name and description. Then he uploads the packaged model files. The KOI-API performs simple checks to determine if the model needs further parameters for instantiating.

To visualize the later collected samples, the user also uploads a visual plugin written in JavaScript. This plugin decodes the data of one sample and presents it. Because the active learning feature is required, an appropriate plugin is also uploaded. This plugin similarly presents the sample as the visual plugin but also provides user interactions translated to additional data for training. For example, a dropbox with class-labels or a text-box for descriptive text.

After finishing all uploads, the user finalizes the model and makes it available for instantiation by himself or other users with sufficient rights.

5.2   First Training

Using the KOI-PWA, the user creates a new instance of the just made model. Again, he can assign a name and description to the instance. If the model provides additional parameters, like hyper-parameters or logging information, the user can specify them at this point. This example assumes the existence of no parameters. Finalizing the new instance makes it available to worker processes to train it and to data sources to add samples.

The user connects the target system through the KOI-Core to the API, to send samples collected in the operation of the target system. At the same time, he uses an already collected set of data and sends it via a simple script, also using the KOI-Core, to the same instance.

The KOI-Worker collects information about the instance and all samples available up to this point from the API. The worker process performs the training as specified in the model files. While training, some samples do not fit the predicted distribution and are flagged for active learning. The system holds all samples with this request flag for later processing.

After finishing the training, the worker saves its current state and registers it with the API for later continuation, while also creating a state of the model optimized for inference and registers this as well.

This process repeats every time new samples arrive, and a worker process is available for training.

5.3   Deployment

After each training epoch, the model registers an inference version of itself with the API. The user collects this state file through the KOI-Core and runs it for inference on the targeted system. The user access rights determine which user account has access to the training data and inference data.

5.4   Label Requests

While training the system, an algorithm flags some labels for active learning. This flag signals that the model requests additional insight into the data, or the given label does not fit the predicted distribution. Through the previously uploaded plugin, the user, or any other user with sufficient access rights, can use the KOI-PWA to visualize the requests and respond to them. When responding to requests, the system removes the active learning flag and sets a new flag alongside the additional information. The new flag marks the sample as renewed and signals the model the completion of the request-response cycle.

The whole response process can also be automated since the needed functions are available through the KOI-Core interface.

6   Conclusion and Future Research

This paper presents a novel framework for machine learning deployment and lifecycle management. We designed the proposed work with academic as well as industrial applications in mind. Key features of the proposed systems are user-management, continual learning mechanisms, and online accumulation of training data. After giving an introduction with an example from the optical inspection domain and how machine learning could improve said applications, the paper described possible stakeholders and requirements of the proposed system. The systems design is given in detail using layered architecture diagrams. Finally, the user experience is described by following a fictional user through the process of model creation and training using the proposed system.

Future research will evaluate the system's performance and acceptance in different inspection and testing domains. New use-cases will emerge, and new requirements will drive the development of the system further. Additional features will broaden the scope of the system towards a more general usage and a wider audience. The system could lay the groundwork for new approaches in the fields of optical inspection and testing of electrical components and appliances. Additionally, the integration of third-party libraries and toolsets to improve the capabilities of KOI, will be considered.

References

1. Abadi, M., Agarwal, A., Barham, P., et al.: TensorFlow: large-scale machine learning on heterogeneous distributed systems. https://arxiv.org/pdf/1603.04467
2. Abadi, M., Barham, P., Chen, J., et al.: TensorFlow: a system for large-scale machine learning. In: Proceedings of OSDI '16: 12th USENIX Symposium on Operating Systems Design and Implementation, pp. 265–283. USENIX Association, Berkeley (2016)
3. Caffe: Convolutional architecture for fast feature embedding (2014)
4. Paszke, A., Gross, S., Massa, F., et al.: PyTorch: an imperative style, high-performance deep learning library. In: Wallach, H., Larochelle, H., Beygelzimer, A., et al. (eds.) Advances in Neural Information Processing Systems, vol. 32, pp. 8026–8037. Curran Associates, Inc. (2019). https://proceedings.neurips.cc/paper/2019/file/bdbca288fee7f92f2bfa9f7012727740-Paper.pdf
5. Chollet, F., et al.: Keras (2015). https://keras.io. Accessed 22 Aug 2020

6. The Theano Development Team, Al-Rfou, R., Alain, G., et al.: Theano: A python framework for fast computation of mathematical expressions. arXiv p. arXiv:1605.02688 (2016)
7. Shultz, T.R., Fahlman, S.E., Craw, S., et al.: Continual learning. In: Sammut, C., Webb, G.I. (eds.) Encyclopedia of Machine Learning, p. 226. Springer, New York (2011). https://doi.org/10.1007/978-0-387-30164-8_171
8. Kakas, A.C., Cohn, D., Dasgupta, S., et al.: Active learning. In: Sammut, C., Webb, G.I. (eds.) Encyclopedia of Machine Learning, pp. 10–14. Springer, New York (2011). https://doi.org/10.1007/978-0-387-30164-8_6
9. GitHub: Nvidia/tensorrt. https://github.com/NVIDIA/TensorRT. Accessed 15 Aug 2020
10. GitHub: onnx/onnx. https://github.com/onnx/onnx. Accessed 15 Aug 2020
11. GitHub: cortexlabs/cortex. https://github.com/cortexlabs/cortex. Accessed 15 Aug 2020
12. TensorFlow: TensorFlow serving with docker TFX. https://www.tensorflow.org/tfx/serving/docker. Accessed 16 Aug 2020
13. Sergeev, A., Del Balso, M.: Horovod: fast and easy distributed deep learning in TensorFlow. https://arxiv.org/pdf/1802.05799
14. GitHub: allegroai/trains. https://github.com/allegroai/trains/. Accessed 15 Aug 2020
15. Simon Brown: The c4 model for visualising software architecture (2020). https://c4model.com/. Accessed 20 July 2020
16. Fielding, R.: Architectural styles and the design of network-based software architectures. https://www.ics.uci.edu/~fielding/pubs/dissertation/fielding_dissertation_2up.pdf. Accessed 21 July 2020
17. Russell, A.: Progressive web apps: escaping tabs without losing our soul - infrequently noted. https://infrequently.org/2015/06/progressive-apps-escaping-tabs-without-losing-our-soul/. Accessed 15 Aug 2020
18. Python Software Foundation: Python 3.8.5 documentation. https://docs.python.org/3/. Accessed 15 Aug 2020
19. Leach, P., Mealling, M., Salz, R.: RFC 4122 - a universally unique identifier UUID URN namespace. https://tools.ietf.org/pdf/rfc4122.pdf. Accessed 15 Aug 2020
20. The Pallets Projects: Flask. https://palletsprojects.com/p/flask/. Accessed 20 Aug 2020
21. The Polymer Project Authors: Litelement (2020). https://lit-element.polymer-project.org/. Accessed 22 Aug 2020
22. Abramov, D.: The Redux Development Team: Redux (2020). https://redux.js.org/. Accessed 22 Aug 2020

Improving Credit Client Classification
by Using Deep Neural Networks?

Klaus B. Schebesch[1]([✉]) and Ralf W. Stecking[2]

[1] Vasile Goldiş Western University, Arad, Romania
kbschebesch@uvvg.ro
[2] Carl von Ossietzky University, Oldenburg, Germany
ralf.stecking@uni-oldenburg.de

Abstract. Credit client classification which is useful for building models to forecast probable defaulting behavior is of obvious practical importance and interest. There are many technical alternatives in order to achieve this goal. A huge variety of *statistical learning* and *nature inspired* black-box techniques where used to search through different classification templates and to apply many kinds of numerical adaptation. Combinatorial rule complexity was traded for massive parameterization. Retraction to computationally less demanding techniques followed and was generally well received by financial practitioners. Explainable modeling also grew in demand. This promised to change with the arrival of deep Neural Networks (dNN), a revival of the repeatedly deprecated classical neural nets, but now with essential improvements in its computational structures at the implementation side and also with some conceptual advances. Using two credit client data sets of different sizes and structure we show which out-of-sample performance measure can be consistently improved by dNN and to what extend manual tuning and modeler's decisions may be delegated to automatic modeling process. We compare these new models to our best performing models published in the past, which were obtained using the same input data.

Keywords: Classification models · Deep Neural Networks ·
ROC-AUC criterion · Client behavior

1 Introduction

1.1 General Overview

Credit client classification which is useful for building models to forecast probable defaulting behavior is of obvious practical importance and interest [1,4]. The general underlying concept is not restricted to financial credit but applies to many similar situations, where client action is conditioned on class or group affiliation determined by some multivariate data analysis. A huge variety of *statistical learning* and *nature inspired* black-box techniques where used to search through different classification templates and to apply many kinds of numerical

© Springer Nature Switzerland AG 2021
D. Simian and L. F. Stoica (Eds.): MDIS 2020, CCIS 1341, pp. 129–148, 2021.
https://doi.org/10.1007/978-3-030-68527-0_9

adaptation. Combinatorial rule complexity was traded for massive parameterization. Retraction to computationally less demanding techniques like Logistic Regression and *moderately expensive* Support Vector Machines (SVM, [19]) followed and was generally well received by financial practitioners. Explainable modeling also grew in demand. Over a broad collection of data sets, at least with regard to primary forecasting goals like out-of-sample hit-rate, accuracy, etc., no clear winner emerged, as the relative superiority of single model classes (like Bagging, Random Forests, SVM, CART, ...) was never truly dramatic and data-set dependent, possibly also as a consequence of the prevailing relative sparseness of the credit client data. Considerable *hand-tuning* of models was still needed. This promised to change with the arrival of deep Neural Networks (dNN), a revival of the repeatedly deprecated classical neural nets, but now with essential improvements in its computational structures at the implementation side and also with some conceptual advances. Using two credit client data sets of different sizes and structure we show which out-of-sample performance measure can be consistently improved by dNN and to what extend manual tuning and modeler's decisions may be delegated to automatic modeling process. We compare these new models to our best performing models published in the past, which were obtained by using the same input data.

1.2   Research Using Deep Machine Learning in Credit Client Modeling

In recent years a few research articles appeared in the literature about credit client classification with deep neural networks. Without claiming completeness, but in attempting to cover a wider spectrum of approaches, we include the following:

1. Credit Risk Analysis Using Machine and Deep Learning Models [2].
2. Predicting mortgage default using convolution neural networks [10].
3. Credit scoring with a feature selection approach based deep learning [6].
4. A novel multistage deep belief network based extreme learning machine ensemble learning paradigm for credit risk assessment [20].
5. A Classification Restricted Boltzmann Machine for a comprehensible credit scoring model [18].

These contributions are mainly directed towards analyzing and improving out-of-sample forecasting performance over that of more traditional, and, more often than not, computationally much cheaper methods. The approaches are experimental, whereby effective practical application in the financial industry often hinges on explainable forecasts (transparent rules, which are difficult or impossible to extract from the dNN models).

In the remaining sections we describe our credit client data using a more general view on applied classification problems, hereby highlighting important features and restrictions on data use and accessibility (Sect. 2). Next, we comment on the more recent revival of neural networks, especially in the context of deep machine learning (Sect. 3). Our new results on credit client classification in the light of deep neural networks is presented in Sect. 4 for a small empirical data set and in Sect. 5 for a large data set, respectively. The results are compared to those obtained in extensive past work using the more traditional approaches. Section 6 further explains some specific results. Finally, in Sect. 7 we conclude and present some outlook concerning future applications.

2   Our Credit Scoring Data for Classification Problems

2.1   Empirical Approach – Credit Data Set Description

Formally, every credit client case contains a number of $N > 0$ credit clients. Each client is described by m_0 raw data features. To every feature vector x_i, $i = 1, ..., N$ has an associated label $y_i \in \{-1, +1\}$, with $y_i = 1$ if the client i was eventually *defaulting* (i.e. could not reimburse the outstanding credit). Hence the raw data set is given by the matrix (x, y) which will be split up into a training and test set by the modeler.

In our computational experiments we use two credit client data sets of different size and structure, a small and a big data set, respectively

Data Set I:

– German Credit Data from UCI repository: information about N=1000 clients
– 300 defaulting credit clients: *default rate is* 30%
– Credit client information: m_0=20 input variables, e.g. age, personal status and sex, property, credit amount, savings account and bonds, etc.

Data Set II:

– N=139951 clients for a building and loan credit
– 3692 defaulting credit clients: *one year default rate is* 2.6%
– Credit client information: m_0=12 input variables, e.g. loan–to–value ratio, repayment rate, amount of credit, house type, etc.

After transforming the raw data set into a standardized numerical format, the effective number of input features m may grow (i.e. $m > m_0$, most often by using a binary representation for categorical variables)

2.2   Experimental Setup

Our experimental setup searches for deep Neural Network models which forecast the probable credit defaulting behavior of new clients. A newly arriving client has data $(x, -)$. Hence, the y-part is not yet known, because it would label his/her future behavior and hence represents itself the required prediction target. This is simulated by means of reserving a test data set $N = N_{train} + N_{test}$. The performance of a model is measured over the test data set by the *Receiver Operating Characteristic* (ROC curve) and the so called AUC (area under curve).

1. Repeated random sub-sampling validation: Divide both credit client data sets **randomly** into *training* and *test* sets with $N_{train} : N_{test} = 2:1$.
2. Repeat ten times:
   (a) Model building for every possible training/test set combination: *Linear* and *RBF-SVM* (SVM using the nonlinear radial basis function kernel) plus Logistic regression additionally built as a benchmark model
   (b) Add **(deep) Neural Networks**
   (c) Report test set results

3   Deep Machine Learning: Renaissance of Messy Neural Networks

Being composed of basically simple, highly repetitive structures neural networks may be viewed as fundamental building blocks of machine learning. In principle this was already known in the 1950s and 60s. Their input-output behavior is determined by training data via adapting a large number parameters by minimization of some error (or energy) function. From a structural point of view these networks are highly scalable (up to millions of free parameters with today's average-level computer technology). Furthermore they usually have a completely differentiable structure making them amenable to adaptation ("learning") by standard gradient descent. Being heavily parameterized seems at first to contradict basic principles of statistical modeling, e.g. that the number of examples should be (much) bigger then the number of parameters, and henceforth of input feature dimensions. However, as whole branches of neural networks can be effectively deactivated by appropriate combined action of parameters and inputs, the true capacity of a network is more difficult to assess. But all these attractive features were almost annihilated by the excessive amount of computation required to train even modest-sized networks and henceforth they were considered as "messy" and lost much of their popularity for data-oriented modeling such as classification. However, as classification was to become evermore important, and as it is a natural application of neural networks, namely by separating input space via hyperplanes and compositions thereof, they were never put completely aside by the data-modeling community.

3.1   Deep Neural Networks

Neural networks resurfaced more recently after revisiting older ideas about Boltzmann machines inspired by statistical physics and ideas of learning in stochastic networks [7] and was subsequently vigorously developed by a wider community from computer science and the (applied) computational sciences ([5,11]). An architecturally important enabler for many dNN is the use of an autoencoder [5] which enables self defined convolutions to act as a massive pattern transformer. However, this is more relevant in a data context with geometrical information; hence we do not use this feature for models of our client data, since they do not contain images or any kind of causally or geometrically chained information (although they have been used occasionally for bankruptcy prediction [10]). More relevant to us are (the revival of) error backpropagation and adapted activation functions, like e.g. ReLUs which do not inhibit backprop errors over arbitrarily many network layers [3].

Around the year 2015 dNNs became increasingly accepted as being surprisingly effective and also efficient, especially on supervised classification problems. Many such real-world problems do have, respectively, do need (a) vast (Terrabyte sized) amounts of data, (b) a large number of hidden layers (so called "deep" neural models) and (c) efficient means for (auto-)regularization as a consequence of the huge numbers of free parameters of the initial (untrained) models. Considerable gains in both efficiency but also classification accuracy were seen in some application domains like visual pattern recognition. However, these gains seem to be obtained primarily by much increased hardware performance, using various types of parallelism and co-processing. Both, graphical and tensor processing units (GPUs and TPUs) also proved to be very useful. Finally, there is also new back- and front-end programming [17] for these model classes.

3.2   Our Deep Neural Networks, Activations and Dropout

By now there exists a broad spectrum of high-performance, open domain Machine Learning software instruments, many of them focusing on deep Neural Networks. Examples are TensorFlow, Theano, Torch, Julia-based Flux, to name a few more widely employed. For our computational experiments we use **R** as a experiment chain driver and **Keras** [9] from inside **R** as model chain driver for deep NN which then in turn calls **TensorFlow** for effective model training (other variants like Python – Tensorflow or Julia – Flux are equally possible).

Our procedure of re-modeling credit client our data models in order to test for possible performance gains over the more traditional ML-techniques employed in past models is the following: We start with our best models in terms of ROC-AUC from past work for two credit scoring data sets of different structure and size. We generate dNN models evaluated over training-test data pairs. The models are of vastly different size and of different potential functional complexity. We then evaluate the out-of-sample results in comparison to the benchmark models and inquire into the role of imposed and automatic regularization as well as modalities of early training stop.

4   Re-Modeling Our Small Data Set with dNN

In order to find out if and also in which way dNN may surpass the more traditional ML-based classification models we first turn to our smaller data set. After some preliminary computational experiments with all our credit client data sets, we note that there is no substantial advantage of dNN – at least in terms of the most direct and popular measures of out-of-sample performance, like hit-rate and accuracy – over the more traditional methods described above and used in previous work. As is detailed in Sect. 3.1 with regard to other more structured and much bigger data sets this may now not come as a surprise. One of the most practically useful additional performance criteria in the context of credit client scoring is the Receiver Operating Characteristic (ROC, which is also used in many other application domain) and its aggregate scalar expression, the area under curve (AUC).

4.1   What to Match and Possibly to Surpass: Data Set 1

Our question is then whether for a credit client data set of size $N \approx 10^2 - 10^3$, (d)NN can beat the more traditional Machine Learning models in terms of AUC. More specifically, we also ask in which way this is achieved. To this end, it seems natural to compare the AUC per model obtained on different data subsets. Hence, our general scheme of running experiments is guided by the following:

- We concentrate on out-of-sample **ROC/AUC** as a widely accepted performance criterion.
- The best out-of-sample AUC performance from our past extensive model comparison in [16] averaged over the ten randomly selected (but henceforth retained) train/test sets is *Logistic Regression* (LogReg) model with

Model	Mean	Std. Dev.	Min	Max
LogReg*	0.782	0.021	0.756	0.826

- A very similar AUC performance is obtained by the *Linear kernel SVM*; in a way, due to **relative sparseness** good results for linear-like models are to be expected as there are not enough data to reliably detect an essentially non-linear class-separation function. In this case there are $m = 20$ recorded input features, then "expanded" into a 61-dim representation by using categorical variables (binary) encoding where appropriate.

4.2   How Do Relate Differently Sized dNN to the Best LogReg?

The structure of deep NN can be described along many architectural as well as numerical mathematics features (Sect. 3.1) which are more or less relevant to modeling the given problem data. For our data which has no feature-based geometric structure (as is rather the case in data about images, time-series, etc.) we refrain in our experiments from using an autoencoder block (which tends to remap the geometry to some useful end) and we directly use stacked layers of conventional NN. The items we put to use from the Keras-Tensorflow library are mainly (a) efficient modeling chains offered by Keras-Tensorflow, (b) within some deeper variants of NN the use of ReLU activation functions in order to strongly limit the otherwise fast decay of the backpropagation signals and to reduce training time (c) layer-wise dropout in order to automatically prune the model (i.e. to potentially avoid over-training) and (c) the training with enhanced solution search of very large models (using the Adams optimizer) still amenable to present personal-grade computer power.

In the computational experiments reported below we use the following convention for a short-hand identification of the dNN model structures:

Inputs:	Raw data input dimension plus categorical
Layers:	number of activation nodes (neurons) layers, etc.
Activation functions:	Sigmoid or ReLU
Outputs:	one, as the goal is forecasting a binary class-label
Validation:	share of training data reserved for validation
Dropout:	share of neurons to de-active or to eliminate

Hence our shorthand model description used in the upcoming tables is for instance $80^s_{.2} - 100^r_{.3} - 30^r_{.1} - 1^s (v = 0.25)$ meaning that starting from a constant number of inputs (not included in the notation) we proceed via first layer using sigmoidal activation functions and a dropout rate of 0.2 to the second layer using ReLU activation functions and a dropout rate of 0.3 and so on until reaching the output node of the four layer network. Also 25% of the training examples are reserved for in-sample cross-validation. The split of all data into a training and a test set is assumed to proceed all modeling. We always use a feedforward, fully connected layer to layer structure as a starting architecture. Models with no dropout or partial dropout and without in-sample validation ($v = 0$ or no annotation) have also been used.

Table 1. Overall out-of-sample AUC for models trained on the small data set

MODEL	Mean	Std. Dev.	Min	Max
LogReg*	0.782	0.021	0.756	0.826
$25^s_{.3} - 1^s$	0.790	0.019	0.766	0.820
Regret AUC	−1.13%	9.18%	−0.89%	0.00%
$100^s_{.3} - 50^s_{.2} - 1^s$	0.785	0.022	0.757	0.826
Regret AUC	−2.96%	21.39%	−3.11%	−0.03%
$100^s_{.3} - 80^s_{.3} - 50^s_{.2} - 1^s$	0.788	0.023	0.757	0.824
Regret AUC	−3.32%	23.42%	−3.00%	−0.36%

The results concerning the NN models on the small data set are displayed in Table 1. They are trained without in-sample validation. Layer-wise **dropout** with rates 0.2–0.3 is used. Hence, no training data is "wasted" for validation sets. Training has been restricted to 80 epochs, which on the basis of running many other test problems form different application domains is judged to be quite extensive for this relatively modest training data volume. The LogReg reference model is taken from previous work [16].

For every new NN model we include a line of measurements which is the exact counterpart of the LogReg-entry. The AUC means, standard-deviations as well as their min and max values are taken over the ten training-test-set pairs, which are exactly the same as those used previously for LogReg. The line added to every new NN model denotes the regret which is produced by stopping the model at epoch 80 instead of (hypothetically) having been able to stop it at the (generally different) epoch which would produce the maximum AUC on the out-of-sample set. Hence negative values denote a standardized measure of loss while positive values indicate a gain. These measurements are for comparative purposes in order to rank the models and should not be interpreted for empirical credit client matters.

This table (Table 1) informs us that the simplest nonlinear neural model, namely $25^s_3 - 1^s$ with $v = 0$ (see the above conventions) is the best regarding mean AUC and with low variation over the training-test data pairs, and it also slightly surpasses the LogReg reference model. The deeper NN models lead to more variation in AUC which means they may adapt better than average to some training-test data set pairs and vice versa. We tested many more model architectures than those reported here. However, the result do not change the overall situation described.

4.3   Pair-Wise Training-Test AUC Curves (80 Training Epochs)

We now report on the ten different deep NN models for every evaluated model architecture, by depicting the AUC-curve on the training set and on the test, respectively.

In Fig. 1 we start with the smallest model from Table 1. i.e. the NN with one hidden layer. In each plot inset we show two curves. The thin curve is the model AUC measured over the training epochs for the training data set. The fat dotted curve is the corresponding evolution of the model AUC over the test set. Obviously, the latter one is the relevant measure of the out-of-sample model performance (the performance practitioners are most interested in). While the AUC grows smoothly over the training epochs as one would expect, the out-of-sample curve is leveling off for most of training-test data pairs. Using the change of slope of the training curve (an information effectively available during training of real life classification models) as a hint for early training stop is however not reliable. Overall, this model is quite robust in that using a wide range of sufficiently large (for more than 15 epochs, say) but otherwise randomly chosen training stops, out-of-sample performance does not fluctuate much.

In contrast to Fig. 1, Fig. 2 depicts the AUC evolution for the deepest or architecturally most complex model from Table 1. The meaning of the curves is the same of that from the previous figure.

This AUC grows smoothly as well but it clearly shows one or more dents over the training epochs. The out-of-sample curve is clearly leveling off for all but one training-test data pair. Using the location of the first dent in the slope of the training curve can be a more reliable hint for early training stop. Overall, this model is slightly over-trained at the last learning epoch (80). In that using a wider range of sufficiently large but otherwise randomly chosen training stops, out-of-sample performance does not fluctuate more than that of the simple NN.

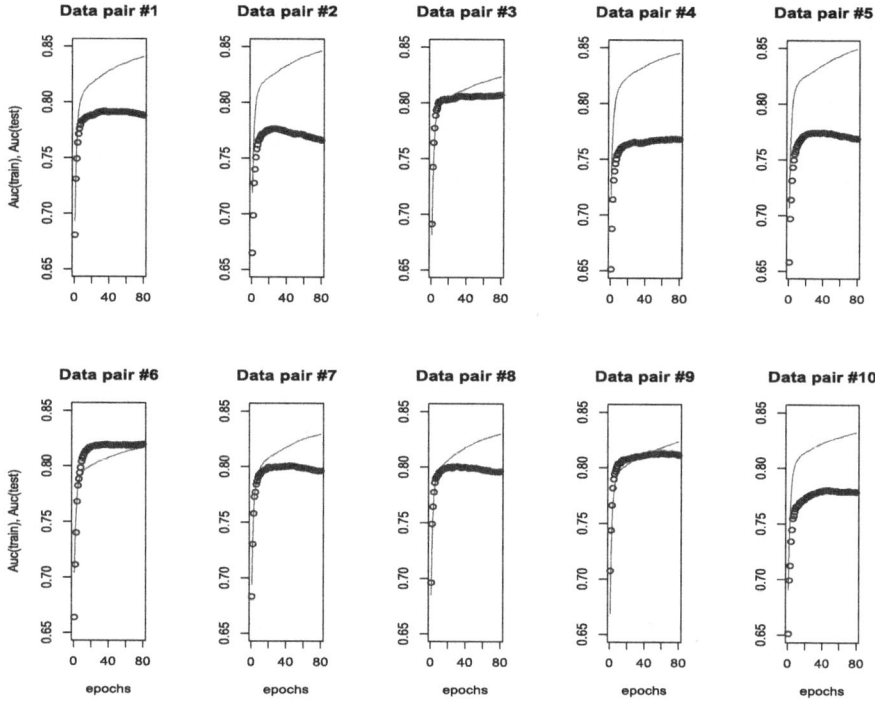

Fig. 1. AUC levels over training and out-of-sample test sets for the $25^s_{.3} - 1^s$ model

5   Re-modeling Our Big Data Set with dNN

Our large credit client data set is taken to represent classification problems in the range $N > 10^5$. This may not be much for technical problems but it is nevertheless a quite substantial data set size for a single financial company. Recall that data pooling is a rather weird but also highly relevant problem in finance owing foremost to various data privacy constraints [16]. Our big data set comprises 12 observed inputs as raw data. Some of the variables are transformed into categorical binary representation leading the 25 effective input variables (client features per case). Furthermore, the data come with highly asymmetric case label frequencies, with non-defaulting credit client vastly outnumbering the defaulting ones. From past extensive experiments the best model turns out to be a Support Vector Machine [14,19] using Radial Basis Function kernels (RBF-SVM). Such types of SVM are in general very powerful at building class separation functions over input space (here client feature space) by using the theoretically appealing and practically very useful concept of support vectors which can be used in many ways to get further insight into the classification model at hand [13].

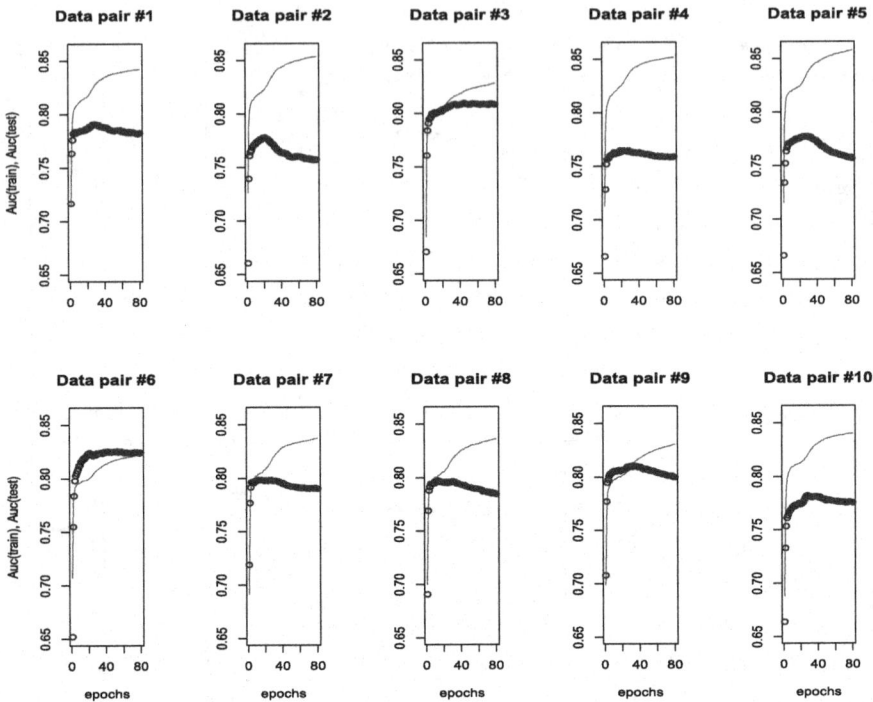

Fig. 2. AUC levels over training and out-of-sample test sets for the $100^s_{.3}-80^s_{.3}-50^s_{.2}-1^s$ model

5.1   What Model to Match and Possibly to Surpass?

Here our question is if deep NN can clearly beat highly principled ML-models such as RBF-SVM. As in the previous sections the question is addressed specifically with regard to our credit client data but now in terms of AUC for $N > 10^5$. Owing to the much bigger data volume all NN models are trained now over 160 epochs. All the other conventions for describing the computational experiments are identical to those from Sect. 4.

In Table 2 we report the results regarding the successively deeper NN models in comparison to the RBF-SVM reference model. After increasing the simple one layer NN architecture to 100 neurons in the hidden layer, and neither using dropout nor in-sample validation, we arrive at a quite remarkable result (at least for these type of data), namely a jump in mean AUC compared to the SVM-model. By using rule-based training stop all outcomes can be further increased (at least marginally). The next model with two hidden layers improve this result further. Note that here too, there is no imposed regularization (no dropout, no in-sample validation). The last two models then behave as one would expect, given the first two: a drop in all measured criteria, especially in the variant without rule-based training stop.

Table 2. Overall out-of-sample AUC for the big data set

Model	Mean	Std. Dev.	Min	Max
RBF-SVM	0.700	0.005	0.692	0.707
$100_0^s : 1^s \ (v=0)$	0.713	0.007	0.698	0.723
TrainStop: W	0.718	0.005	0.711	0.727
$100_0^s : 50_0^s : 1^s \ (v=0)$	0.707	0.008	0.695	0.723
TrainStop: F	**0.722**	0.006	**0.713**	**0.732**
$100_{.1}^s : 50_{.1}^s : 1^s \ (v=0)$	0.566	0.008	0.549	0.576
TrainStop: G	0.716	0.006	0.705	0.725
$100_{.1}^s : 50_{.1}^s : 1^s \ (v = 0.2)$	0.636	0.048	0.572	0.698
TrainStop: G	0.711	0.005	0.704	0.718

5.2   Pair-Wise Training-Test AUC Curves (160 Training Epochs)

In Fig. 3 we depict the AUC evolution over the ten training-test data pairs for the first NN model from Table 2. Owing to the size of the data set, we train the models over 160 epochs. The expectation is that such a comparatively large number of training epoch will be sufficient for any of the training-test data pairs. Here too, the meaning of the curves is the same with that from the previous figures.

In this case the AUC over the training set grows less smoothly and it is now showing globally a uni-modal (locally a multi-modal) evolution. The out-of-sample curve is clearly leveling off at later epochs and its evolution is locally irregular, although if follows globally approximately the AUC curve over the training set. In this case a useful candidate for a early stopping rule would be the onset of more pronounced irregularity in the training set curve. Overall, this model is over-trained at the last learning epoch (160). Randomly chosen training stops over epochs leads to quite strong fluctuations of out-of-sample performance. Note that in this case there is no imposed regularization of any kind.

In the case of the model with two hidden layers depicted in Fig. 4 we use both types of regularization, dropout with a rate of 10% (of neurons to delete) and a in-sample validation set with 20% of retained cases (out of all available training cases or observations). Here the evolution of the AUC over training epochs is again much smoother, with the shapes of both curves matching to a high degree. While the peak AUC performance is out-of-sample quite good, there may be a

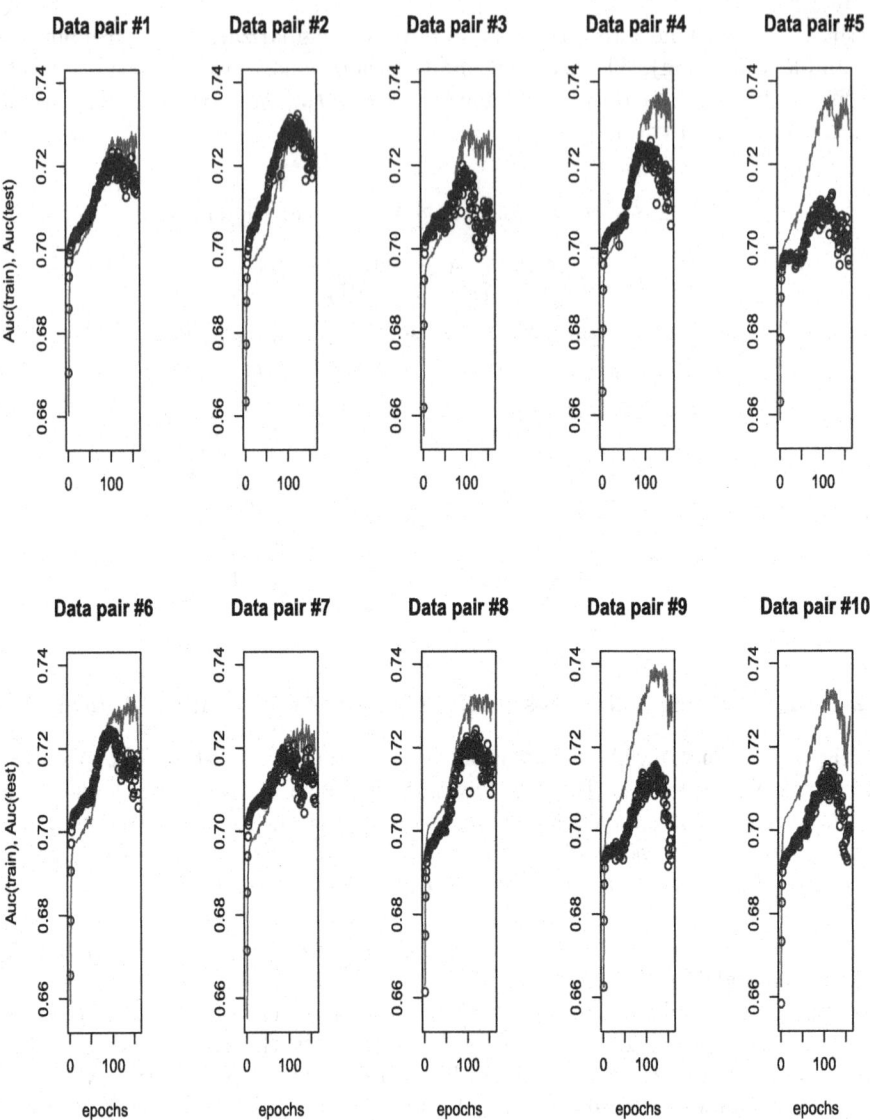

Fig. 3. AUC levels over training and out-of-sample test sets: the $100^s : 1^s$ model (d = 0, v = 0)

sharp drop of performance toward the end of the training epochs. Here a random training stop in later epochs is risk-laden while it is completely feasible over an interval starting at, say, 1/2 and reaching 2/3–3/4 of the maximum epochs.

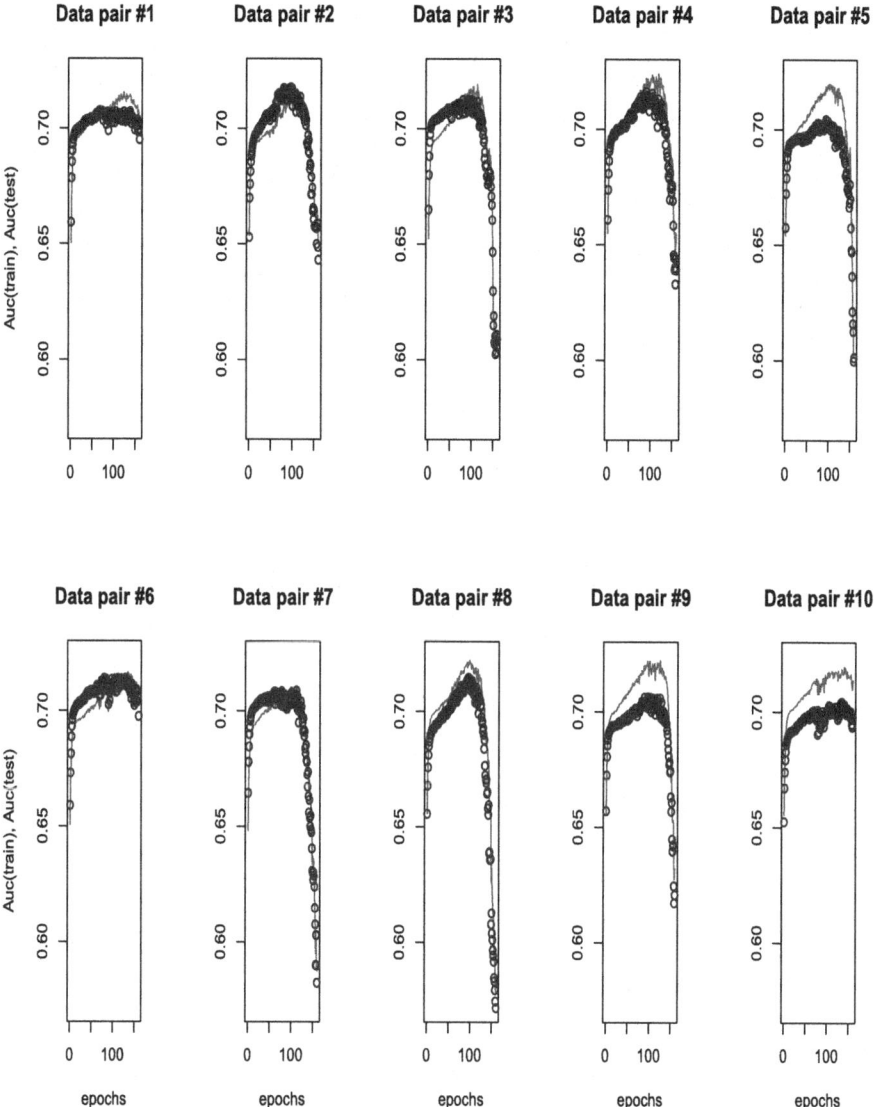

Fig. 4. AUC levels over training and out-of-sample test sets: the $100^s : 50^s : 1^s$ model $(d = 0.1, v = 0.2)$

5.3 What About Deeper NN Models? Again, for Data Set 2

Should one detect progress in out-of-sample performance by using successively deeper models (i.e. NN models with more layers) this hints at the possible presence of some non-linear dependencies, hitherto unseen by simpler models. If one does not find such progress, this does not preclude the existence of such dependencies in principle, but possibly, that they can't be detected owing to the relative sparseness of data (i.e. relatively few data points N compared to input feature dimension m). Hence, In general, a big number of training/validation examples is in itself not enough to produce more intricate, reliably detectable input-output relations. Data may also be very redundant and/or the number of different class-members may be extremely asymmetrical. The latter is certainly the case in our credit client data. Hence, questions pertaining to this problem area can't be conclusively answered except by explicitly evaluating the resulting models on the application-specific data instances. In our case this means adding more layers to the models (making the NN deeper). Hence, in what follows, we report on using NN with – at least potentially – much more complicated behavior.

Table 3. Overall out-of-sample AUC for increasingly deep NN models

Model	Mean	Std. Dev.	Min	Max
RBF-SVM*	0.700	0.005	0.692	0.707
$400^s_2 : 400^s_1 : 1^s$ ($v = 0.2$)	0.710	0.006	0.701	0.719
Regret AUC	−0.37%	11.89%	−0.68%	0.00%
$100^s_0 : 30^r_0 : 30^r_0 : 30^r_0 : 1^s$	0.701	0.012	0.678	0.720
TrainStop: G+	0.712	0.005	0.704	0.720
$100^s_0 : 30^s_0 : 30^s_0 : 30^s_0 : 1^s$	0.640	0.050	0.548	0.704
TrainStop: G	0.710	0.006	0.702	0.719
$100^s_0 : 50^s_0 : 30^r_0 : 30^r_0 : 30^r_0 : 1^s$	0.680	0.046	0.555	0.710
TrainStop: G+	**0.716**	0.007	**0.705**	**0.728**

The meaning of the entries of Table 3 is the same as for the previous tables. The regularized model with two large hidden layers beats the reference RBF-SVM slightly in terms of mean, min and max AUC. Regularization does work by smoothing the otherwise irregular training curve (thin continuous line). There are models for 5 out of 10 training-test data pair where out-of-sample AUC is above training AUC. To some degree this was also observed in runs shown in previous figures over the first epochs. As this recurs, we interpret it as a peculiarity of the respective training data sets. After increasing the number of hidden layers similar or better AUC performance can only be obtained by (rule-based) training stop. Note that the use of ReLU activation functions leads to improvements over the models which use sigmoidal activations only: the second and the fourth model are better then the third one. Overall, compared to RBF-SVM, all our deep NN models lead to a higher variance of AUC over our ten training-test data sets. The only exception is the second NN, provided one succeeds in stopping the training appropriately.

5.4   Pair-Wise Training-Test AUC Curves of Deeper NN

For the first NN with two layers containing 400 neurons respectively only the first 40 epochs are shown in Fig. 5. For this model we use both dropout and in-sample validation. For the "deeper" models (Figs. 6 and 7) all 160 epochs are shown. Here we do not use any dropout or in-sample validation, leaving all the work to the parameter adaptions by the error-backprop algorithm.

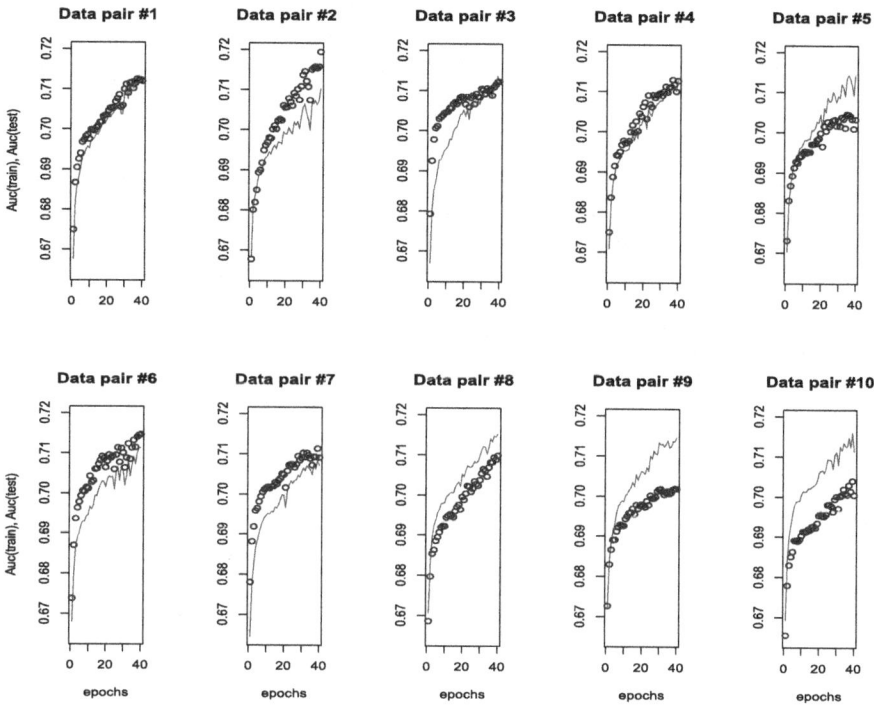

Fig. 5. AUC levels in training and out-of-sample test: the $400^s_{\cdot 2} : 400^s_{\cdot 1} : 1^s$ model (val = 0.2)

In Fig. 5 we display the first 40 training epochs of the 400 neurons per layer, two-layer NN. In this epoch range, an early-stopping moving-average rule over epoch-wise AUC evolution would work reliably.

In Fig. 6 we display all 160 training epochs of the four hidden layer network (the third NN model from Table 3) which uses no imposed regularization but sigmoidal activations in the higher layers. Provided one could find an adequate early-stopping rules for training (moving-average seems to work here as well) this model would beat the RBF-SVM reference model. A random training stop, however, would produce much inferior results. Note that the model is degrading over the later training epochs, but is also able to recover. It is unclear if, e.g. by using very long training, one could find still better models.

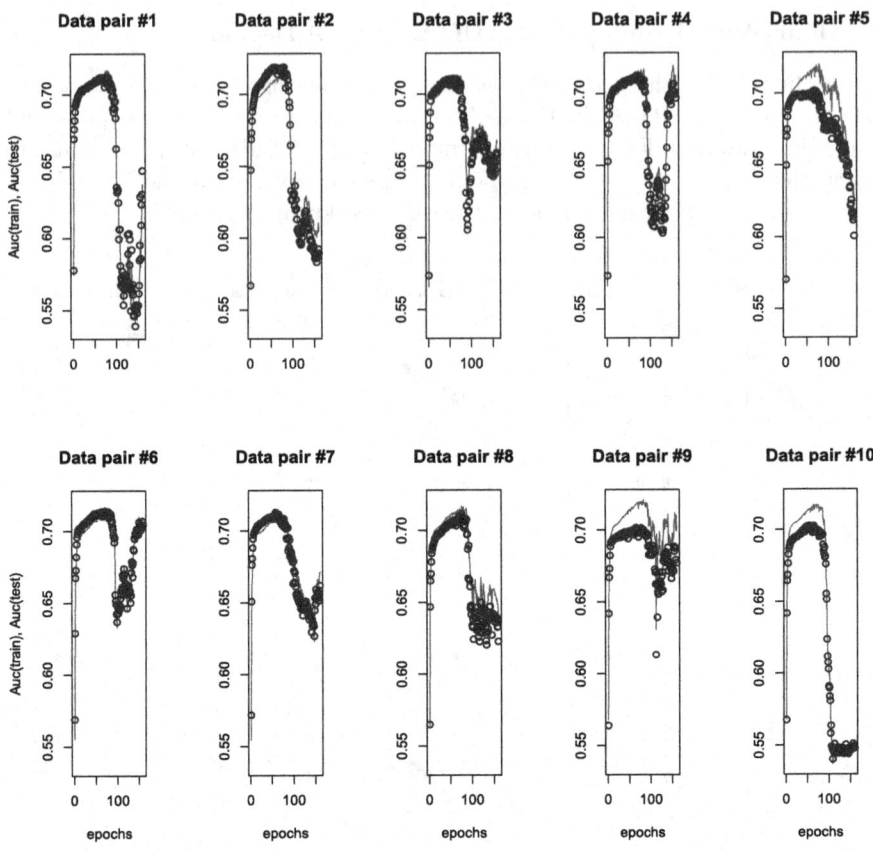

Fig. 6. AUC levels in training and out-of-sample test: the $100^s : 30^s : 30^s : 30^s : 1^s$ model

Finally, in Fig. 7 we also display all 160 training epochs of the five hidden layer network (the fourth NN model from Table 3) which also does not use imposed regularization but ReLU activations in all higher layers. Compared to the last described model (in Figure 6) this model does degrade much less over later training epochs. Hence, using adequate early-stopping rules for training (moving-average seems to work here quite well) are less risky to apply. This model would beat the RBF-SVM reference model (together with the second NN model from Table 3 which also uses ReLU activations) by the largest margins.

A recommendation for further progress in finding the appropriate (as well as robustly good performing) models is to further study possible rule-based stopping criteria for deep NN training. This may proceed for instance by starting with simple moving-average based rules and may employ more elaborate shape analysis of learning curves on training sets as true out-of-sample sets are not available in real world applications.

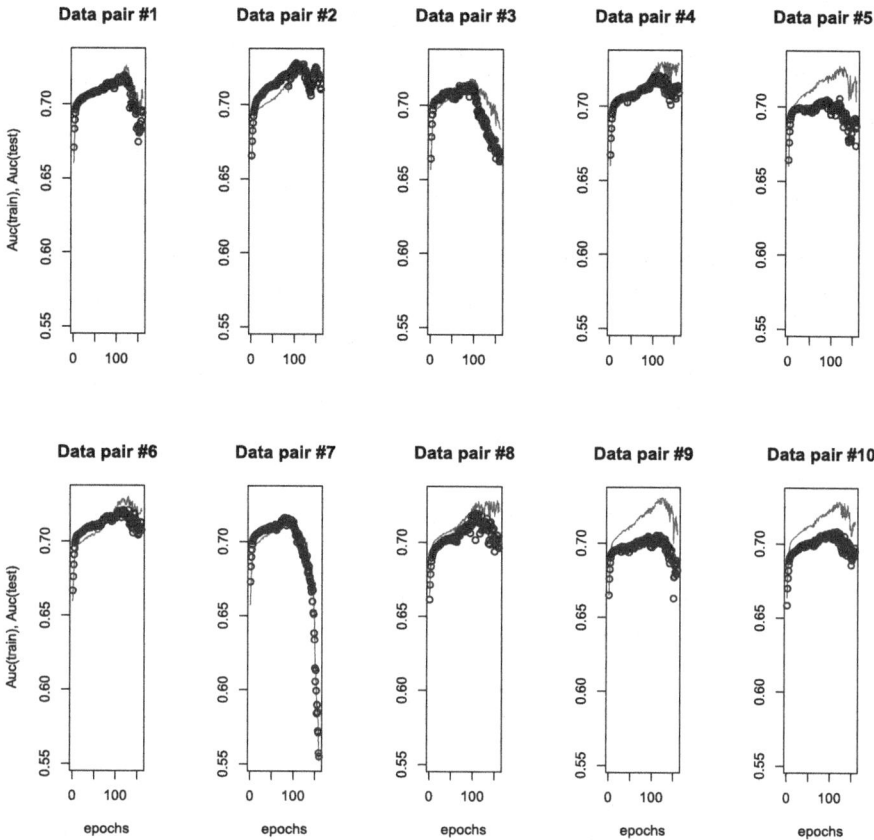

Fig. 7. AUC levels in training and out-of-sample test: the $100^s : 50^s : 30^r : 30^r : 30^r : 1^s$ model

6 Some Attempts at Explaining Our Results

The gain of using deep neural networks for our credit client data is significant – but not dramatically so, as one may expect. Especially for the "deeper" models, ROC-AUC is more volatile and is conditioned on the ability of the modeler to be able to make a good guess for the training epoch for which a model will produce a higher (or perhaps maximal) ROC-AUC value over a test set.

6.1 About Computational Cost of Our Models

For the models on our small data set, Logistic Regression is competitive. Compared to any other more complex modeling technique (SVM, dNN, ...) the computational costs are negligible. For the models on our big data set, we have to compare RBF-SVM against dNN. Due to the use of the RBF-kernel and due to the need to perform grid search for both hyperparameters "capacity" and "kernel-width" the cost of the SVM is not significantly lower compared to that of the highly refined and optimized dNN model-building via Tensorflow (time consumption on an Intel i5 4-core computer is around 2–3 h for the whole computational cycle, including 10-fold train-test splits per model).

6.2   Why Are dNN Superior with Regard to ROC-AUC?

ROC curves are calculated by confronting the vector of the binary, e.g. (0, 1)-reference output for every data example (case) with the vector of continuous probabilities of choice (the effective model output) for the cases by varying a real valued "cutoff". As all the measures based on discrete information (hit-rates, etc.) do not differ very much between our models, only the particular fine-structures of the output-probabilities acts differently on the results as the cut-off is gradually shifted. In past modeling [15] we observed – especially in capacity constrained SVM-models, which intentionally allow for a certain amount of mis-classification in order to avoid overtraining – the formation of substantially different vectors of output probabilities for the same data when using just slightly different nonlinear kernels (e.g. of Coloumb-type instead of RBF, see [8,15]. A similar effect may hold in the present model population, as dNN are capable of more diverse input-output mappings.

6.3   Concerning the Prediction of the Best Moment for Training Stop

To a certain extend we can predict the best stopping moment (in terms of number of epochs) by fitting piecewise (local) linear regressions to re-scaled learning curves resulting from minimizing the binary cross-entropy which is used as the effective objective. This is useful if the models with the maximum out-of-sample AUC value tend to be near inflection points between local (linear) approximations. This is true in a majority of cases but there are notable exceptions. Combining such models may yield still superior out-of-sample AUC values. Other variants are to use the AUC learning curve (from training) to predict best stopping moments for out-of-sample AUC levels. Furthermore, in order to improve overall AUC levels, one may seek to optimize directly with regard to AUC instead of reading it off as a co-metric (personal communications by Frank-Michael Schleif). This, however, may require non-differentiable methods for model parameter search.

7   Conclusion and Outlook

Credit scoring data come in different input shapes and volumes. Their associated classification problems face both relative data sparseness and substantial noise. However, owing to various reasons, for instance data privacy, data availability of such applications will and cannot be increased at will. Hence, besides the our long-lasting aim of increasing forecasting performance (e.g. [13]), more recently

incorporating privacy constraints [16], and to better understand the data themselves [12], here we are concerned with what aspects of modeling can be further
improved. Our extensive tests using deep neural networks (dNN) for two very
different credit client data sets do find that training and validating classification
models leads to limited improvement over the more traditional out-of-sample
performance – as is measured by hit-rate and classifications accuracy. However
using these neural models with AUC as a criterion for training stop, proves to be
significant in that it is often reliable and the resulting (superior) out-of-sample
AUC is in itself a robust measure of overall model performance and stability. As
truly dramatic improvements over base models are very unlikely for our data,
we nevertheless conclude that (1) dNN may produce significantly better out-of-
sample forecasts than standard ML or statistical learning models; and even the
grossly over-sized dNN model variants still produce highly competitive results
with limited training effort. (2) Dropout as a regularization method of dNN may
produce more stability w.r.t. advantageous training stops; it hence may replace
the widely practiced in-sample validation. In general, it may be stated that such
and related modeling successes may be mainly attributed to the ability of today's
computationally advanced dNN to successfully auto-regularize. (3) For our bigger data set, which is more difficult to classify and in some sense degenerate,
dNN clearly show their potential. Again such dNN are highly robust against
over-specification. These findings concern our implicit goal of maximizing out-
of-sample ROC/AUC. However, as far as using dropout in place of in-sample validation is concerned, the message carries over to other modeling problems and
for other performance measures (e.g. recurrent dNN and MSE-error, as other
own work finds). Finally, (4) the chances of obtaining valuable models by using
dNN increases if the financial industry manages someday (against all odds) to
merge their client data bases. The source for training data would become truly
huge, and perhaps more structured, including contexts, news, client attitudes
and sentiment, all conveniently time-stamped. This may then lead to additional
and interesting modeling targets.

References

1. Abdou, H., Pointon, J.: Credit scoring, statistical techniques and evaluation criteria: a review of the literature. Intell. Syst. Account. Finance Manag. **18**, 59–88
   (2011)
2. Addo, P.M., Guègan, D., Hassani, B.: Credit risk analysis using machine and deep
   learning models. Risks **6**(38), 1–13 (2018). https://doi.org/10.3390/risks6020038
3. Bengio, Y.: Practical recommendations for gradient-based training of deep architectures. arXiv:1206.5533 [cs.LG], pp. 1–20 (2012). https://arxiv.org/pdf/1206.5533.
   pdf
4. Experian, The credit reference agency explained, a guide for consumer
   advisers, experian (2013). http://www.experian.co.uk/downloads/consumer/
   creditRefAgencyExplained.pdf
5. Goodfellow, I., Bengio, Y., Courville, A.: Deep Learning. The MIT Press, Cambridge (2016)

6. Ha, V.-S., Nguyen, H.-N.: Credit scoring with a feature selection approach based deep learning. In: MATEC Web of Conferences 54, 7th International Conference on Mechanical, Industrial, and Manufacturing Technologies, pp. 1–5 (2016). https://doi.org/10.1051/matecconf/20165405004

7. Hinton, G.: A practical guide to training restricted Boltzmann machines. Working Paper UTML TR 2010–003, pp. 1–20, University of Toronto (2010). http://www.cs.toronto.edu/~hinton/absps/guideTR.pdf

8. Hochreiter, S., Mozer, M.C., Obermayer, K.: Coulomb classifiers: generalizing support vector machines via an analogy to electrostatic systems. In: Advances in Neural Information Processing Systems 15, pp. 561–568, MIT Press (2003)

9. Keras, called from R. https://cran.r-project.org/web/packages/keras/index.html. Accessed 19 Sept 2020

10. Kvamme, H., Sellereite, N., Aas, K., Sjursen, S.: Predicting mortgage default using convolutional neural networks. Expert Syst. Appl. pp. 1–43 (2018). https://doi.org/10.1016/j.eswa.2018.02.029. Accepted for publication

11. Rackauckas, C., et al.: Universal differential equations for scientific machine learning. arXiv:2001.04385v3, pp. 1–45 (2020)

12. Schebesch, K.B., Stecking, R.W.: Topological data analysis for extracting hidden features of client data. In: Doerner, K.F., Ljubic, I., Pflug, G., Tragler, G. (eds.) Operations Research Proceedings 2015. ORP, pp. 483–489. Springer, Cham (2017). https://doi.org/10.1007/978-3-319-42902-1_65

13. Schebesch, K.B., Stecking, R.W.: Support vector machines for credit applicants: detecting typical and critical regions. J. Oper. Res. Soc. **56**(9), 1082–1088 (2005)

14. Schölkopf, B., Smola, A.: Learning with Kernels. The MIT Press, Cambridge (2002)

15. Stecking, R., Schebesch, K.B.: Comparing and selecting SVM-kernels for credit scoring. In: Spiliopoulou, M., Kruse, R., Borgelt, C., Nürnberger, A., Gaul, W. (eds.) From Data and Information Analysis to Knowledge Engineering. Studies in Classification, Data Analysis, and Knowledge Organization. Springer, Berlin (2006). https://doi.org/10.1007/3-540-31314-1_66

16. Stecking, R.W., Schebesch, K.B.: Classification of credit scoring data with privacy constraints. Intell. Data Anal. **19**(s1), S3–S18 (2015)

17. Song, M., Hu, Y., Chen, H., Li, T.: Towards pervasive and user satisfactory CNN across GPU microarchitectures. In: International Symposium on High Performance Computer Architecture (HPCA), pp. 1–12 (2017)

18. Tomczak, J.M., Zieba, M.: Classification restricted Boltzmann machine for comprehensible credit scoring model. Expert Syst. Appl. **42**(2), 1789–1796 (2015). https://doi.org/10.1016/j.eswa.2014.10.016

19. Vapnik, V.: Statistical Learning Theory. Wiley, New York (1998)

20. Yu, L., Yang, Z., Tang, L.: A novel multistage deep belief network based extreme learning machine ensemble learning paradigm for credit risk assessment. Flex. Serv. Manuf. J. **28**(4), 576–592 (2015). https://doi.org/10.1007/s10696-015-9226-2

Integrated Tool for Assisted Predictive Analytics

Florin Stoica$^{(\boxtimes)}$ ⓘ and Laura Florentina Stoica ⓘ

Research Center in Informatics and InformationTechnology, Lucian Blaga
University of Sibiu, Sibiu, Romania
{florin.stoica,laura.cacovean}@ulbsibiu.ro

Abstract. Organizations use predictive analysis in CRM (customer relationship
management) applications for marketing campaigns, sales, and customer services, in manufacturing to predict the location and rate of machine failures, in
financial services to forecast financial market trends, predict the impact of new
policies, laws and regulations on businesses and markets, etc. Predictive analytics is a business process which consists of collecting the data, developing
accurate predictive model and making the analytics available to the business
users through a data visualization application. The reliability of a business
process can be increased by modeling the process and formally verifying its
correctness. Formal verification of business process models aims checking for
process correctness and business compliance. Typically, data warehouses are
usually used to build mathematical models that capture important trends. Predictive models are the foundation of predictive analytics and involve advanced
machine learning techniques to dig into data and allow analysts to make predictions. We propose to extend the capability of the Oracle database with the
automatic verification of business processes by adapting and embedding our
Alternating-time Temporal Logic (ATL) model checking tool. The ATL model
checker tool will be used to guide the business users in the process of data
preparation (build, test, and scoring data).

Keywords: Predictive analytics · ATL · Oracle

1 Introduction

Predictive analysis has developed a lot in recent years benefiting from advances in
support technologies, especially in the fields of machine learning and big data.

With the help of predictive analytics, organizations can determine, build, and
exploit models contained in their own data warehouses to detect opportunities, risks, or
to make analysis projections (financial forecast, sales volumes forecast, etc.).

Models can be designed, for instance, to discover relationships between various
monitored variables, behaviour factors and to track specified indicators.

Predictive analytics is aimed at making predictions about future outcomes based on
historical data and analytics techniques such as statistical modelling and data mining.
Historical data is used to build a mathematical model (predictive model) that is used on

© Springer Nature Switzerland AG 2021
D. Simian and L. F. Stoica (Eds.): MDIS 2020, CCIS 1341, pp. 149–166, 2021.
https://doi.org/10.1007/978-3-030-68527-0_10

current data to predict what will happen next, or to suggest actions to take for optimal outcomes.

Organizations can use predictive analytics to exploit current and historical data to detect trends, forecast events and situations that should occur at a specific time, based on the provided parameters and variables.

We intend to design and implement an Integrated Tool to provide Assisted Predictive Analytics (ITAPA) which will operate in the field of financial forecast to provide assisted predictive modelling capability by guiding business users within the predictive analytics workflow.

The tool will be based on custom services embedded in an extensible database system. These services will be integrated into a Web application, developed as a component-based tool using the low-code development platform APEX.

One of the reasons why we chose the Oracle Database as the target of our developments is that Oracle Database provides support for developing, storing, and deploying Java classes, which facilitates the embedding of new services. The new services provided at the database level will be exploited at the business level within the organization through applications developed in Oracle Application Express (APEX) - a low-code development platform which can be used to deliver solutions to real business problems and to build scalable, secure enterprise apps. We will exploit also the Oracle Advanced Analytics, which provides a broad range of in-database, parallelized implementations of machine learning algorithms to solve many types of business problems.

The focus of the ITAPA tool will be on financial forecast, in order to maintain the financial balance of the company, based on the anticipation of positive and negative flows in the medium term (2–5 years).

The predictive analytics will used to establish the proportions in the activity of financial exploitation, strategic and financial policy of the company, with positive or negative influences on the production process and the economic-financial phenomena, in order to achieve the profitability, balance and prevention of bankruptcy risk.

The main objectives of the financial forecast are:

- Ensuring the financial balance and an optimal financial structure;
- Rational allocation of the money resources that requires the evaluation of the expenses for new investments/modernizations and the measurement of their efficiency, in order to avoid capital assets;
- Assessment of the company's ability to generate cash flow in the medium and long term in order to cover the debt service.

Analysis projections will be made, using one or more variables:

- Exchange rate fluctuation – increase/decrease: EUR, USD;
- Interest rate changes – EURIBOR, ROBOR;
- New investments loan;
- Change of wage costs based on government laws;
- Sales volumes forecast – based on marketing research.

The ITAPA software will rely on Oracle database which will embed its core functionality (custom services embedded in the Oracle database and made accessible as

Java stored procedures). The Web interface of ITAPA will be made using Oracle APEX, a component of Oracle RAD Stack included by default in the Oracle database. The ITAPA tool will include a data visualization component, to make the analytics available to the business users.

The rest of this paper is organized as follows. In Sect. 2 is briefly described the supervised learning model used for regression analysis in the ITAPA tool. In Sect. 3 we present a methodology based on model checking to control predictive analytics workflow. Section 4 contains the architecture of the proposed tool for Assisted Predictive Analytics, which is based on embedded custom services within Oracle database. In Sect. 5 are presented technical details about the process of embedding the ATL model checker into Oracle database. Some representative tools for analytical predictions are mentioned in Sect. 6. Conclusions and further directions of study are formulated in Sect. 7.

2 Building Predictive Models Using Support Vector Machines

In the recent years, Support Vector Machines (SVMs) have become a very popular tool for machine learning tasks which can be used for both classification and regression analysis. Many applications of SVM have been done in various fields: face detection, handwriting recognition, image classification, bioinformatics, prediction of financial time series. Each instance of data contains one target values and several attributes. The goal of SVM is to produce a model which predicts target value of data instances in the testing set for which only the attributes are given. SVM training involves optimization of a convex cost function, to find a hyperplane in an N-dimensional space (N - the number of features) that distinctly classifies the data points. Hyperplanes are decision boundaries that help classify the data points.

The objective of the SVM algorithm is to find an optimal separating hyperplane with a maximal margin, i.e. the maximum distance between data points of both classes.

In the case of non-separable data the SVM algorithm uses a technique called the kernel trick. The SVM kernel is a function that takes low dimensional input space and transforms it to a higher dimensional space i.e. the data are projected in a space with higher dimension in which they are separable by a hyperplane [4]. Kernel functions can be linear or nonlinear.

SVM uses an epsilon-insensitive loss function to solve regression problems. Oracle Data Mining (ODM) has its own proprietary implementation of Support Vector Machines (SVM), which exploits the many benefits of the algorithm while compensating for some of the limitations inherent in the SVM framework [7, 13]. The standard implementation of the Support Vector Machine algorithm in the Oracle database uses a Linear or Gaussian (RBF) kernel.

SVM regression tries to find a continuous function such that the maximum number of data points lie within the epsilon-wide insensitivity tube. Predictions falling within epsilon distance of the true target value are not interpreted as errors. Because SVR is the name used to denote a version of a SVM for regression, we will note with ODM-SVR its specific implementation by Oracle.

3   Using Model Checking to Verify and Assist the Predictive Analytics Process

Predictive analytics starts with a business goal: to shape the strategic and financial policy of organizations to their unique socio-economic conditions. The process capitalizes on heterogeneous, often massive, data sets in models that can generate clear, actionable results to support the achievement of business objectives.

To prove that all errors have been found, we will use a formal method, namely model checking, to detect and eliminate bugs in the design of a predictive analysis process.

Design validation involves checking whether the business process design satisfies the system requirements.

The verification process involves translating the business process model given in the BPMN 2.0 (Business Process Model and Notation) to a formal specification in the form of concurrent game structures [11]. This transformation makes possible formal and automatic verification of the functional and qualitative requirements imposed to a given business process using an ATL model checker.

Our ATL model checker will be also used to guide (auto-recommendations, auto-suggestions, wizards) the user in data preparation (build, test, and scoring data) used by the predictive model.

3.1   Verification of Business Processes Using ATL Model Checking

Alur et al. [1] introduced Alternating-time Temporal Logic (ATL), a temporal logic based on a computational model appropriate to describe compositions of open systems, called concurrent game structure (CGS).

ATL models generalize turn-based transition trees from game theory, includes notions of agents, their abilities and strategies (conditional plans) explicitly in its models and thus it is relatively simple to encode turn-based game in the formalism of concurrent game structures (imposing that only one agent makes a move at any given time step). The semantics of ATL is formalized by defining games such that the satisfaction of an ATL formula corresponds to the existence of a winning strategy.

In [12] we presented a method based on our ATL model checker for verification of a business processes described using the BPMN 2.0.

Automated verification of a business process by ATL model checking is the formal process through which a given specification expressed by an ATL formula and representing a desired behavioural property is verified to hold for the ATL model obtained by parsing the BPMN specification of the business process.

3.2   Alternating-Time Temporal Logic

The model checking problem for ATL is to determine the states of a model in which a given ATL formula is satisfied.

ATL path quantifiers defines "cooperation modalities" of the form $\langle\langle A\rangle\rangle\varphi$, where A is a set of agents. The ATL formula $\langle\langle A\rangle\rangle\varphi$ express the fact that the agents A can cooperate to ensure that φ holds against any behavior of the other agents (equivalently, that A have a winning strategy for φ) [6].

In the following we will present the computational model used to describe the ATL semantics, called Concurrent Game Structure (CGS). Mode details about ATL syntax and semantics can be found in the paper [11], where the own implementation of an ATL model checker is described.

A concurrent game structure is defined as a tuple $S = \langle \Lambda, Q, \Gamma, \gamma, M, d, \delta \rangle$ with the following components:

- $\Lambda = \{1, \ldots, k\}$ is a nonempty and finite set of all considered agents;
- Q is the finite set of *states*;
- Γ represents the finite set of *propositions* with which the states of the model will be labeled;
- $\gamma : Q \to 2^\Gamma$ is called the *labelling* function, defined as follows: for $\forall q \in Q, \gamma(q)$ is the set of propositions *true* at q;
- M denotes the nonempty finite *set of moves*;
- the function $d: \Lambda \times Q \to 2^M$ associates each player $a \in \Lambda$ and each state $q \in Q$ with the set of available moves of agent a at state q. In the following, we will use the notation $d(a, q) = d_a(q)$. For $\forall q \in Q$, a tuple $<j_1,\ldots,j_k>$ such that $j_a \in d_a(q)$ for $\forall a \in \Lambda$, represents a *move vector* at q.
- the transition function $\delta(q, \langle j_1, \ldots, j_k \rangle)$, associates to each state $q \in Q$ and each move vector $\langle j_1, \ldots, j_k \rangle$ at q the new state in which the transition from q is made if every player $a \in \Lambda$ chooses the move j_a.

A *computation* of S is an infinite sequence of states $\lambda = q_0, q_1, \ldots$ such that q_{i+1} is the successor of q_i, $\forall i \geq 0$. A computation starting at state q is called a *q-computation*.

3.3   ATL Syntax and Semantics

Given a concurrent game structure S and a set of agents $A \subseteq \Lambda$, an ATL formula is defined by the following syntax:

$$f ::= p \mid \neg \varphi \mid \varphi_1 \vee \varphi_2 \mid \langle\langle A \rangle\rangle \bigcirc \varphi \mid \langle\langle A \rangle\rangle \square \varphi \mid \langle\langle A \rangle\rangle \Diamond \varphi \mid \langle\langle A \rangle\rangle \varphi_1 U \varphi_2.$$

where $p \in \Gamma$ and $\neg \varphi, \varphi_1, \varphi_2$ are well-formed ATL formulas.

A *strategy* for the player $a \in \Lambda$ determines for every finite prefix λ of a computation a move for the player a in the last state of λ.

For a game structure S, we write $q \models \varphi$ to indicate that the formula φ is satisfied in the state q of the structure S. For each state q of S, the satisfaction relation $\models$ is defined inductively as follows:

- for $p \in \Gamma$, $q \models p \Leftrightarrow p \in \gamma(q)$.
- $q \models \neg \varphi \Leftrightarrow q \nvDash \varphi$
- $q \models \varphi_1 \vee \varphi_2 \Leftrightarrow q \models \varphi_1$ or $q \models \varphi_2$.
- $q \models \langle\langle A \rangle\rangle \bigcirc \varphi \Leftrightarrow$ there exists a strategy for each player in A, such that for all q-computations λ following these strategies, we have $\lambda [1] \models \varphi$ (the formula φ is satisfied in the successor of q within computation λ).

- $q \vDash \langle\langle A \rangle\rangle \; \square \; \varphi \Leftrightarrow$ there exists a strategy for each player in A, such that for all q-computations λ following these strategies, and all positions i $\geq$ 0, we have $\lambda[i] \vDash \varphi$ (the formula φ is satisfied in all states of computation λ).
- $q \vDash \langle\langle A \rangle\rangle \diamond \; \varphi \Leftrightarrow$ there exists a strategy for each player in A, such that for all q-computations λ following these strategies, there exists a position i $\geq$ 0 such that $\lambda[i] \vDash \varphi$ (the formula φ is satisfied in at least one state of computation λ).
- $q \vDash \langle\langle A \rangle\rangle \; \varphi_1 \; U \; \varphi_2 \Leftrightarrow$ there exists a strategy for each player in A, such that for all q-computations λ following these strategies, there exists a position i $\geq$ 0 such that $\lambda[i] \vDash \varphi_2$ and for all positions $0 \leq j < i$, we have $\lambda[j] \vDash \varphi_1$.

3.4   Formal Verification of a Business Process

Business Process Model and Notation represents the de-facto standard for business processes diagrams that allows the building a visual model which expresses a work-flow's business requirements graphically.

A simplified BPMN diagram is composed by two basic categories of elements: Flow Objects and Connecting Objects. Flow Objects are the main graphical elements from the diagram of a business process used to define its behavior. There are three types of Flow Objects: Tasks, Gateways and Events [3] and three types of Connecting Objects:

1. *Uncontrolled flow*: a flow that is not affected by any conditions or does not pass through a Gateway;
2. *Conditional flow*: has an associated condition to determine whether or not the flow will be used;
3. *Default flow*: this flow will be used only if all the other outgoing conditional flow is not true.

We note with $F = T \cup G \cup E$ the set of Flow Objects, composed by Tasks (*T*), Gateways (*G*) and Events (*E*).

The set of Connecting Objects is denoted by $C = C_U \cup C_C \cup C_D$ and is composed by Uncontrolled flows (C_U), Conditional flows (C_C) and Default flows (C_D).

Each graphic element of the diagram will be labelled with the function:

$\pi: F \cup C \rightarrow \Pi \cup \{none, default\}$ defined as follows:

$$\pi(o) = \begin{cases} none \; if \; o \in C_U \\ default \; if \; o \in C_D \\ name_o \in \Pi \; if \; o \in F \cup C_C \end{cases}$$

where Π denotes the finite set of names of Flow Objects (*F*) and Conditional flows (C_C).

We denote by *ce* the restriction of π to *C*:

$ce = \pi_{|C}: C \rightarrow \Pi \cup \{none, default\}$ where:

$$ce(c) = \begin{cases} none \ if \ c \in C_U \\ default \ if \ c \in C_D \\ \pi(c) \ if \ c \in C_C \end{cases}$$

For simplicity, we will consider that ce function assigns to each connecting object from C its *condition expression*.

To complete the formal definition of a business process we define the transition function $\delta_{BP}(f_o, c_o)$ which associates to each flow object $f_o \in F$ and each outgoing connecting object $c_o \in C$ the flow object connected to f_o by c_o.

For a business process model defined above, we build an equivalent concurrent game structure $S = \langle \Lambda, Q, \Gamma, \gamma, M, d, \delta \rangle$ following the rules:

- $\Lambda = \{1\}$;
- $Q = F$;
- $\Gamma = \pi_{|F}(F) \cup \{p, c\}$; for each task $t \in T \subseteq F = Q$ we introduce two additional labels: p for task in progress (or not started) and c for task completed.
- The *labelling* function $\gamma : Q \rightarrow 2^{\Gamma}$ is defined as follows:

$$\gamma(q) = \begin{cases} \pi_{|F}(q) \ if \ q \in G \cup E \ (gateway \ or \ event) \\ \pi_{|F}(q) \cup \{p\} \ if \ q \in T, \ q \ is \ in \ progress \ or \ not \ started \\ \pi_{|F}(q) \cup \{c\} \ if \ q \in T, \ q \ is \ completed \end{cases}$$

- $M = \bigcup_{c \in C} ce(c) \cup \{w\}$; the special move w (wait) aims to wait for a task to complete;
- The function $d : \Lambda \times Q \rightarrow 2^{M}$ is defined by:

$$d(1, q) = \begin{cases} \bigcup_{c \in C_q} ce(c) \ \forall q \in G \cup E \\ \bigcup_{c \in C_q} ce(c) \ if \ q \in T, \ q \ is \ completed \\ \{w\} \ if \ q \in T, \ q \ is \ in \ progress \ or \ has \ not \ started \end{cases}$$

where C_q represents the set of connecting objects outgoing from q.
- The transition function δ is defines as follows:

$\delta(q, <j>) = \delta_{BP}(q, j), \forall q \in Q$ for which $p \notin \gamma(q)$ and $\forall j \in \bigcup_{c \in C_q} ce(c)$ and
$\delta(q, <w>) = q, \forall q \in T$ for which $p \in \gamma(q)$ (*task in progress*).

The formal verification of a given business process, using the ATL model checking, is done in two steps:

1. The equivalent ATL model is constructed following rules described above;
2. The ATL formula representing a desired behavioural property is verified to hold for the model obtained at step 1.

In this way, the user is guided to perform the tasks which will be presented in Sect. 3.5. The accomplishment of these tasks is necessary for performing the transitions (moves) in the ATL concurrent game structures, for the assistant agent to win the ATL games.

The process of verifying a business process using ATL model checking is shown in Fig. 1.

3.5 Control Predictive Analytics Workflow with Winning Game Strategies

Predictive analytics is sequenced as following steps – Model the Data, Data preparation for ODM-SVR, Training and Testing the Model, Scoring/Predict/Forecast the outcome.

Data preparation can be done manually but for automation we will prefer Oracle Automatic Data Preparation (ADP).

We will consider each step as a (sub)process which will be modelled through an ATL concurrent game structure (CGS) with two agents, the System (the agent 1) and the business user (the agent 2).

After designing all the business processes that define the predictive analytics workflow, the properties to be verified are described as goals to reach or conditions to be met, and are expressed by ATL formulas. The ATL model checker takes as input for each step the ATL formula φ (the property to be checked) and the concurrent game structure which corresponds to the respective business process.

The game is modelled rather cooperatively, the business user being guided to favour the winning of the game by the assistant agent (system), following his winning strategy.

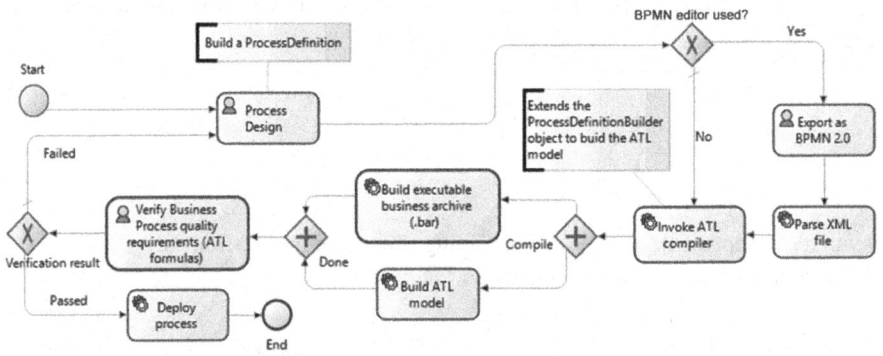

Fig. 1. Formal verification of a business process using ATL model checking [12].

For each process mentioned above, will be implemented an algorithm which looks for infallible conditional plans to achieve a winning strategy that is defined via the ATL formulae φ which specifies the successful completion of the respective step.

We will impose as requirement for player 1 (the Assistant Agent) to win the game: the formula $q_0 \vDash \langle\langle\{1\}\rangle\rangle \diamond \varphi$ states that there exists a set of strategies of agent 1 such that for all q_0-computations λ following these strategies, the formula φ is satisfied in at least one state of the computation λ (regardless of the other agent's moves), where q_0 is the state from CGS corresponding to the initial state of the process.

The business user (the agent 2) will be "guided" through the processes of the predictive analytics workflow forcing him to choose at next move a state from the set (of states in which the formula) $\langle\langle\{1\}\rangle\rangle \circ \varphi$ (is satisfied) or from the set $\langle\langle\{1\}\rangle\rangle \Diamond \varphi$, in this order.

Due to the fact that there is no competition between agents, the implementation will consider only player 1 (according to the Sect. 3.4) and the execution of tasks will be done according to its strategy.

Implementation details are encapsulated in the ITAPA tool and do not involve the business users, they only have to execute the tasks proposed by the system (which follows the Assistant Agent's strategy).

Details on using ATL in business process verification can be found in [10, 12]. In [12] a case study is presented and in [10] the ATL model checker was successfully used in verification of JADE agents with Finite State Machine (FSM) behaviours.

4 Embedding Custom Services Within Oracle Database

Oracle Database provides support for developing, storing, and deploying Java applications [5] on the strength of a complete implementation of the standard Java programming language and a fully compliant JVM.

In Fig. 2 and Fig. 3 are depicted the classical architecture of our ATL model checker tool and the new architecture, respectively. In the new architecture, the model checker tool is embedded within Oracle database and provided as custom service through Java stored procedure technology.

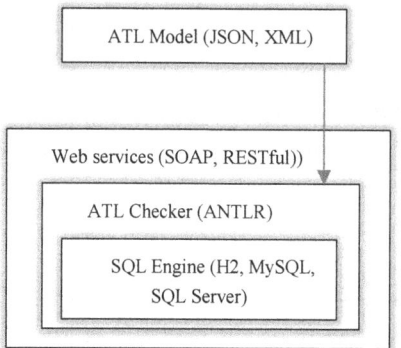 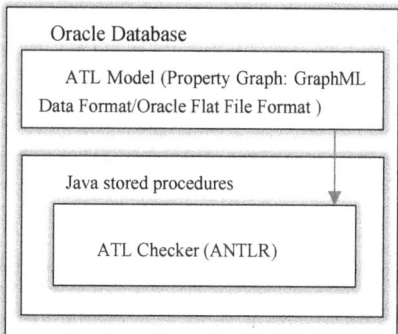

Fig. 2. Current architecture of ATL model checker (Web service)

Fig. 3. Embedded ATL model checker (Java stored procedure)

A property graph consists of a set of objects or vertices, and a set of arrows or edges connecting the objects. Vertices and edges can have multiple properties, which are represented as key-value pairs and thus a property graph is very suitable to store an ATL concurrent game structure.

Property graphs are supported in Oracle Database by of a set of PL/SQL packages, a data access layer, and an analytics layer [8].

The data access layer provides a set of Java APIs that can be used to create and drop property graphs, add and remove vertices and edges, search for vertices and edges using key-value pairs and perform other manipulations which can be used to build ATL models.

Property graphs are stored in Oracle Database in tables which are used internally to model the vertices and edges. To improve query performance, some indexes are created by default in the database.

Property graphs provide very good support for storing ATL models as concurrent game structures (CGS). The GraphML Data Format is very similar with the current XML format used in our ATL model checker and our JSON fromat can be easily translated to the Oracle Flat File Format.

Using Oracle data access layer for property graphs, we intend to eliminate the need to use internal (Java) data structures in our ATL model checker.

Storing the entire ATL model in the database will greatly contribute to increasing scalability of the ATL model checker tool.

Figure 4 illustrates how our ATL model checker resides on top of the Java core class libraries, which reside on top of the Oracle Database JVM. Because the Oracle Java support system is located within the database, the JVM interacts with Oracle Database libraries, and does not directly access the operating system [5].

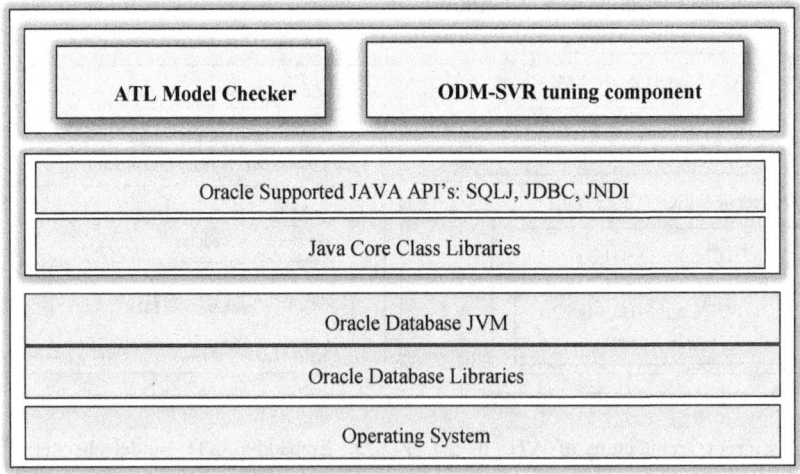

Fig. 4. Embed custom Java applications in Oracle Database Java Component Structure (adapted from [5])

The combination of Java and Oracle Database allows creating component-based, network-centric or data-centric applications that can be easily developed and updated when business needs change. The Oracle JVM takes advantage of the session architecture of Oracle database to concurrently run Java applications for hundreds to thousands of users.

Figure 5 shows the architecture of our tool that will exploit the services embedded in the database. The figure also shows how Oracle Net Services Connection Manager can combine many network connections into a single database connection. This enables Oracle Database to support a large number of concurrent users [5].

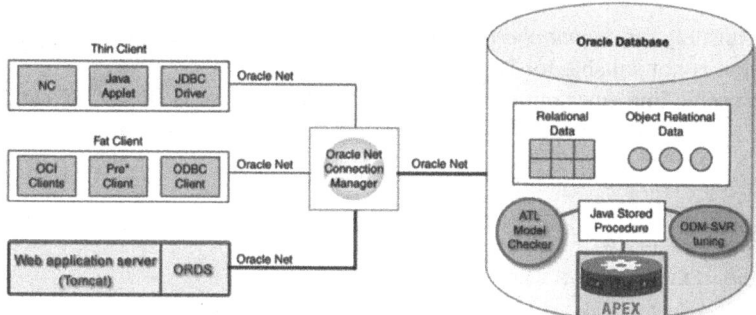

Fig. 5. The architecture of the proposed tool for Assisted Predictive Analytics (adapted from [5]).

We will use 3-tier architecture, client/server configuration, in which the clients (remote or from within the database) call Java stored procedures in the same way they call PL/SQL stored procedures.

A web request from the web browser is sent to Oracle REST Data Services (ORDS) where it is handed to Oracle Database to be treated. Within the database, the request is processed by Oracle APEX, which allows the direct invocation of our custom services by calling the Java procedures that encapsulate them. Once the processing is complete, the result is sent back through ORDS to the browser.

This architecture is based on the Oracle RAD Stack, an inclusive technology (composed by three core components: Oracle REST Data Services (ORDS), Oracle APEX, and Oracle Database) for the rapid development of scalable, secure apps, with enterprise-class features.

5   Embedding the ATL Model Checker into Oracle Database

An ATL model checker must implement the function $Pre(A, \Theta) = \{q \in Q \mid \forall a \in A \; \exists\; j_a \in d_a(q)$ such that for every $b \in \Lambda \setminus A$ and $j_b \in d_b(q)$, $\delta(q, <j_1,..., j_k>) \in \Theta\}$. Therefore $Pre(A, \Theta)$ (with $A \subseteq \Lambda$ and $\Theta \subseteq Q$) represents the set of states from which players A can enforce the system transition into some state from Θ in one move.

Our ATL model checker relies to implement this function on the database engine and thus uses optimizations specific to the respective database.

We consider the states in Q to be integers, stored into a *HashSet* object. Implementing the function *Pre(A, Θ)* involves efficient insertion of the states from Θ into a table.

For Oracle, a quick solution is:

```
insert into R select * FROM sys.odcinumberlist(…)
```

but unfortunately *sys.odcinumberlist*() accepts a maximum of 1000 arguments and thus the solution is not suitable for large models.

The specific implementation code for the Oracle database is shown below:

```
HashSet Pre(HashSet r) {
    int longArray[] = new int[r.size()];
    Iterator it = r.iterator();
    while (it.hasNext()) {
          i++;
          longArray[i] = (int)it.next();
    }
    ArrayDescriptor desc = ArayDescriptor.createDescriptor(
                        "NUMBERTABLE", conn);
    ARRAY array_to_pass = new ARRAY(desc, conn, longArray);
    OraclePreparedStatement pstat =
    (OraclePreparedStatement) conn.prepareStatement(
     "INSERT INTO R (SELECT distinct * FROM TABLE (
            SELECT CAST(? AS NUMBERTABLE) FROM DUAL))");
    pstat.setARRAY(1, array_to_pass);
    pstat.executeUpdate();
    pstat.close();
    …
}
```

where NUMBERTABLE is a custom type created with the statement:

```
CREATE TYPE NUMBERTABLE AS TABLE OF NUMBER;
```

The Oracle database limits the number of characters in a string to 32767, insufficient to store large models. To avoid this limitation, the method of the ATLOracle class by which it is invoked the model checker uses CLOB (Character Large OBject) parameters:

```
public static java.lang.String checkModel(
  oracle.sql.CLOB model, int size,
  oracle.sql.CLOB[] response)
{ … }
```

A CLOB value is defaulted to 2 GB characters. In the first version of the implementation, both the model and the formula to be verified are encoded in JSON format and then encapsulated in CLOB format.

The loading of the necessary libraries in the database is done with the script:

```
loadjava -thin -user system/passwd@//localhost:1521/
xepdb1 -verbose .\antlrworks-1.4.jar .\javax.json.jar
```

The Java source files of our ATL model checker are loaded into database with the following command:

```
loadjava -thin -user system/passwd@//localhost:1521/
xepdb1 -verbose .\ATLLexer.java .\ATLParser.java
.\ATLJson.java .\ATLOracle.java
```

The PL/SQL wrapper to invoke the newly loaded Java class is:

```
CREATE OR REPLACE FUNCTION ATL_ORACLE (model IN CLOB,
                                length IN NUMBER,
                                response IN OUT CLOB)
RETURN VARCHAR2 AS
LANGUAGE JAVA NAME 'ATLOracle.checkModel(oracle.sql.CLOB,
int, oracle.sql.CLOB[]) return java.lang.String';
```

Invoking the ATL model checker from SQL*Plus or SQL Developer can be done in the following way:

```
declare shortResponse VARCHAR2(32767);
        model CLOB;
        response CLOB;
begin
  model := ' your ATL model here…';
  DBMS_OUTPUT.enable();

  DBMS_LOB.CREATETEMPORARY(response, TRUE);
  shortResponse :=
  ATL_ORACLE(model,DBMS_LOB.GETLENGTH(model),response);
  dbms_output.Put_line('1:'||shortResponse);
  dbms_output.Put_line('2:'||response);
  dbms_lob.freetemporary( response );
end;
```

Details about coding a model in JSON format for its forwarding to the Java stored procedure are provided by ATLDesigner, the client component of our ATL model checker, available at https://use-it.ro/ (select the button *Web Service* from the main toolbar and then click on button *Show model* from the window that opens).

For testing purposes, we extended ATLDesigner with the facility to invoke the stored procedures in the Oracle database, in order to call the embedded version of the ATL model checker. Using Oracle Data Provider for .NET, the process of verification of an ATL model is described below:

```
OracleConnection conn = new OracleConnection(
                            STRING_CONNECTION);
conn.Open();
var clob = new OracleClob(conn);

OracleCommand cmd = new OracleCommand();
cmd.Connection = conn;
cmd.CommandText = "ATL_ORACLE";
cmd.CommandType = CommandType.StoredProcedure;

OracleParameter states = new OracleParameter(
"return_value", OracleDbType.Varchar2, 32000);
states.Direction = ParameterDirection.ReturnValue;
cmd.Parameters.Add(states);

OracleParameter model = new OracleParameter(
"model", OracleDbType.Clob,int.MaxValue);
model.Direction = ParameterDirection.Input;
model.Value = request;
cmd.Parameters.Add(model);

OracleParameter size = new OracleParameter(
"length", OracleDbType.Int32);
size.Direction = ParameterDirection.Input;
size.Value = request.Length;
cmd.Parameters.Add(size);

OracleParameter response = new OracleParameter(
"response", OracleDbType.Clob, int.MaxValue);
response.Direction = ParameterDirection.InputOutput;
response.Value = clob;
cmd.Parameters.Add(response);

cmd.ExecuteNonQuery();

byte[] buffer = new byte[clob.Length];
clob.Read(buffer, 0, (int)clob.Length);
string rez = Encoding.Unicode.GetString(buffer, 0,
                                        buffer.Length);
```

By setting the Oracle option in the Play Game window, the embedded ATL model-checker is used to design a game strategy when playing Tic-Tac-Toe. In the actual stage of the development, experimental results are encouraging, showing that our tool is able to handle large systems efficiently. The tests were performed on an Oracle Database Express Edition (XE) which has a memory limitation of 2 GB (Fig. 6). We expect better performance when we can make the most of the in-memory database facility, using the enterprise version of the database.

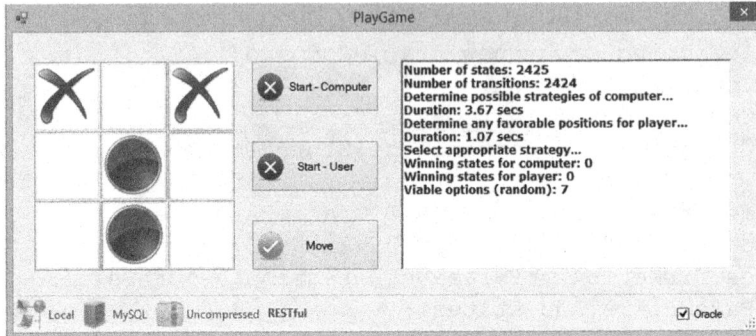

Fig. 6. Use the embedded ATL model checker to determine a winning strategy in Tic-Tac-Toe game

6   Related Work

In [2] the most representative tools for analytical predictions are presented. However, some tools qualify as a business intelligence tool for data analysis but not for data analytics.

Microsoft R Open, IBM SPSS Predictive Analytics Enterprise, RapidMiner Studio and SAS Advanced Analytics are suitable solutions for Predictive Analytics and offers flexibility, but requires technical knowledge to end-uses (connect to databases, script programming, etc.).

The ITAPA tool intends to offer a higher-level perspective, being addressed to business users without technical knowledge of databases or prediction models, easily adaptable to the specific of the company.

7   Conclusions and Future Work

As presented in the previous sections, the ITAPA tool will rely on Oracle database. We will use Oracle Database In-Memory – a suite of features that greatly improves performance for real-time analytics and mixed workloads [9]. Using the In-Memory Column Store (IM column store) we will store tables used by ODM-SVR prediction model in memory, using a columnar format optimized for rapid scans. Also, the internal tables used by ATL model checker will use In-Memory Compression format, optimized for access speed.

ITAPA tool will incorporate results obtained from previous research in distinct domains (model checking, machine learning) that will be used synergistically to generate predictions with superior accuracy to existing approaches, in an environment controlled by formal methods, with the intention of being exploited in the economic environment.

The next steps for full implementation of the ITAPA tool are:

- Use the data access layer for Oracle Property graphs to handle and store the CGS structures into database (GraphML format or Oracle Flat File Format);
- Design Processes for Automatic Data Transformations (PADT) - transformations applied to data attributes to generate the model attributes – BPMN format;
- Design Processes for Training/Testing (PTT) the prediction models used by ODM-SVR method – BPMN format;
- Design Processes for ODM-SRV tuning – BPMN format;
- Translate the BPMN formats to CGS structures and store them as Property Graphs in database;
- Formal verification of business processes using the embedded ATL model checker to verify that desired behavioural properties expressed by an ATL formulas are verified to hold for the corresponding ATL models;
- Design and develop Web user interface (APEX) incorporating wizards for user guidance to follow winning strategies (using assistant agents).

The SVR is often been viewed as a tool for experts. Using Oracle Automatic Data Preparation provided by Oracle and having custom ODM-SRV tuning, usability is a major enhancement targeted by the ITAPA tool.

Acknowledgement. The authors were supported from the project financed from Lucian Blaga University of Sibiu & Hasso Plattner Foundation research action LBUS-RRC-2020-01.

References

1. Alur, R., Henzinger, T.A., Kupferman, O.: Alternating-time temporal logic. J. ACM **49**(5), 672–713 (2002)
2. Best Predictive Analysis Software of 2019. https://financesonline.com/predictive-analysis/. Accessed 22 Sep 2020
3. Business process model and notation™ (BPMN™) Version 2.0.2 [Online]. The Object Management Group (OMG). https://www.omg.org/spec/BPMN/. Accessed 21 Sep 2020
4. Chang, C.-C., Lin, C.-J.: LIBSVM: a library for support vector machines (2019). https://www.csie.ntu.edu.tw/~cjlin/libsvm. Accessed 21 Aug 2020
5. Das, T.: Oracle Database Java Developer's Guide, Release 18c (2018). https://docs.oracle.com/en/database/oracle/oracle-database/18/jjdev/. Accessed 24 July 2020
6. Kacprzak, M., Penczek, W.: Fully symbolic unbounded model checking for alternating-time temporal logic. J. Auton. Agents Multi-Agent Syst. **11**(1), 69–89 (2005)
7. Oracle Advanced Analytics' Machine Learning Algorithms. https://www.oracle.com/technetwork/database/options/advanced-analytics/odm/odm-techniques-algorithms-097163.html. Accessed 10 Aug 2020
8. Oracle Property Graph Developer's Guide 18c. https://docs.oracle.com/en/database/oracle/oracle-database/18/spgdg/index.html. Accessed 14 Aug 2020

9. Oracle Database In-Memory Guide 12c Release 2 (2019). https://docs.oracle.com/en/database/oracle/oracle-database/12.2/inmem/index.html. Accessed 18 June 2020

10. Stoica, L.F., Stoica, F., Boian, F.M.: Verification of JADE agents using ATL model checking. Int. J. Comput. Commun. Control **10**(5), 718–731 (2015). ISSN 1841-9836

11. Stoica, F., Stoica, F.L.: Implementing an ATL model checker tool using relational algebra concepts. In: Proceeding of the 22th International Conference on Software, Telecommunications and Computer Networks (SoftCOM), Split-Primosten, Croatia, pp. 361–366 (2014)

12. Stoica, F.: Formal verification of business processes using model checking. In: Proceeding of the 27th International Business Information Management Association (IBIMA) Conference, Milan, Italy, pp. 2563–2575 (2016)

13. Surampudi, S.: Oracle Data Mining, Concepts (2020). https://docs.oracle.com/en/database/oracle/oracle-database/19/dmcon/index.html. Accessed 01 Sep 2020

Creating Web Decision-Making Modules on the Basis of Decision Tables Transformations

Aleksandr Yu. Yurin$^{(\boxtimes)}$ and Nikita O. Dorodnykh

Matrosov Institute for System Dynamics and Control Theory,
Siberian Branch of the Russian Academy of Sciences,
134, Lermontov Street, Irkutsk 664033, Russia
`iskander@icc.ru`

Abstract. Creating embedded decision-making modules for web applications that implement artificial intelligence methods in the form of knowledge bases is quite an interesting task. Specialized methodologies and software are being developed to solve them. At the same time, the use of generative and visual programming principles, as well as model transformations, can provide better results. In our previous works, we proposed to apply these principles combined with the model-driven approach for the automated creation of expert systems and knowledge bases. In this paper, we extend the previously developed method with new platforms, in particular: PHP (Hypertext Preprocessor) and Drools, as well as we add the possibility to use the decision tables formalism and Microsoft Excel tools for their construction. The modified (extended) method allows one to effectively create knowledge bases with a large number of logical rules and generate the source code for web embedded decision-making modules. This extension is implemented as a plugin for an expert system prototyping system, namely, Personal Knowledge Base Designer. This paper describes the extended method and examples of its application for the development of web application modules: for making decisions when detecting banned messages and identifying customers who violate rules of using the SMS notification service ("Detector"), and interpreting signs of emotions within the HR-Robot application ("EmSi-Interpreter"). The proposed method was also evaluated in solving educational (test) tasks.

Keywords: Model transformations · Decision tables · Knowledge bases · Rules · Web applications

1 Introduction

Knowledge bases engineering remains the most difficult stage in the development of intelligent decision-making modules, in particular, in the case of web applications. The principles of generative and visual programming, as well as model

© Springer Nature Switzerland AG 2021
D. Simian and L. F. Stoica (Eds.): MDIS 2020, CCIS 1341, pp. 167–184, 2021.
https://doi.org/10.1007/978-3-030-68527-0_11

transformations, can be used to increase the efficiency of activities of this stage related to conceptualization, formalization, and code generation.

These principles are implemented separately in various methods and tools, in particular, in [1,2]. However, resulting solutions do not always support the target software platforms and meet the expectations and requirements of experts and those users who do not have programming skills. Principles under consideration are used together in such areas as Model-Driven Engineering (MDE) [3], Model-Driven Development (MDD), and Model-Driven Architecture (MDA). In our previous works [4,5], we proposed a method for creating expert systems and knowledge bases using MDA and sequential transformations of conceptual models. Our method had some limitations:

1. By platforms, in particular, knowledge bases and expert systems created only for CLIPS (C Language Integrated Production System) that can be integrated mainly into desktop applications developed using C, C++, C#, Delphi, etc.
2. By the size of created knowledge bases: the usage practice has shown a high complexity of using conceptual models with the number of concepts and relationships more than 50 items.
3. By the sources of information that can be used for developing models: concept maps (Cmap Tools, XMind, FreeMind), UML class diagrams (StarUML, IBM Rational Rose).

In this paper, we propose to extend this method to automate the creation of embedded decision-making modules for web applications based on model transformations. Also, it is proposed to use the well-known formalism of decision tables [6–9] to describe logical rules. This formalism allows one to create large knowledge bases. Tables themselves are formed using the widely used Microsoft Excel tool and converted to logical rules using the plugin for the Personal Knowledge Base Designer (PKBD) [5].

The proposed method was tested while developing modules for solving the following case studies: decision-making when detecting banned messages and identifying customers who violate rules of using the SMS notification service within the "SMS-Organizer" platform [10]; interpretation of signs of emotions in the "HR-Robot" application [11].

The paper contribution consists of the following results:

1. An extended transformation-based method for automated creation of decision-making modules that use logical rules; the method is based on model transformations and uses conceptual models and the decision tables formalism to obtain source codes and specifications.
2. A software component (a plugin) for PKBD [5] that implements transformation of decision tables; this component supports the generation of source codes for PHP (Hypertext Preprocessor) and Drools. PHP is a general-purpose language that focused only on web application development. Drools a specialized language for creating knowledge bases in the form of business rules, it can be used in Java applications (whether desktop or web).

3. Two embedding decision-making modules for PHP-based web applications ("Detector" for the "SMS-Organizer" platform, "EmSi-Interpreter" for the "HR-Robot" application), as well as, results of an experimental evaluation on educational (test) tasks.

The paper is organized as follows. Section 2 presents the preliminaries of this work. Section 3 explains our extended five-step method. Section 4 describes case studies. Section 5 considers discussion and concluding remarks.

2   Preliminaries

2.1   Model-Driven Approach and Model Transformation

Generative and visual programming principles, as well as, model transformations are widely used in the area known as the Model-Driven Engineering (MDE) [3]. MDE is a software engineering approach that uses transformations and interpretations of information (in particular, conceptual) models for the creation of software. The most standardized version of MDE is the Model-Driven Architecture (MDA) based on the complex use of OMG standards. MDA implies the use of UML as the main mean for formalization and visualization of models while following a certain sequence (chain) of model transformations: from models with greater abstraction to models with less abstraction and specific program codes. In this case, the transformation chain has the form [4]: $CIM \rightarrow PIM \rightarrow PSM \rightarrow \{software, codes, specifications\}$. Where CIM is a computation-independent model, PIM is a platform-independent model, and PSM is a platform-specific model. In addition to transformation chains, MDA as MDE variation implements a four-level metamodel scheme. According to this scheme, there can be following transformation types [12]: horizontal and vertical, endogenous and exogenous, unidirectional and bidirectional. As means of implementing these transformations, both specialized model transformation languages (e.g., ATL, QVT, etc.) and general-purpose programming languages can be used.

In our previous works [4], we proposed to use basic MDA principles in the context of creating intelligent decision-making modules. In this paper, we extend our method to creating modules for web applications.

2.2   Design of Knowledge Bases and Intelligent Systems Modules Based on Model Transformation

There are available several examples of knowledge bases and intelligent systems engineering based on MDD/MDA principles. However, they can be classified according to the following criteria:

1. Source models and transformation chains that they use. Following trends can be determined under this criterion: using own methodological principles without a clear definition of standard models and transformation chains [13–17]; implementing transformation chains and defining models [4,18,19].

2. Target software platforms. According to this criterion, we can determine the following trends of using: general-purpose languages such as Perl or C#, .NET [13,19]; specialized languages for creating knowledge bases such as Jess, Drools, CLIPS [4,15,18,20]; own original language means, in particular, PRISMA [19].

3. Information sources. Following trends can be determined in accordance with this criterion: using ontologies [14,15,18], UML models [4], trees [16] or XML-like structures [13].

4. Focus on programmers [13,15,18,20] or non-programmers [4,14,17,19].

Based on this brief analysis, we concluded that non-programmers should be more involved in the developing process, and as a result, we propose to use various conceptual models as information sources, including the tabular form and employ chains of a transformation of these models with the possibility of synthesizing source codes, both in general-purpose languages and specialized programming languages for knowledge bases engineering.

Thus, in this paper, we use MDA specialization that had previously developed in [4], which utilizes sequential horizontal endogenous transformations implemented using a general-purpose language in the form of a tool (PKBD) [5].

2.3   Techniques for Describing Logical Rules

Despite the popularity of semantic technologies, in particular, ontologies as main knowledge representation formalism, knowledge representation languages based on logical rules are still used in intelligent systems engineering. The attractiveness of this knowledge representation model is caused by its simplicity and clarity for domain experts, high modularity, and transparency of a mechanism of logical inference.

The task of automating the creation of knowledge bases with the aid of specialized software that facilitates obtaining, conceptualizing, and formalizing domain knowledge remains relevant despite the existence of a variety of special languages for coding logical rules: CLIPS, Jess, Drools, RuleML, SWRL, etc. and standards, for example, RIF (Rule Interchange Format). Based on the classification [21] and its subsequent modification, we can distinguish the following main groups of techniques designed to solve this problem:

1. Textual techniques. These techniques provide manual manipulation of constructs of programming languages. Approaches of this group designed for programmers and are implemented in the form of specialized editors. There are extensions, for example, VISUAL JESS [22] for Jess (Java expert system shell) that provides color highlighting of syntactic structures and interactive definition of knowledge base elements.

2. Graphical techniques [23,24]. These techniques provide the creation of visual abstractions corresponding to elements of logical rules with their subsequent translation into source code for a specific programming language.

3. Tabular techniques. It is based on decision tables and their transformation into source code. Both the standard table formalisms [6–9] and its specializations, such as eXtended Tabular Trees (XTT2) [25], are used in this area.

The last two groups are the most promising, since they not only provide an effective representation of logical rules for non-programmers, but also allow one to use the non-specialized CASE tools, and the general-purpose tools such as Microsoft Excel.

In this paper, we propose to use tabular and graphical techniques together and to employ an original domain-specific language, namely, the Rule Visual Modeling Language (RVML) [4].

2.4 Rule Visual Modeling Language

RVML [4] is a domain-specific notation designed for modeling logical rules. The RVML main elements are derived from the UML concepts "class" and "association". RVML uses the separate graphical elements to visualize all components of logical rules (Fig. 1), rather than stereotypes or typed classes as in UML. RVML users can assign subjective probabilities to facts and rules (Fig. 1b,c,d) in the form of the certainty factors (CF); indicate the type of actions performed with facts (Fig. 1e), such as adding (+) or deleting (−).

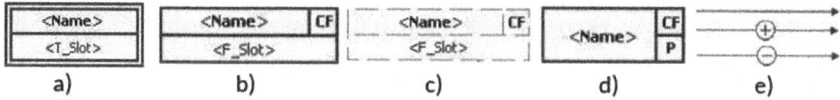

Fig. 1. The main RVML elements: a) a fact template; b) a fact; c) a condition; d) a rule node; e) connectors of elements without and with the indication of actions

RVML elements can be used to describe rule templates, fact templates, specific rules, and specific facts. Examples of using RVML are given below in the paper.

2.5 Background

The method considered in this paper is an extension of the previously proposed one [4]. This method is a specialization of MDA for creating a certain type of software such as rule-based expert systems and knowledge bases. This specialization required redefining the models used in the standard transformation chain, as well as developing specialized tools.

Formally, the method can be represented as follows:

$$M = \langle MOF, L^{ES}, CIM^{ES}, PIM^{ES}, PSM^{ES}, PDM^{ES}, T \rangle , \tag{1}$$

$$T = \langle T^{ES}_{CIM-to-PIM}, T^{ES}_{PIM-to-PSM}, T^{ES}_{PSM-to-CODE} \rangle , \tag{2}$$

where MOF (Meta Object Facility) is an abstract language for describing models (a metamodel description language); L^{ES} is a set of languages and formalisms used for building models, that will be transformed; in our case $L^{ES} = \{UML, CM, ET, RVML\}$ where UML is a Unified Modelling Language; CM is a concept or mind maps formalism; ET is an event trees formalism; $RVML$ is a Rule Visual Modeling Language.

CIM^{ES} is a computation-independent model in the form of a domain model represented with the aid of L^{ES}; this model describes the main concepts and relationships of the domain, the main architectural elements of the software is created, and business logic. Since the method is intended for creating a specific type of system, all the logic and description of software elements are integrated into the toolkit.

PIM^{ES} is a platform-independent model, in our case this model represents logical rules in the RVML notation. The method uses this model only to describe the knowledge base structures.

PSM^{ES} is a platform-specific model, in our case this model takes into account the features of the programming language with the use of RVML. This model is used to refine the knowledge base structures for further code generation.

PDM^{ES} is a set platform description models, $PDM^{ES} = \{CLIPS, PKBD\}$. The model is used when implementing code generation.

$T^{ES}_{CIM-to-PIM}, T^{ES}_{PIM-to-PSM}, T^{ES}_{PSM-to-CODE}$ are the rules for model transformations implemented in the toolkit.

Our method [4] is a sequential chain of model transformations and consists of the five main steps: creating a domain (CIM) model, creating a platform-independent model, creating a platform-specific model, generating source codes and specifications, testing. These steps are described in more detail in the next section.

The main difference between the proposed extension described in this paper is the addition of new formats for input models due to tabular representation (decision tables, canonical tables of a special type), and the addition of support for new platforms (Drools, PHP). The benefit of this extension is the ability to work with large data presented in tabular form, as well as integrate the results of code generation into web applications.

3 Method

The extended method consists of same five steps as the earlier method. Let's describe these steps in detail.

Step 1: Creating a domain model. A domain model for a description of the decision-making process is created in this step. The resulting model is considered as a computation-independent model (CIM) and can be represented in the form of conceptual models, such as OWL (Web Ontology Language) ontology, UML class diagrams (see Fig. 2), or mind and concept maps.

Ways to form these models are not specified: users can use existing conceptual models created by the software for ontological and conceptual modeling (e.g., OntoStudio, CmapTools, IBM Rational Rose Enterprise, StarUML, etc.). From the viewpoint of the classical knowledge base engineering, this step can be considered as conceptualization.

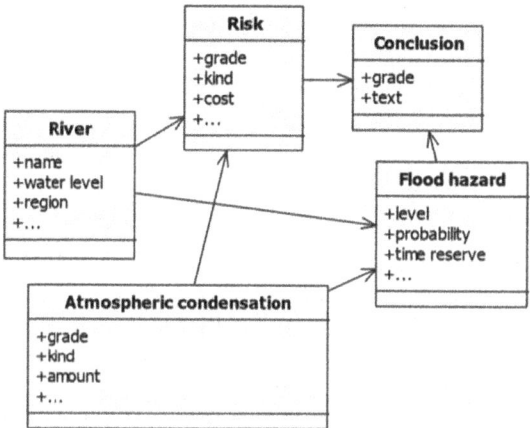

Fig. 2. An example of a domain model fragment used in the educational process for the creation of knowledge bases (this UML class diagram is used for the task of river flood hazard prognosis)

Step 2: Creating a platform-independent model. A model of this step is the result of the direct or indirect transformation of CIM.

There are many ways to implement such transformations [4]. In this work, we use PKBD plugins that provide interaction with IBM Rational Rose, FreeMind, StarUML, XMind, CmapTools, and TabbyXL. The general model of logical rules is used as a platform-independent model (PIM) that can be represented in the form of RVML-schemas for visualization and correction.

Most transformations can be shown as tables (Table 1).

Table 1. Examples of correspondences of model elements for Fig. 2

CIM elements (for UML class diagrams)	PIM elements (for a general model of logical rules)
Class	Template
Attribute	Slot (Property)
Expression	Default property value
Association	Cause-and-effect relationship

Table 1 presents transformations in a simplified form for understanding their principles. In practice, there are several types of element correspondences, for example, one-to-one correspondence (equivalence), as in the table; ambiguous correspondence (synonymy); indistinguishable correspondence (homonymy), etc., as well as extreme cases: redundancy and lack of expressive ability. Special languages and approaches are created for processing these special cases and describe transformations, for example, Transformation Model Representation Language (TMRL) [26]. In our case transformations are implemented with the aid of general-purpose language.

In terms of classical knowledge base engineering, this step corresponds to the formalization stage.

We propose to formalize the PIM data as decision tables made in the standard table editor (Microsoft Excel) in the CSV format. Let's introduce some constraints for naming column headers in decision tables for a PIM:

1. A first column has the header as "Rule Name" and should contain names of rules.
2. Headers of dependent columns are marked with the symbol "#".
3. A column header name can be compound, indicating an entity name and its property name separated by "::" string.

A decision table fragment for Fig. 2 (River flood hazard prognosis) is presented in Fig. 3.

	1	2	3	4	5	6	7	8	9
1	RuleName	Risk::grade	Risk::kind	...	Flood-haz	...	#Conclusion::grade	#Conclusion::text	#Conclusion::cf
2	Risk+Flood hazard->low	natural	low		low		safe situation	0.9	

Fig. 3. A decision table fragment for Fig. 2

The decision table structure from Fig. 3 corresponds to a general rule (or a rule template) of the "if-then" type, a corresponding rule template represented as an RVML scheme (Fig. 4a).

This rule template is interpreted as follows: if there is a "Risk" of a certain grade and kind, and a "Flood hazard" of a certain level and probability, then some "Conclusion" is made. Since this is only a rule template that describes the logical structure of some knowledge, slots (properties) values specify the default values or possible ranges of values only.

Table rows of the decision table correspond to specific rules (Fig. 4b). The structure of this rule corresponds to a previously defined rule template filled with specific content, i.e. the slot values are defined. Thus, the rule is interpreted as follows: if the "Risk" grade is low and the "Flood hazard" level is also low, then a "Conclusion" is made about a safe situation and indicated that this fact is added to the working memory. In the presented example the sections for the certainty factors (CF) are not filled (marked as "-"), and the rule priority is marked as 0 (the normal priority level).

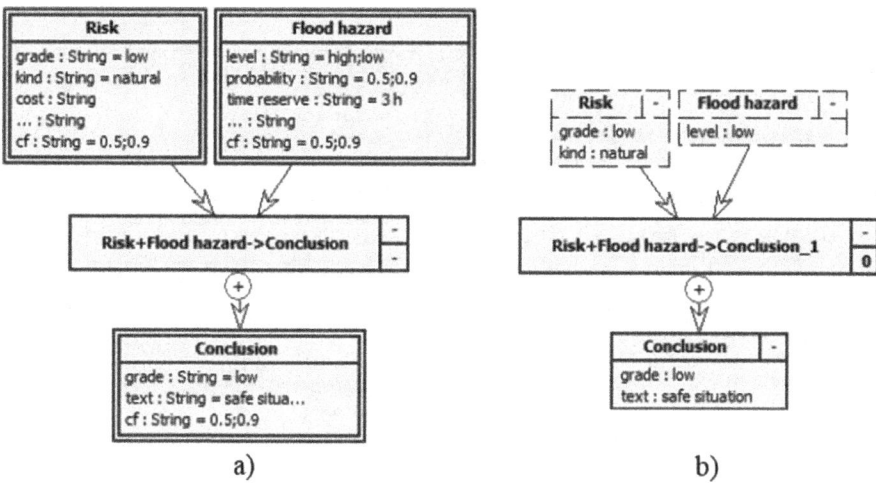

Fig. 4. Examples of a rule template (a) and a specific rule (b) in the form of RVML schemas

Step 3: Creating a platform-specific model. A model of this step is the result of automatic transformations of a PIM. By the method, the developer should refine the RVML model taking into account the characteristics of a target software platform. For example, in the case of PHP, rule priorities and default slot values are specified.

Step 4: Generating source codes and specifications for a knowledge base and an intelligence system. At this step, RVML schemas are direct-mapped to constructs of programming languages. Correspondences with model elements and constructs can be shown as a table (Table 2).

Table 2. Examples of correspondences of model elements

PSM elements (for a general model of logical rules)	PHP elements	Drools elements
Template	class	declare
Slot (Property)	var	"<Property>"
Default property value	"<value>"	"<value>"
Cause-and-effect relationship	if (...) {...}	rule ...when ... then ... end

The resulted source codes (Fig. 5 and 6) can be integrated into the application.

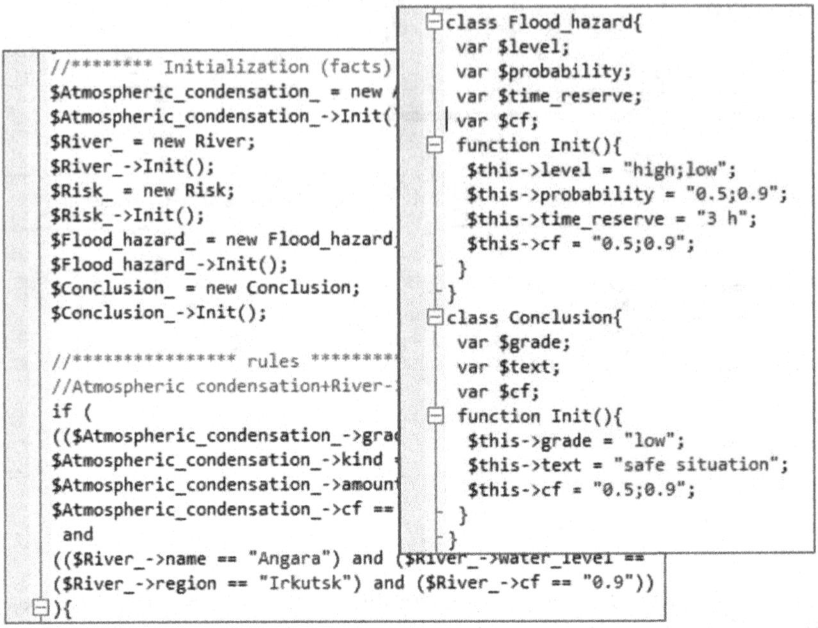

```
//******** Initialization (facts)
$Atmospheric_condensation_ = new /
$Atmospheric_condensation_->Init(
$River_ = new River;
$River_->Init();
$Risk_ = new Risk;
$Risk_->Init();
$Flood_hazard_ = new Flood_hazard
$Flood_hazard_->Init();
$Conclusion_ = new Conclusion;
$Conclusion_->Init();


//*************** rules ********
//Atmospheric condensation+River-
if (
(($Atmospheric_condensation_->gra
$Atmospheric_condensation_->kind
$Atmospheric_condensation_->amoun
$Atmospheric_condensation_->cf ==
 and
(($River_->name == "Angara") and ($River_->water_level ==
($River_->region == "Irkutsk") and ($River_->cf == "0.9"))
){
```

```
class Flood_hazard{
    var $level;
    var $probability;
    var $time_reserve;
    var $cf;
    function Init(){
        $this->level = "high;low";
        $this->probability = "0.5;0.9";
        $this->time_reserve = "3 h";
        $this->cf = "0.5;0.9";
    }
}
class Conclusion{
    var $grade;
    var $text;
    var $cf;
    function Init(){
        $this->grade = "low";
        $this->text = "safe situation";
        $this->cf = "0.5;0.9";
    }
}
```

Fig. 5. Examples of generated PHP source codes

```
declare Conclusion
    grade : String
    text : String
    cf : String
end

//*********************** Rules ***************************
rule Atmospheric_condensation+River_>Flood_hazard_1 "Description of the
rule: Atmospheric condensation+River->Flood hazard_1" salience 0
dialect "mvel"
when
    Atmospheric_condensation(grade == "low", kind == "snow", amount == "15
mm", cf == "1")
    and River(name == "angara", water_level == "low", region == "irkutsk", cf
== "0.9")
then
    $Flood_hazard = new Flood_hazard();
    $Flood_hazard.level = "low";
    $Flood_hazard.probability = "0.9";
    $Flood_hazard.cf = "0.9";
    insert($Flood_hazard);
end
```

Fig. 6. An example of generated Drools source codes

Step 5: Testing. At this step, obtained source codes are tested both in the special software (for example, in the PKBD interpreter) and in the web application after the integration of generated codes.

As practice has shown, generated codes require improvement for its successful integration.

In this paper, we use UML class diagrams (designed in StarUML) and decision tables (designed in Microsoft Excel) as information sources for transformations.

Thus, the chain of model transformations is the following:

$$UML \rightarrow DecisionTables \rightarrow RVML \rightarrow \{PHP, Drools\}$$

4   Case Studies

We use our method to solve the following two practical tasks:

1. Creation of a web-based embedded decision-making module, namely, "Detector" for detecting banned SMS messages and customers who violate rules of using the SMS notification service within the "SMS-Organizer" platform [10].
2. Creation of a prototype of a knowledge base for emotion signs interpretation module, namely, "EmSi-Interpreter" for the HR-Robot application [11].

4.1   Detector: A Decision-Making Module for Detecting Banned Messages

Banned messages fall into the category of SPAM messages that violate provisions of federal laws. For timely and automatic detection of such messages in large traffic, as well as, for determining customers who send them, a specialized subsystem, namely, "Detector" was developed for a telecommunication company specializing in providing SMS mailing services for legal entities. "Detector" is a part of the "SMS-Organizer" platform [10] and includes an intelligent decision-making module with a knowledge base.

Next, we consider in detail the development of this module using the method proposed.

1. Creating domain models. At this step we analyzed the domain and defined the following main concepts (Fig. 7, step 1): "Message", "Sender" and "Keywords" (or phrases), so-called "SPAM markers". We analyzed a database containing 1366490 messages and selected 829 messages of customers who previously had been blocked by the moderators, 46 messages form them were excluded from further consideration because they did not contain any signs of banned messages. Selected 783 messages formed a training corpus, which contained four main messages groups: propaganda of prohibited substances, unfair advertising, threats and insults. For each group, the possible response of the decision-making module is determined, in particular, blocking sending messages (changing its status) and blocking customers.

2. Selected concepts and "SPAM markers" were used in the creation of PIM in the form of a decision table storing data about 487 unique "SPAM markers" sets (Fig. 7, step 2). The table was developed by an automated analysis of the database of messages. The table structure is formed by listing properties of all concepts from the domain model, where each property is a column of the table.

3. PSM was built using PKBD by importing the obtained decision table and then refining it in the form of RVML (Fig. 7, step 3), including the priority of rules and default values.

4. The generator based on RVML synthesized 6871 lines of PHP code describing 487 rules (Fig. 7, step 4).

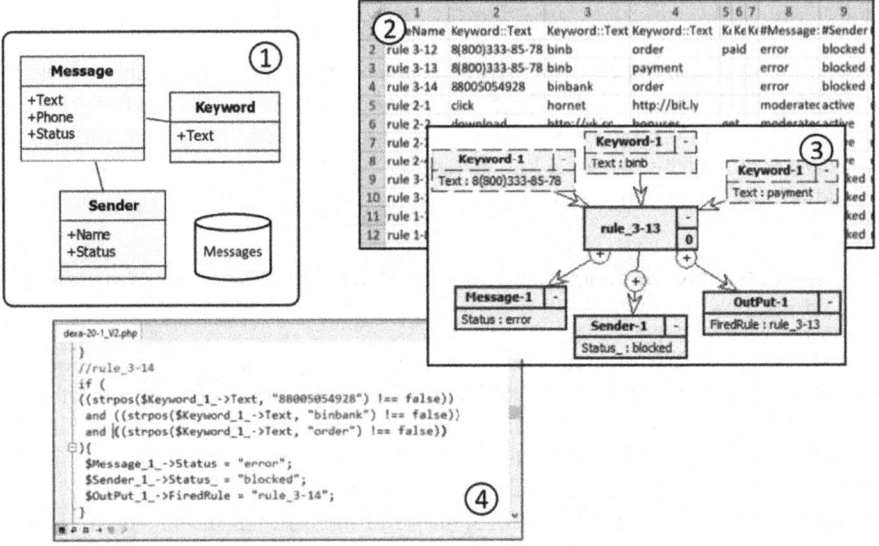

Fig. 7. Steps of development of the "Detector" decision-making module

Rule priorities are taken into account by sorting atomic conditional operators within the computing block (a function). Currently, this mechanism is not used in the Detector module (this is not necessary), i.e. all rules have the same priorities, and all matching rules are "executed" since it is enough to execute at least one of any rules to detect spam.

Later, the generated code was tested, evaluated and integrated into "SMS-Organizer". The knowledge base for the "Detector" module after modification includes 498 rules that provided detection of 653 banned messages from the training corpus. In our case for the training corpus the accuracy was calculated using the standard formula:

$$Accuracy = \frac{(TP - TN)}{TP + TN + FP + FN} = \frac{DetectedBannedMessages}{AllMessages} \qquad (3)$$

So, the accuracy was 0,83. The average execution time was 0,00026 s. We detected 1145 spam messages and 25 new suspected customers while checking the database of 1366490 messages with the aid of the developed module, 8 messages of the additionally detected ones were false positives.

4.2   EmSi-Interpreter: A HR-Robot Feature Interpretation Module

The proposed method was used when prototyping one of the modules of the HR-Robot application [11]. The main purpose of HR-Robot is to support decision-making when selecting candidates for vacancies and checking staff for motivation (research of the psychological situation in the team) based on analyzing emotions. The system consists of the following main modules: video processing, emotions signs detection, and emotions signs interpretation. This section deals with the development of a knowledge base prototype for emotions signs interpretation module called "EmSi-Interpreter".

Now, let's consider the main steps in detail:

1. Conceptual models describing parts of the face and its main elements, which would be tracked in the process of determining emotions, were created as a domain model. A fragment of one of the face models is shown in Fig. 8, step 1.
2. Decision tables describing the structural aspect of the domain (domain models), as well as, the knowledge of psychologists, were developed. These tables contain information about combinations of signs describing emotions, for example, "fear" (Fig. 8, step 2). Each row of the table is a logical rule. Next, we used PKBD, which provided the import of decision tables and their representation in the form of logical rules. In this example, the knowledge base segment for the "fear" emotion includes 5 fact templates, 1 rule template, and 11 specific rules.
3. Imported decision tables were refined in the RVML form (Fig. 8, step 3).
4. For this segment of the knowledge base, 250 lines of code were generated for PHP, and 453 lines were generated for Drools (Fig. 8, step 4).

HR-robot is currently under development: compressors and conditions for video shooting (shooting angle, light, etc.), image stabilization algorithms, and feature detection are improving. For this reason, only the principal testing of "EmSi-Interpreter" is conducted on the correct test data (honest test persons and actors with a clear expression of emotions) and there are no objective assessments of its accuracy.

In the future, it is planned to extend the knowledge base "EmSi-Interpreter" for working with unknown people or different ethnics/cultures.

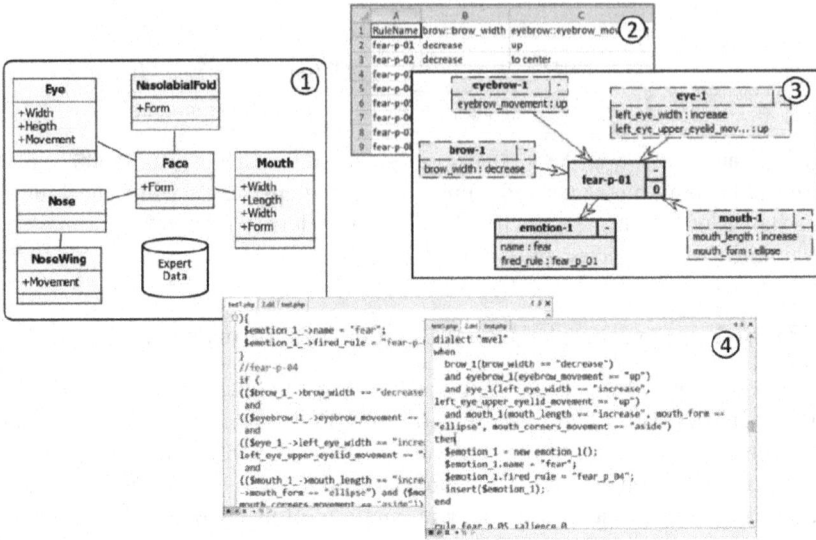

Fig. 8. Steps of development of the "EmSi-Interpreter" decision-making module

5   Discussion and Conclusions

Test tasks from the educational process were used to evaluate our method. In this evaluation 36 students of the Institute of information technology and data analysis of the Irkutsk National Research Technical University (IrNRTU) took part.

Our evaluation objective was to assess the time required to perform the development of knowledge bases (intelligent decision-making modules) using various methods: our previous method (only conceptual modeling and PKBD, let's denote this approach as A1); our extended method (conceptual modeling, decision tables (Microsoft Excel) and PKBD, let's denote this approach as A2); hand (manual) coding (hereby denoted as A3).

All the students who participated in the evaluation of the method had programming skills, but they did not use them for A1 and A2, because they only built graphical models and formed decision tables, code generation was carried out automatically.

For our evaluation, we used 12 educational tasks for the creation of diagnostic or prognostic decision-making modules in various domains (Table 3). The following constraints were imposed for tasks: 5–10 entities in the models; up to 3 properties of domain entities; 5–10 connections between domain entities; 3–4 cause-effect relationships; 10–15 instances of cause-effect relationships (possible specific rules).

The assessment of a time required for structuring, formalizing, and programming knowledge was made. The average value of the three results for each task is shown in Table 4.

Table 3. The description of tasks

#	Domain	Number of domain entities	Number of connections	Number of cause-effect relationships	Number of instances of cause-effect relationships
1	Car diagnosing	6	5	3	10
2	Mushrooms identification	5	6	3	10
3	Computers diagnosing	8	5	3	10
4	Flu diagnosing	5	8	4	11
5	Harvest prognosis	8	7	3	12
6	Electric kettle diagnosing	9	5	3	10
7	Public opinion prognosis	5	6	3	14
8	Toaster diagnosing	8	7	4	14
9	Iron diagnosing	6	5	3	15
10	Weather prognosis	7	10	3	12
11	River flood hazard prognosis	5	6	3	11
12	Forest fires prognosis	8	7	3	12

Table 4. Evaluation results

#	A1 (min.)	A2 (min.)	A3 (min.)	A2 vs. A1 (%)	A3 vs. A2 (%)
1	4,17	8,44	25,33	50,6	66,7
2	10	20,5	21,5	51,2	4,7
3	6,99	11,6	21	39,7	44,8
4	5,85	8,14	13,86	28,1	41,2
5	9,95	16,7	47	40,4	64,5
6	3,75	6,06	26	38,1	76,7
7	16,1	27,3	56,7	41	51,9
8	10	16,6	28,98	39,7	42,8
9	7,12	9,56	34,05	25,5	71,9
10	9,5	15,6	22,2	39,2	29,6
11	4,87	7,97	27,5	38,6	71
12	8	12	29,04	33,3	58,7
Avg	8	13,4	29,4	38,8	52

The comparison showed (Fig. 9) that the use of decision tables can reduce the time spent compared to hand-coding (an average of 52%), however, it is less effective an average of 38,8% (Fig. 9) than to use special PKBD wizards or manipulating graphical primitives (for example, RVML).

The use of decision tables when developing knowledge bases is a well-known practice [6–9], especially when we deal with a large number of logical rules. It should be noted that in this experiment, decision tables were created manually (with the use of Microsoft Excel) and the test knowledge bases contain up to 15

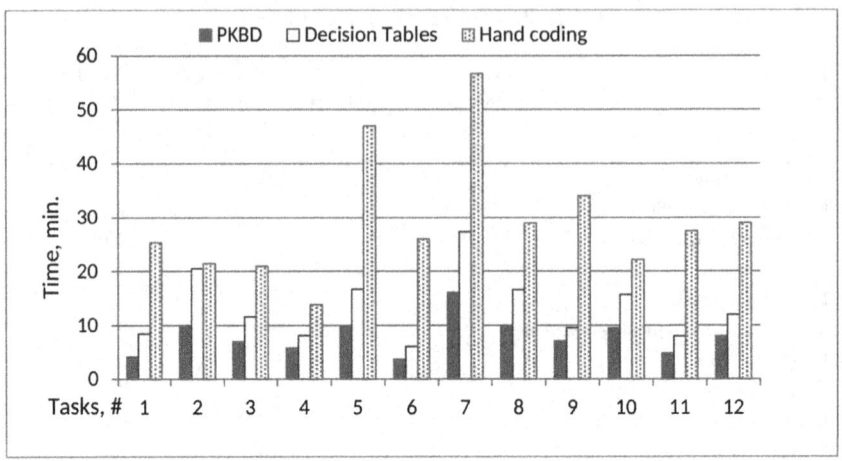

Fig. 9. Results of comparison of A1 (PKBD), A2 (Decision Tables), and A3 (Hand coding) by the time criterion

rules. However, the efficiency can be even higher with the automated generation of decision tables containing a large number of records (487, as in the above case study). Herewith, the use of model transformations and specialized tools reduce the codification complexity of rules.

In general, in this paper, we propose a method for developing decision-making modules and knowledge bases. This method uses principles of model transformations, as well as, UML class diagrams and decision tables as inputs. Our method is not focused only on creating modules and knowledge bases for web applications. However, discussed in the paper case studies are devoted to the web applications, and source codes are generated for Drools and PHP, while Drools can be used in Java applications (whether desktop or web), PHP is focused only on web development.

The proposed method is focused on non-programmers (e.g., domain experts and analysts) and is realized as software called PKBD [5]. PKBD was used when prototyping the expert system for identifying causes of damages and destruction of construction materials in petrochemistry [27]. The proposed method and software were also used in the educational process at the IrNRTU.

Acknowledgement. This work was supported by the Council for Grants of the President of Russia (grant No. MK-1647.2020.9).

References

1. Schreiber, G., et al.: Knowledge Engineering and Management. The CommonKADS Methodology. The MIT Press, Cambridge (2000)
2. Stokes, M.: Managing Engineering Knowledge: MOKA: Methodology for Knowledge Based Engineering Applications, 6th edn. ASME Press, New York (2001)

3. Silva, A.R.D.: Model-driven engineering: a survey supported by the unified conceptual model. Comput. Lang. Syst. Struct. **43**, 139–155 (2015). https://doi.org/10.1016/j.cl.2015.06.001
4. Yurin, A.Yu., Dorodnykh, N.O., Nikolaychuk, O.A., Grishenko, M.A.: Designing rule-based expert systems with the aid of the model-driven development approach. Expert Syst. **35**(5), 1–23 (2018). https://doi.org/10.1111/exsy.12291
5. Yurin, A.Yu., Dorodnykh, N.O.: Personal knowledge base designer: software for expert systems prototyping. SoftwareX **11**, 100411 (2020). https://doi.org/10.1016/j.softx.2020.100411
6. Pollack, S.L., Hicks Jr., H.T., Harrison, W.J.: Decision Tables: Theory and Practice. Wiley Interscience, Hoboken (1974)
7. Santos-Gomez, L., Darnell, M.J.: Empirical evaluation of decision tables for constructing and comprehending expert system rules. Knowl. Acquis. **4**(4), 427–444 (1992). https://doi.org/10.1016/1042-8143(92)90004-K
8. Vanthienen, J., Wets, G.: From decision tables to expert system shells. Data Knowl. Eng. **13**(3), 265–282 (1994). https://doi.org/10.1016/0169-023X(94)00020-4
9. Seagle, J.P., Duchessi, P.: Acquiring expert rules with the aid of decision tables. Eur. J. Oper. Res. **84**(1), 150–162 (1995). https://doi.org/10.1016/0377-2217(94)00323-5
10. SMS-Organizer Home. http://centrasib.ru/index.php?p=smso. Accessed 16 Oct 2020
11. Personnel Evaluation Home. http://www.ocenkakadrov.ru/. Accessed 16 Oct 2020
12. Mens, T., Gorp, P.V.: A taxonomy of model transformations. Electron. Notes Theoret. Comput. Sci. **152**, 125–142 (2006). https://doi.org/10.1016/j.entcs.2005.10.021
13. Dunstan, N.: Generating domain-specific web-based expert systems. Expert Syst. Appl. **35**, 686–690 (2008). https://doi.org/10.1016/j.eswa.2007.07.048
14. Nofal, M.A., Fouad, K.M.: Developing web-based semantic and fuzzy expert systems using proposed tool. Int. J. Comput. Appl. **112**, 38–45 (2015). https://doi.org/10.5120/19682-1414
15. Shue, L., Chen, C., Shiue, W.: The development of an ontology-based expert system for corporate financial rating. Expert Syst. Appl. **36**, 2130–2142 (2009). https://doi.org/10.1016/j.eswa.2007.12.044
16. Ruiz-Mezcua, B., Garcia-Crespo, A., Lopez-Cuadrado, J., Gonzalez-Carrasco, I.: An expert system development tool for non AI experts. Expert Syst. Appl. **38**, 597–609 (2011). https://doi.org/10.1016/j.eswa.2010.07.009
17. Kadhim, M.A., Alam, M.A., Kaur, H.: Design and implementation of intelligent agent and diagnosis domain tool for rule-based expert system. In: Proceedings of the International Conference on Machine Intelligence Research and Advancement, pp. 619–622. IEEE Xplore Press, Katra (2013). https://doi.org/10.1109/ICMIRA.2013.129
18. Canadas, J., Palma, J., Tunez, S.: InSCo-Gen: a MDD tool for web rule-based applications. Web Eng. **5648**, 523–526 (2009). https://doi.org/10.1007/978-3-642-02818-2_53
19. Cabello, M.E., Ramos, I., Gomez, A., Limon, R.: Baseline-oriented modeling: an MDA approach based on software product lines for the expert systems development. In: Proceedings of the 1st Asian Conference on Intelligent Information and Database Systems, pp. 208–213. IEEE Xplore Press, Dong Hoi (2009). https://doi.org/10.1109/ACIIDS.2009.15

20. Chaur, G.W.: Modeling rule-based systems with EMF. Eclipse Corner articles. http://www.eclipse.org/articles/Article-Rule%20Modeling%20With%20EMF/article.html. Accessed 16 Oct 2020
21. Gavrilova, T.A., Gulyakina, N.A.: Visual knowledge processing techniques: a brief review. Sci. Tech. Inf. Process. **38**, 403–408 (2011). https://doi.org/10.3103/S0147688211050042
22. Grissa-Touzi, A., Ounally, H., Boulila, A.: VISUAL JESS: an expandable visual generator of oriented object expert systems. Int. J. Comput. Inf. Eng. **1**(11), 1668–1671 (2007). https://doi.org/10.5281/zenodo.1057263
23. Visual Rules BRM. https://www.bosch-si.com/bpm-and-brm/visual-rules/business-rules-management.html. Accessed 16 Oct 2020
24. VisiRule. Logic Programming Associates. http://www.lpa.co.uk/ind_hom.htm. Accessed 16 Oct 2020
25. Nalepa, G.J., Kluza, K.: UML representation for rule-based application models with XTT2-based business rules. Int. J. Softw. Eng. Knowl. Eng. **22**(4), 485–524 (2012). https://doi.org/10.1142/S021819401250012X
26. Dorodnykh, N.O., Yurin, A.Yu.: A domain-specific language for transformation models. In: CEUR Workshop Proceedings (ITAMS 2018), vol. 2221, pp. 70–75 (2018)
27. Yurin, A.Yu., Berman, A.F., Nikolaychuk, O.A., Dorodnykh, N.O.: Knowledge base engineering for industrial safety expertise: a model-driven development approach. Stud. Syst. Decis. Control **199**, 112–124 (2019). https://doi.org/10.1007/978-3-030-12072-6_11

Machine Learning

Taking a Close Look at Twitter Communities and Clusters

Kowshik Bhowmik[(⊠)] and Anca Ralescu

University of Cincinnati, Cincinnati, OH 45219, USA
bhowmikk@mail.uc.edu.com

Abstract. The rise in the popularity of Social Networking Sites has made Community Detection in such networks a major research interest. The edges connecting the entities in the network are the principal foci in graphical community detection. At the same time, large volume of data is produced on these Social Networking Sites, a large portion of which being text data. Document Clustering methods utilize the textual properties of text documents to cluster similar documents together while separating dissimilar documents. This paper treats text data collected from Twitter as a set of documents. The clusters produced by the document clustering methods are associated with the respective users. These clusters are then compared with the communities detected in the graphical representation of the network generated from the users and the relationships between them. NodeXL was used to collect data from Twitter while Gephi was used for visualizing the collected dataset. Different feature representation and clustering methods were applied for clustering the tweets(documents) and in turn the users associated with them.

Keywords: Community detection · Document clustering · Social media mining

1   Introduction

The internet, with billions of pages and hyperlinks that act as edges among nodes in a graph, is the largest web network [1]. The widespread use of social networking sites is contributing to this ever growing network. As social media generated networks are the center of this analysis, in this paper they are collectively referred to as Social Media Networks.

Social Media Networks can be an important source of intelligence in the modern world because the activities and feedback of millions of social media participants are encoded within them. One can extract actionable insights from these networks that can help navigate different social phenomena. Since Social Media Networks are huge in scale and dynamic by nature, analyzing them is a challenging task.

The emergence of Social Media Networks has added a new framework for the word 'community' which already has many a connotations based on the context

© Springer Nature Switzerland AG 2021
D. Simian and L. F. Stoica (Eds.): MDIS 2020, CCIS 1341, pp. 187–201, 2021.
https://doi.org/10.1007/978-3-030-68527-0_12

in which it is used. In the social media ecosystem we can see a wide range of entities interacting with each other. These entities can be represented as nodes in a network while the numerous kinds of relationship and connections they have with each other can be thought of as edges connecting those nodes, thereby forming the Social media Networks.

Social media networks can be represented using the same notation as a traditional graph, $G = (V, E)$, where G represents the entire network made up of numerous nodes and the edges between them, V stands for the set of all vertices, and E represents the edges between nodes in V. In a Social Media Network, vertices can represent entities that cover the range of users, the content they create and even the metadata objects related to the users of the content. The edges, based on their type (plain, weighted, guided, multi-way) can represent different kinds of relationship between the entities. Community Detection is a technique that helps us recognize groups of vertices in a network that are closer together than the rest of the network. It is an important tool in analyzing networks that are large and complex, such as a Social Media Network.

A large portion of the content generated on Social Networking Sites are text data. Twitter, a popular micro-blogging and social networking service enables the generation of vast amount of text data through their features such as tweets, comments, mentions and sharing of hyperlinks [2]. Network clustering problem is related to the graph partitioning problem. Its goal is to isolate, within a network, groups of nodes that are closely connected together. The nodes in a Social Media Network can be clustered based on the textual content they produced using the different text clustering techniques available. The question this paper addresses is whether the communities detected in a graph by utilizing its linkage properties would correlate to the clusters of users that can be formed based on clustering the textual contents that they produced.

2   Related Works

Twitter is a social networking and microblogging service released in 2008. Users could originally publish a 140 character post [3] – a limit now increased to allow 280 characters [4]. Twitter, unlike other social networking sites, does not require mutual acceptance between users to facilitate communication between them. Unless explicitly specified, whatever a user posts is available for other users to read, reply-to, retweet and favorite. These functionalities enable a free and swift spreading of information as well as public opinions and sentiments.

Twitter has a widespread use in fields as diverse as politics, entertainment, marketing, emergency events etc. [5] Tumasjan et al. [6] showed that Twitter is a tool of political deliberation and also that political sentiments offline is reflected in related twitter messages. In their work, Sakaki et al. [7] treated users on Twitter as a geographically-distributed sensor network and approximated earthquake epicenters in Japan successfully. Lerman and Ghosh's study [8] on the transmission of information on Twitter found that Twitter is an important medium of information diffusion and the flow of information is influenced by the structure of the network.

Many previous approaches of discovering communities on twitter were based solely on the users' social structure on the network [9]. Among these, many considered Twitter users as nodes and the relationship between them as an edge. Here, relationship between users were dictated by whether they followed each other on Twitter. Huberman et al. [10] modeled 'relationship' between users based on interaction between them. Here there is an edge between two users if there are two or more replies from one of them to the other. The school of researchers who based the discovery of communities only on the social structure among the users basically ignored the other available properties of nodes and the contents of the interaction. These methods were based on agglomerative clustering, min-cut based graph partitioning, centrality based and Clique percolation methods [11,12].

Semantic contents of the social media networks have been utilized to discover communities within the larger network by some researchers. Zhou et al. [13] propose the CUT (Community-User-Topic) model which leverages the semantic contents of the interaction among users in order to discover communities. CUT models a community as a random mixture who have, associated with them, a distribution of topics or interests. This model does not leverage the structural or linkage information of the graph from the communities are to be detected. Models such as CUT assume that people who interact on shared interests are interconnected while other models assume that users who are already connected with each other are interested in the same topics. However, these assumptions do not always hold true in social media networks. On Twitter, there may be users who form connections by following but never interact or contribute in similar conversations. There may also be users who take part in an ongoing conversation but do not follow one another. The work described in this paper considered interaction between users as the edges between two users and found communities based on those connections. Furthermore, users were clustered based on the textual content they produced. Finally, possible relations among the communities was investigated.

3 Proposed Method

3.1 Data Collection

Twitter enables users/ researchers to access its data through two different APIs: the Rest API and the Streaming API. The Rest API lets the users access data from the recent past while the Streaming API makes it possible to access and work with the data in real time. Both APIs let users fulfill their specific criteria by using keywords, username and their combinations as the search term.

NodeXL, an application developed by Social Media Research Foundation [14], simplifies the tasks associated with network analysis to a great extent and also facilitates collection and analysis of online Social Media Networks data. Built on top of the familiar spreadsheet framework, it has additional features that support collection, storage, analysis, visualization and publishing of datasets.

NodeXL has features to connect to online social media networks such as Facebook, Twitter, flickr etc. and also lets users to export the collected network data in different formats facilitating interaction with other social network analysis programs such as Gephi.

The research reported here used the Twitter Data importer extension provided by NodeXL. The selection of 'Import from Twitter Search Network' causes a separate window to pop-up. The data was collected post the second round of Democratic National Debate which took place on July 30–31, 2019. The search was done using the popular hashtags associated with the topic, '#DemDebate OR #DemDebate2020' (the disjunction-the use of OR- makes this query more flexible). A choice was made to import a 'Basic network' which means a network that will contain tweets that were made as replies to tweets that fulfilled the search criteria as well as those that mentioned them. The other import option allowed adding some of the collected users' friends in the network. As this choice would slow down the process, it was not selected. The number of tweets to be collected was limited. Twitter's rate limit policy that restricts users from fetching more than fifteen user information every fifteen minutes is a major roadblock in collecting data from the Social Networking site. NodeXL has a mechanism in place that deals with the rate-limiting imposed by Twitter by pausing and caching the collected data.

NodeXL efficiently stores the collected data in various Worksheets of an Excel Workbook. It can be opened and manipulated like any Excel workbook. It was planned to carry out the network visualization & community detection tasks in Gephi. To that end, the workbook was exported in GraphML format which can be worked on by Gephi. Again, to perform Natural Language Processing & Machine Learning techniques on the collected tweets, the 'Edge' worksheet was exported in 'xlsx' format which can be imported and converted to a Pandas dataframe.

3.2   Visualization

Gephi is a popular, open source network visualization software [15]. It helps the users visualize their respective network data on a two or three dimensional space. Through its different features, Gephi enables its users to analyze the patterns and overall structures of the networks and thus retrieve actionable insights from them. Gephi can be used to calculate in-degree, out degree of nodes, betweenness centrality and then to have those metrics reflected in the visualization. The vector file format ensures that the graphics produced my Gephi are readable and of high quality.

Gephi is well-suited to support the task of data visualization which is often a non-linear process. It not only helps users visualize the data but also enables them to interact with it. With the use of size, shapes, color Gephi creates a visual language for the network data with which users can intuitively interpret the network and communicate their findings (Fig. 1).

The visual representation of our network data corroborates some well-known facts. The participants of the second Democratic Party presidential debate made

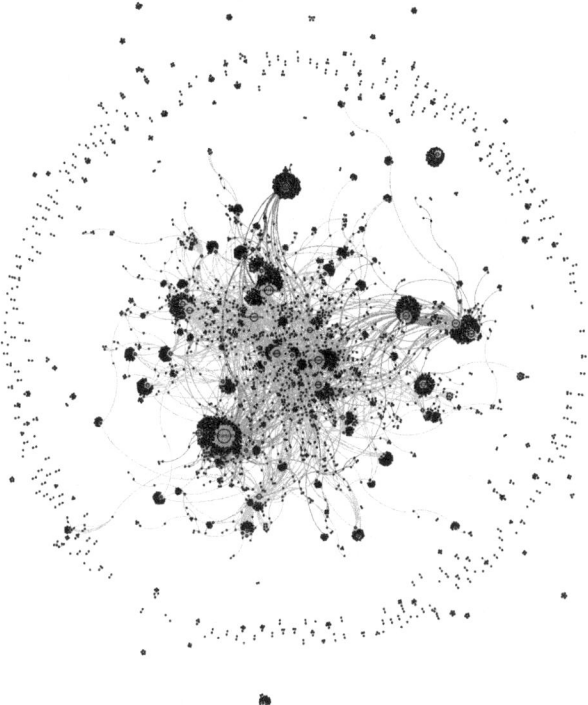

Fig. 1. Force Atlas layout applied on the network data

an appearance in the visualization. Since in-degree was selected as the deciding factor for a node's size in the visualization, the nodes depicting the major candidates are clearly visible. Bernie Sanders, Joe Biden, Kamala Harris, Elizabeth Warren, Pete Buttigieg, Julián Castro, Andrew Yang all had prominent presence in the visualization whereas participants Amy Klobucher, Beto O'Rourke, Cory Brooker were comparatively less prominent. Apart from these 10 candidates the nodes that are most prominent are those of President Donald Trump (personal), Alexandria Ocasio-Cortez (@aoc), news portal ABC, podcast host Scott Dworkin (@funder), Times Editor Anand Giridharadas (@AnandWrites), AARP Advocates, Official twitter handle of the Democratic National Committee (@TheDemocrats), the official twitter handle for the Trump campaign (@TeamTrump) and the personal handle of the Governor of Louisiana, Bobby Jindal (@BobbyJindal).

The Modularity report generated by Gephi shows that there are as many as 385 communities found in the network data set. But upon close inspection, it is revealed that most of these communities contain less than 1% of the total users in the dataset. The Force Atlas, Yifan Hu, Fruchterman-Reingold, OpenOrd and Force Atlas 2 layouts have respectively 14, 15, 15, 13, 14 communities that contain more than 2% members of the dataset. From visual inspection also it is apparent that there are around 15 well defined communities in the graphical representation.

3.3   Text Mining of Social Media Data

Text is the most prevalent means of communication in social networking sites such as Facebook and Twitter [16]. English remains the lingua franca on these sites. But English is, by no means the first language of all the people expressing their views and opinions on these sites. English speakers, especially the younger generation, are paying less attention to correct use of language and use a lot of slang words. Overall, much of the text data found on online social networking sites on Twitter contain spelling or grammatical errors or both [17]. On top of it all there is heavy use of emoticons. These factors make the analysis of text data in social media a complex process. Furthermore, some tweets feature code-switching, which means the users composed the tweets using words or phrases belonging to multiple languages [18]. These tweets may then come up in search results since they contain the search terms, thereby adding more complexities to the problem.

Text Preprocessing. Standard preprocessing steps were first applied on the network data. The 'Edges' sheet was loaded into a pandas dataframe. Since linguistic data are of interest in this part of our experiment, 'Tweet' column along with the 'Source' and 'Target' columns were retained. The goal is to eventually cluster the users based on the document clusters that they are a part of. After that, preprocessing steps are performed on the collected tweets.

A brief visual inspection revealed that there are a number of duplicate tweets in the dataset. This is due to the way the data is stored. A tweet occupies multiple rows if it appears under different 'Source'-'Target' pairing. For example, if a user tweets something and another retweets it, the same tweet will appear in two rows. In the first row, the person who tweeted it will occupy both the spaces as source and target, while another row will depict the retweet relationship where the source is the user retweeting the original tweet and the target is the user whose tweet is being retweeted. The same is true for the 'Mention' relationship: if a user mentions multiple different users in his tweet, it will be repeated with the source user being the one who mentioned the other users while the mentioned users will occupy different rows in the 'Target' column. Since the presence of identical tweets has the potential to influence the document clustering step, it was decided to remove all the duplicate tweets retaining only their first occurrence in the dataset. This decision causes the loss of many users, but keeping them caused only identical tweets to be clustered together which gives us no information.

After that, tweets were lower-cased and usernames were removed from the tweets so that they cannot influence the document clustering stage. Non-English tweets were removed from the dataset along with stopwords listed in the NLTK English stop-words list. Then Data Visualization technique provided by the Yellow-Brick libraries was used to visualize the most frequent tokens [19]. This

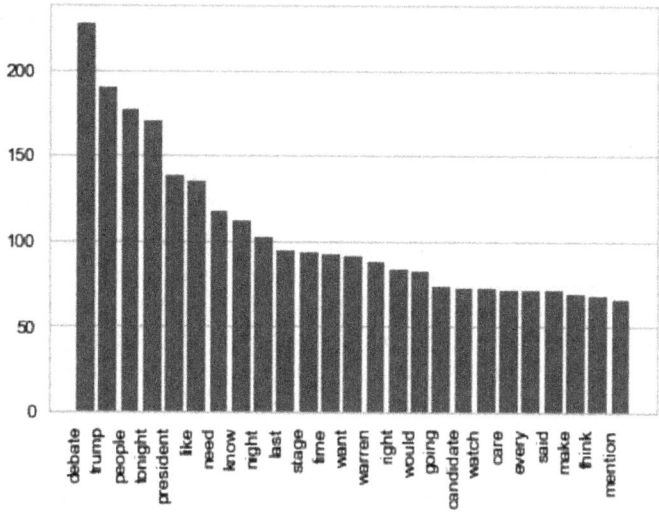

Fig. 2. List of unigram tokens after preprocessing

visualization revealed the existence of some more words that are stop words in the context of our dataset. These words and the two search hashtags are then removed from the dataset. A second visualization shows an improvement as now most of the frequent tokens are also domain specific (Fig. 2).

Feature Representation. Machine Learning algorithms do not take text data as input. Text data need to be transformed into numeric values for Machine Learning algorithms to work on them. The numerical or vector representation of text data can take many forms. They can be in the form of Bag of N-Grams, a term-frequency inverse document frequency or TF-IDF matrix, word-embeddings or document embeddings.

Once the pre-processing steps were completed, 1426 tweets were left. The tweets were represented in four different ways: Bag of Words, tf-ids, Topic Models and Document Embeddings.

For the Bag-of-Words (N-gram) Model, scikit-learn's CountVectorizer method was used. For the selection of n-gram range, they were re-visualized along with their frequencies. From this visual exploration, the decision was made to use a combination of bigrams and trigrams. It is possible to set the range of n-grams through the parameter $ngram\_range$, (2,3) for bigrams and trigrams. While many of the parameters retain their default values, min_df was set to 5. It means that for a bigram or trigram to appear as a feature in the feature matrix it must at least appear in five documents. This was done to keep the feature matrix from being too sparse.

The second feature representation approach was Tfidf Model. scikit-learn makes it possible to transform text data set into a Tfidf representation so as to apply machine learning techniques such as the clustering and classification algorithms. Since the basic preprocessing has already been done, parameters such as *lowercase, preprocessor, tokenizer, stop_words* etc. are set to the default 'None'. Just as it has been done with the CountVectorization, the *ngram_range* was set to (2,3). No *min_df* was set for extracting the Tfidf features.

Representing a set of documents as the distribution of a prespecified number of topics is not very common. For this one first needs to train an Latent Dirichlet Allocation (LDA) model, for which gensim's LDA library was used. The model takes *id2word*, a corpus and *num_topics* as parameters among others. id2word is a mapping between the tokens in the document to their IDs. It is common to use individual words to create this mapping. But upon visualization of the models created by using individual words as tokens, it was decided to use a similar combination of bigrams and trigram as was used to fit our CountVectorizer and Tfidf models. The parameter *passes* was set to 100. This means that there will be 100 passes through the corpus during the training process.

After training the LDA model, the topic distribution of each of the tweets in the dataset is obtained. This distribution is in a matrix form where each row depicts a document while each column depicts the topics. An entry (i, j) in this matrix is the probability of document i belonging to topic j. The graphical representation of the network data found 15 major communities. Since the clusters formed from these features will be compared to the communities identified in the previous section, cluster size was set to 15- equal to the number of communities. The number of topics was net to 14 since there are documents in the data set that do not concretely fall under any topic and are equally distributed among all the topics. Once document-topic matrix is obtained, it can be used as input to the clustering algorithms.

A pre-trained doc2vec model was used for document representation. This particular model was pre-trained on Wikipedia data. The pre-trained embeddings were used to infer the vectors for the documents (tweets) in our data set. Parameters *start_alpha*, *end_alpha* and *epochs* were set at 0.002, −0.016 and 1000 respectively. Training resulted in an output file that contained vector representations for the tweets in the dataset to be used as input to machine learning algorithms.

Document Clustering. Social Networking Platforms have become a way to disseminate information about major events and products. How the information spread and how users engage with the content generate is difficult to capture since the data generated are neither structured or annotated. Automatic topic annotation can be one way to make better sense of the vast amount of data being generated every second. Users these days are getting access to highly curated contents that are tailored to the users' preferences and thus highly biased. This is a result of phenomena such as echo chambers and filter bubbles [20]. Automatic topic annotation, thus, has become a more important challenge.

In Machine Learning terminology, Clustering is the process of dividing a dataset into different groups in such a way that data belonging to a group

Table 1. Summary of clusters formed by KMeans clustering algorithm on different dataset representations

Prominent Users	N-Gram	Tfidf	LDA	Doc2Vec
AARP Advocates (@AARPAdvocates)	0	0	0	0
abc (@abc)	0	0	0	0
Anand Giridharas (@AnandWrites)	0	0	0	0
Andrew Yang (@AndrewYang)	0	0	0	0
Alexandria Ocasio-Cortez (@aoc)	0	0	0	0
Bernie Sanders (@BernieSanders)	0	0	0	0
Elizabeth Warren (@ewarren)	2	0	1	0
Scott Dworkin (@funder)	0	0	0	0
Joe Biden (@JoeBiden)	1	0	0	0
Julián Castro (@JulianCastro)	0	0	0	0
Kamala Harris (@KamalaHarris)	1	0	1	0
Pete Buttigieg (@PeteButtigieg)	1	0	1	0
Donald Trump (@realDonaldtrump)	0	0	0	0
Team Trump (@TeamTrump)	0	0	0	0
The Democrats (@TheDemocrats)	0	0	0	0

is similar to one another while being different from data belonging to other groups. Similarly Document Clustering is the collective name of Machine Learning techniques that aim to group similar documents together while keeping dissimilar documents in different groups or clusters. The clustering algorithms can be broadly divided into three categories: hierarchical clustering, partitional clustering and semantic-based clustering. The methods proposed for document clustering usually involve use of a feature matrix like a bag of words or tf-idf to represent the corpus in a numeric form. But these representations of text can often be sparse. In a more recent developments, text are being represented by neural word embeddings that retains the semantic nature of the text and require less manual preprocessing.

Recall that the previous section mentioned four different ways. They were Bag-of-Words model, Tfidf model, LDA document-topic matrix and document embeddings trained from our data set using a pre-trained model. The feature engineering was done with the intent of applying clustering algorithms on the data set which would then be projected on the target users of those tweets which would give us a glimpse into the nature of the clusters formed in the data set by the linguistic properties of the tweets alone. Two clustering algorithms on each of the four feature representation of our data set were used: KMeans Clustering algorithm and Agglomerative Clustering algorithm.

In order to compare communities found in the visual representation of the data set using Gephi the initial number of clusters was set as close as possible to

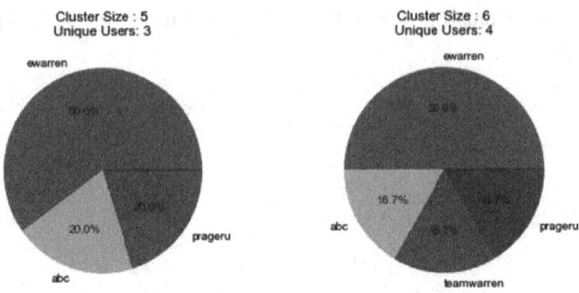

Fig. 3. Similar clusters formed from N-Grams feature representation

the number of notable communities found the graphical representation, in this case, 15. KMeans implementation found in the scikit-learn library requires as to set the numbers of clusters to be formed by the algorithm beforehand, thus the *n_cluster* parameter of the implementation was set to 15. The algorithm was initialized at 'kmeans++' which means that the implementation will select the initial cluster centers as a means to speed up the convergence process. Maximum number of iterations was set at 150 which means that the implementation will iterate at most 150 times in a single run. The other parameters of the KMeans algorithm were kept at their default value.

The implementation of Agglomerative clustering is also found in the scikit-learn library. As in KMeans algorithm, the number of clusters was also set to 15 in this. The *affinity* parameter, which decides the metric that is used to compute the linkage between the clusters, was set to Manhattan metric. After visually analyzing various options, 'average' was selected as the value for the parameter *linkage*. The 'average' linkage takes the average between each pair of observation between two sets.

3.4   Analysis and Comparison

The graphical community detection revealed 15 major communities with 7 of the 10 debate participants forming notable communities along with personal handle of President Donald Trump, Representative Alexandria Ocasio-Cortez (@aoc), handles of news and commentary providers ABC, Dworkin(@funder), Anand Giridharadas (@AnandWrites), AARP Advocates, Official twitter handle of the Democratic National Committee (@TheDemocrats), the Trump campaign (@TeamTrump) and the as well as the Governor of Louisiana, Bobby Jindal (@BobbyJindal). This research checks for any similarities or dissimilarities between the communities formed in the graphical representation in Gephi and the clustering of target users based solely on the linguistic contents of the tweets they posted or were replied to or mentioned in. As discussed above, four types of feature representation were used coupled with two types of clustering algorithms. Both clustering algorithms were set to form 15 clusters which is approximately

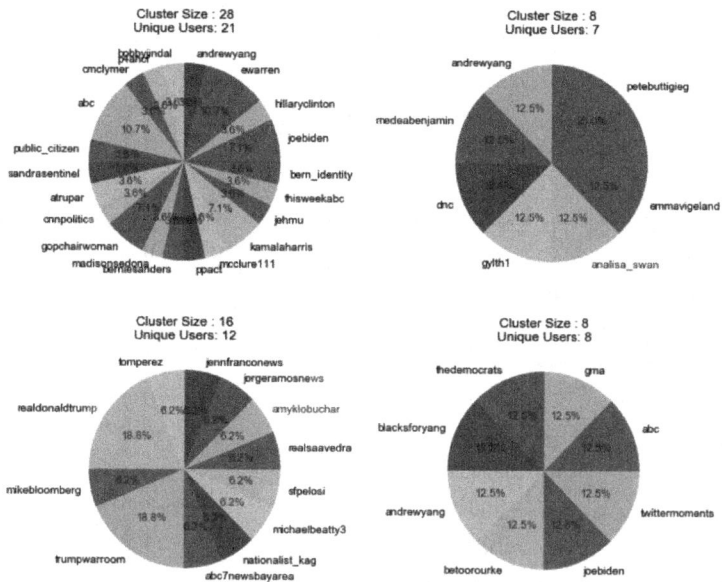

Fig. 4. Clusters formed by KMeans clustering algorithm from N-Grams features (4 of 15)

the number of notable communities visible in the Gephi representation. This choice was made in order to compare the two methods on a similar ground.

The intention is to take a look at the clusters formed, see whether the nodes around which communities have grown in case of the graphical representation also appear prominently in the clusters formed by the documents they are associated with. In the graphical representation, size of a node was chosen to reflect its in-degree. In the document/ user clustering part of the experiment also, only the target nodes were taken into account. For a target node to be considered prominent in the clusters formed it has to fulfill two criteria:

1. Its appearance in a respective cluster must be greater than 20%
2. The size of the clusters and the number of unique Target users can not be equal to each other.

KMeans and Agglomerative clustering algorithms were performed on four feature representations: Bag of N-grams, Tfidf, Document-Topic Matrix and Document Embeddings. This means that documents were clustered in 8 different ways. Each of these 8 combinations of 4 feature representation and 2 clustering algorithms has 15 clusters each. The goal of the study is to study the nature of these clusters and investigate any possible correlation between them and the Graphical Communities detected by Gephi.

The 8 resulting set of clusters show that, applied on the same feature representation, the clustering algorithms tend to form similar clusters. While some of the clusters formed by the two clustering algorithms are identical, there are

Table 2. Summary of clusters formed by agglomerative clustering algorithm on different dataset representations

Prominent Users	N-Gram	Tfidf	LDA	Doc2Vec
AARP Advocates (@AARPAdvocates)	0	0	0	0
abc (@abc)	0	1	0	0
Anand Giridharas (@AnandWrites)	0	0	0	0
Andrew Yang (@AndrewYang)	0	0	0	0
Alexandria Ocasio-Cortez (@aoc)	0	0	0	0
Bernie Sanders (@BernieSanders)	0	0	0	0
Elizabeth Warren (@ewarren)	3	1	1	1
Scott Dworkin (@funder)	0	0	0	0
Joe Biden (@JoeBiden)	2	1	0	0
Julián Castro (@JulianCastro)	0	0	0	0
Kamala Harris (@KamalaHarris)	0	1	1	0
Pete Buttigieg (@PeteButtigieg)	0	0	1	0
Donald Trump (@realDonaldtrump)	0	0	0	0
Team Trump (@TeamTrump)	0	0	0	0
The Democrats (@TheDemocrats)	0	0	0	0

some differences as well (Fig. 3). The combination of Bag of N-Gram feature representation with KMeans (Fig. 4) and Agglomerative clustering algorithms (Fig. 5) have identified 5 clusters each that have a prominent user among them. The first combination identifies three clusters with Joe Biden, Kamala Harris and Pete Buttigieg as prominent users and two clusters with Elizabeth Warren as a Prominent cluster. On the other hand, the latter combination produced the same number of clusters with prominent users (5) but has two of them attributed to Joe Biden and three of them to Elizabeth Warren.

The clusters derived from the LDA Document-Topic matrix also exhibit similar behavior. The Document-Topic Kmeans combination produces 3 clusters with prominent users (Table 1). These users are Elizabeth Wareen, Kamala Harrs and Pete Buttigieg. On the other hand, Agglomerative Clustering applied on the same feature representation produces two clusters with prominent target users with Elizabeth Wareen and Kamala Harris as the prominent users.

On the other hand, the Doc2Vec feature representation produces one cluster with a prominent user when applied Agglomerative clustering algorithm but none in the case of KMeans clustering algorithm. The clusters formed by the Agglomerative clustering algorithm from the Tfidf feature representation contain three clusters with prominent users while KMeans clustering algorithm does not produce any cluster with prominent users from the same feature representation.

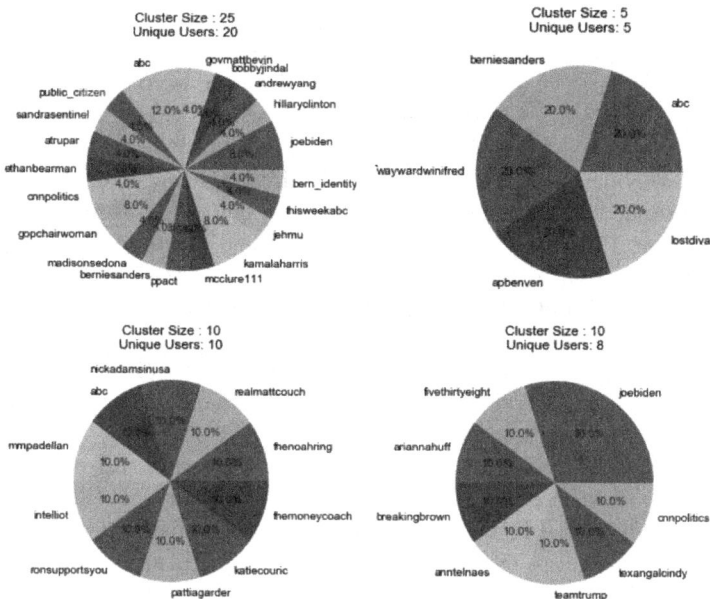

Fig. 5. Clusters formed by agglomerative clustering algorithm from N-Grams features (4 of 15)

Apart from the combination of Doc2Vec feature representation and Agglomerative clustering algorithm, the Tfidf-Kmeans does not produce any cluster with a prominent user. The same representation along with Agglomerative clustering produced 3 clusters with prominent target users (Table 2). Elizabeth Warren and Kamala Harris appear as prominent target users in two of these clusters. The third cluster identified 'abc' as the prominent user. The only combination to have done so out of the four.

Both Bag of N-Grams and Tfidf representation of the data produce one large clusters and fourteen small clusters upon the application of either of the two clustering algorithms. On the other hand, the clusters formed from the LDA Document Topic matrix as well as the Doc2Vec embeddings are more uniform in size.

4 Conclusions

This research aimed to look for any possible correlation between linkage-based community detection in twitter network data and user clusters formed from the textual content of the same network data. Although many factors affect the forming of linkage-based communities, we wanted to investigate the influence of context semantics in this. Conclusive results would have been able give us insights into interaction patterns of users on social networking sites and how much of it is dictated by the textual contents they produce.

The two clustering algorithms, coupled with the four feature representation did not produce many clusters with prominent users. The clusters that were formed were not large enough to be considered as counterparts of the communities formed by Gephi. Considering these factors, it cannot be claimed that the user clusters formed based on their linguistic features have a correlation to the communities based on modularity values. Hence, we cannot claim that the text data produced by the users are the strongest factor in forming the different communities on Twitter. The nature of the dataset can be one of the reasons behind this result. Since the tweets were mostly about the Democratic Debate and most people tweeted on similar topics, the text data may not have been topically diverse enough for the features and the clustering algorithms to capture the finer communities. This is why, one of the future direction can be to repeat the experiment on a different and more topically diverse dataset and compare the results with the results of this thesis.

Another future direction of this work can be the use of different metrics to compare the two methods. This research used the authorities of the communities or in other words, nodes in communities that have the highest in-degree in the most visible communities as the basis of comparison. In future, other methods of calculating the network authorities, e.g. the HITS algorithm can be used. Comparison of hubs and bridges of networks and clusters can be incorporated in the future works.

References

1. Papadopoulos, S., Kompatsiaris, Y., Vakali, A.: Community detection in social media. Data Min. Knowl. Disc. **24**(3), 515–554 (2012). https://doi.org/10.1007/s10618-011-0224-z
2. Aggarwal, C.C., Wang, H.: Text mining in social networks. In: Aggarwal, C. (ed.) Social Network Data Analytics, pp. 353–378. Springer, Boston (2011). https://doi.org/10.1007/978-1-4419-8462-3_13
3. Kim, Y.H., Seo, S., Ha, Y.H., Lim, S., Yoon, Y.: Two applications of clustering techniques to Twitter: community detection and issue extraction. Discrete Dyn. Nat. Soc. (2013)
4. Gligorić, K., Anderson, A., West, R.: How constraints affect content: the case of Twitter's switch from 140 to 280 characters. arXiv preprint arXiv:1804.02318 (2018)
5. Zhang, Y., Wu, Y., Yang, Q.: Community discovery in Twitter based on user interests. J. Comput. Inf. Syst. **8**(3), 991–1000 (2012)
6. Tumasjan, A., Sprenger, T.O., Sandner, P.G., Welpe, I.M.: Predicting elections with twitter: what 140 characters reveal about political sentiment. In: Fourth International AAAI Conference on Weblogs and Social Media (2010)
7. Sakaki, T., Okazaki, M., Matsuo, Y.: Earthquake shakes Twitter users: real-time event detection by social sensors. In: Proceedings of the 19th International Conference on World Wide Web, pp. 851–860 (2010)
8. Lerman, K., Ghosh, R.: Information contagion: an empirical study of the spread of news on digg and Twitter social networks. arXiv preprint arXiv:1003.2664 (2010)

9. Sachan, M., Contractor, D., Faruquie, T. A., Subramaniam, L. V.: Using content and interactions for discovering communities in social networks. In: Proceedings of the 21st International Conference on World Wide Web, pp. 331–340 (2012)
10. Huberman, B.A., Romero, D.M., Wu, F.: Social networks that matter: Twitter under the microscope. arXiv preprint arXiv:0812.1045 (2008)
11. Mucha, P.J., Onnela, J., Porter, M.A.: Communities in networks. Not. Am. Math. Soc. **56**, 1082–1097 (2009)
12. Fortunato, S.: Community detection in graphs. Phys. Rep. **486**(3–5), 75–174 (2010)
13. Zhou, D., Manavoglu, E., Li, J., Giles, C.L., Zha, H.: Probabilistic models for discovering e-communities. In: Proceedings of the 15th International Conference on World Wide Web, pp. 173–182 (2006)
14. Smith, M., et al.: NodeXL: a free and open network overview, discovery and exploration add-in for Excel 2007/2010/2013/2016. http://nodexl.codeplex.com/ from the Social Media Research Foundation. http://www.smrfoundation.org. Accessed 7 Dec 2020
15. Bastian, M., Heymann, S., Jacomy, M.: Gephi: an open source software for exploring and manipulating networks. In: Icwsm, vol. 8, no. 2009, pp. 361–362 (2009)
16. Salloum, S.A., Al-Emran, M., Monem, A.A., Shaalan, K.: A survey of text mining in social media: Facebook and Twitter perspectives. Adv. Sci. Technol. Eng. Syst. J. **2**(1), 127–133 (2017)
17. Bergsma, S., McNamee, P., Bagdouri, M., Fink, C., Wilson, T.: Language identification for creating language-specific Twitter collections. In: Proceedings of the Second Workshop on Language in Social Media, pp. 65–74 (2012)
18. Chang, J.C., Lin, C.C.: Recurrent-neural-network for language detection on Twitter code-switching corpus. arXiv preprint arXiv:1412.4314 (2014)
19. Bengfort, B., Bilbro, R.: Yellowbrick: visualizing the scikit-learn model selection process. J. Open Source Softw. **4**(35), 1075 (2019)
20. Grossetti, Q., du Mouza, C., Travers, N.: Community-based recommendations on Twitter: avoiding the filter bubble. In: Cheng, R., Mamoulis, N., Sun, Y., Huang, X. (eds.) WISE 2020. LNCS, vol. 11881, pp. 212–227. Springer, Cham (2019). https://doi.org/10.1007/978-3-030-34223-4_14

Fake News Detection Without External Knowledge

Zalán Bodó[(✉)] [ID]

Faculty of Mathematics and Computer Science, Babeş–Bolyai University,
Street Kogălniceanu, nr. 1., 400084 Cluj-Napoca, Romania
zbodo@cs.ubbcluj.ro

Abstract. Although written deception is not a new invention, the emergence and progress of electronic media—and more recently social media—has changed the speed and extent of access to information in a good way, but at the same time facilitating the proliferation of disinformation as well. Automatic veracity determination, therefore, became a widely studied problem in the last years. We claim that without using a knowledge base and fact-checking, that is based solely on textual content features one cannot truly fight this phenomenon, nevertheless, such a deception detection system can be used beneficially in certain situations. In the present study we apply text categorization methods to detect fake news without involving any external knowledge base (e.g. lexicons, unlabeled corpora, pre-trained word vectors, etc.). We employ traditional bag-of-words and more recent end-to-end neural network models, and evaluate them on eight—five smaller and three larger—fake news datasets. The experimental results show that one can attain considerably precise detection performance, in some cases even in the very close vicinity of the perfect F_1 score, using solely the labeled data. We also strive to explain why some of these approaches imply a better performance than others.

Keywords: Fake news detection · Text categorization · Machine learning

1 Introduction

The emergence and progression of social media has drastically changed the speed and extent of access to news, providing us with extremely fast information acquisition, on the other hand making it also possible to be flooded with a huge amount of information. The loads of information facing us today greatly encumbers checking the legitimacy of every piece. By the 2018 survey of the Pew Research Center [51], 68% of USA's adult population gets news from social media platforms, while an earlier research revealed that 43% of social media users are

Parts of this paper have been presented at the Pannonian Conference on Advances in Information Technology (PCIT 2019), 31 May–1 June 2019, Veszprém, Hungary.

not even aware of the publication source [36]. The radical change that happened in the last decades in how media communicates information towards people can also distort reality: providing equal time to tell both sides of the story—even on factual matters—can make issues look controversial. The problem of climate change is a good example: there is strong consensus in the scientific community about anthropogenic global warming [6], nevertheless, since the average person acquires information from news, social media, etc. rather than scholarly publications, society is divided on this question [25,26].

Whilst targeted propaganda can manipulate our views, the human brain is also prone to cognitive biases. Trying to avoid psychic discomfort in every situation, the resolution it arrives to is not always rational; often we more easily accept comforting lies than unpleasant truths [31]. Social groups individuals belong to can also affect the behavior and identity of a person, making her/him choose socially safe options [58]. On the other hand, due to the online era, our present choices are likely to influence later ones, for traditional recommendation systems base their suggestions on previous activity, often isolating us in a filter bubble [39].

Finding fake news and reporting them might also backfire [31], that is corrective information might only strengthen the initial belief, therefore such an automated system will likely not provide us the ultimate solution for fighting this complex phenomenon. Nonetheless, a high-accuracy fake news detection tool can be definitely useful: collecting legitimate facts for feeding it into another (machine learning) system, for example, could be carried out automatically with it.

In the present paper we experiment with text categorization methods for detecting fake news without involving any external knowledge base. Our goal is not to develop a better method than previous research, nor to once and for all solve the problem of spreading false information, but to investigate the accuracy by which fake news can be recognized learning only from the training corpus of fake and legitimate sample news articles, and try to explain the whys. We claim that without actual fact-checking it is difficult to fight disinformation effectively; if linguistic features only are used in the decision process it can be relatively simple to deceive these detectors [68].

The paper is structured as follows: Sect. 2 defines the concept of fake news and gives a brief overview of the existing approaches for automatic deception detection. Machine learning approaches from text categorization perspective are enumerated in Sect. 3, focusing on the simple yet efficient bag-of-words models and more recent neural network-based solutions. In Sect. 4 some of the commonly utilized fake news datasets are presented—five smaller and three larger text corpora—used later in the experiments. The experiments are described in Sect. 5, discussed and analyzed in Sect. 6, while Sect. 7 concludes the present work. The appendix in front of the references enumerates the URLs referenced in the paper.

2   Fake News and Deception Detection

2.1   Definition and Categorization of Fake News

In this paper we follow the definition of [54]: "Fake news is a news article that is intentionally and verifiably false". Both deliberateness and verifiability are important here: while an article can contain misinformation due to ignorance or lack of investigation, intentional deception and persuasion might induce a different writing style and progression. Verifiability, in turn, is required to have an *objectively* decidable problem to tackle, to be able to assign labels to the news stories.

McIntyre [31] categorizes falsity as falsehood (unintended deceit made by mistake), willful ignorance (avoidance of becoming informed about something for various reasons) and lying (intentional deception), while the Global Disinformation Index [32] differentiates between politically and financially motivated disinformation. Furthermore, based on knowledge, false information can be labeled as either opinion (e.g. fake reviews on e-commerce websites) or fact-based (e.g. hoaxes) [23]. Further categorizations of fake news into more refined subcategories can be found in the literature; Rubin et al. [45], for example, distinguish between serious fabrications, hoaxes and humorous fakes.

While some might argue that there is no such thing as objective truth, we support *naive realism*, believing that there exist facts that are independent of people's ideology [54].

2.2   Automatic Checking of Veracity

The present section enumerates a few surveys and papers examining different types of features, machine learning methods for deception detection, and mentions some related studies.

Somewhat surprisingly, a relatively large number of surveys on automatic fake news detection can already be found in the literature. The work of [5] surveys the existing fake news detection approaches providing a typology of the assessment methods, proposing also guidelines for building a detection system. Presenting the problem of fake news from a data mining perspective, [54] considers definitions of fake news, psychological and social foundations, and fake news on social media; features, detection models, datasets, evaluation measures, related areas and open issues are also discussed. In [23] diverse aspects of the fake news phenomenon are examined, including the actors involved, rationale of successful deception, impact and characteristics of false information, while it also reviews opinion and fact-based detection algorithms. Existing detection methods, diffusion models and mitigation techniques are analyzed in [43], focusing on 2017 and 2018 works and social networks Twitter, Facebook, Sina Weibo, as well as collaborative platforms Wikipedia and Yelp.

Using traditional text categorization methods for automatic deception detection enriching content-based features with other linguistic ones (psycholinguistic, readability-related and syntactic features) is studied in [42]. The FakeNewsNet data repository is introduced in [53], compiled by the help of fact-checking websites. The authors perform classification experiments as well, using content-based and social context-based features. In [1] building automated detection systems for fake reviews and news is studied, employing n-gram features and traditional supervised learning methods.

In [46] a deep neural network model called CSI (Capture, Score, Integrate) is proposed for fake news detection that takes into account the textual content, temporal user engagements and the tendency of promoting a source. By the empirical investigation of [57], hoaxes and non-hoaxes can be accurately identified based solely on user interactions (i.e. likes), without considering the actual content of the posts. Attention-based LSTMs incorporating speaker profiles are proposed for fake news detection in [28]. Machine learning methods for deception detection—traditional techniques and different neural network architectures—are benchmarked on three datasets in [18].

A glimpse into the ever-growing fake news detection literature instantly shows a diverse and intriguing set of related problems. The vulnerability of natural language processing-based fake news detection systems is examined in [68], using three types of adversarial attacks: fact distortion, subject-object exchange and cause confounding. In [35] open issues in fake news analysis are discussed, emphasizing the importance of interpretability of the results as well as providing supporting evidence. Automatically generating fake news articles given the article title or the headline is studied in [63]; performance evaluation is done by the means of Amazon Mechanical Turk (AMT/MTurk), showing that, in general, humans can easily be fooled by machine-written propaganda.

Our approach focuses on using only the textual content of an article in predicting its veracity, invoking no external knowledge during the classification process. Experimenting with different features, weightings and neural network models we also strive to elucidate the findings.

3   Machine Learning in Text Categorization

Text categorization is the problem of automatically assigning documents written in some natural language to predefined categories (or vice versa); it is a supervised learning task, for the mapping is learned through labeled examples [60].

When dealing with natural language texts, the first and probably most important question is the one concerning representation. One of the most popular and successful approaches is the bag-of-words model [48], representing a document as a vector of term occurrences, assuming independence between these. Since the order of words also carries important information, the more generalized version uses word n-grams as dimensions of the representational space [8]. Instead of using plain occurrences, a weighting scheme is sometimes applied to emphasize the relevance of the words, the most notable one being tf-idf (*term frequency × inverse document frequency*) [48].

In order to better represent dependencies, more advanced methods were later introduced: latent semantic analysis (LSA) [7] linearly combines initial dimensions to find major associative patterns in the data, while novel methods learn continuous vector representations of words based on co-occurrence statistics around the word in question. Prominent models of this approach are word2vec [33], GloVe [41] and fastText [3]. End-to-end learning techniques often use the same representations as traditional methods, namely bag-of-words, one-hot encoding and continuous vector representations [11], however it is also common practice to learn the embeddings during training the classifier.

In this paper we investigate statistical text categorization methods as we consider fake news detection a binary text classification problem.

3.1 GOFTeC: Good Old-Fashioned Text Categorization

By GOFTeC (Good Old-Fashioned Text Categorization) methods we refer to traditional approaches based on bag-of-words (bag-of-n-grams) representation that optionally employ a feature selection (dimensionality reduction) method and a supervised learning algorithm. Despite the simplicity and naivety of the representational model, these algorithms often yield surprisingly good performance.

Preferred methods include document frequency thresholding, mutual information and χ^2 statistics-based feature selection [61], while some of the most successful machine learning algorithms here are support vector machines [16], naive Bayes [44] and logistic regression [64]. Sebastiani [49] gives an excellent overview of these approaches.

3.2 Neural Networks

Despite the many existing neural network models for text classification, we only consider here the classes of convolutional and recurrent networks, briefly presenting the concrete methods that will be used in the experiments.

Convolutional Neural Networks. Convolutional neural networks (CNN) [24] are famous primarily for their unprecedented performance in image classification and analysis, however, they also perform well in diverse natural language processing tasks. The method described in [19] uses pre-trained word2vec vectors and employs 1-dimensional convolutions with different kernel sizes, concatenating the outputs together and passing it to a fully connected softmax layer to perform the classification. The proposed method was tested on multiple datasets and compared to other models, outperforming state-of-the-art ones on 4 out of 7 sentiment analysis and question classification datasets. Zhang et al. [65] introduce a character-based deep neural network architecture for text classification, using one-hot input encoding, 6 convolutional, 3 pooling and 3 fully connected layers. Compared to some state-of-the-art approaches, the new models achieved lower testing errors on 4 out of 8 datasets.

Kipf and Welling [22] introduced graph convolutional networks (GCN) to propagate information from neighboring nodes in a network using spectral graph convolutions. Later, GCNs were applied to text categorization [62], where a single large graph was built to represent word–word and word–document relations in a corpus, jointly learning the word and document embeddings. The two-layer GCN significantly outperformed the baseline models on 4 out of 5 datasets.

Recurrent Neural Networks. Since text can be viewed as a sequence of basic building blocks, such as characters or words, sequence models, e.g. long short-term memory (LSTM), gated recurrent unit networks (GRU), bidirectional LSTM (BLSTM), etc. [11] can be considered a good option for document classification. Nowak et al. [37] compares LSTM, GRU and BLSTM networks for text classification on three datasets. To improve upon the performance of BLSTMs, [66] introduces an attention layer to focus on important words in classification settings. The model proposed by the authors outperformed all the other methods it was compared to on a relation classification task, without using other NLP features.

4   Fake News Datasets

As we consider fake news detection a supervised learning problem, news data along with its veracity-based labeling is needed. The eight datasets presented below—five smaller and three larger corpora—are used to perform the classification experiments; in six of these the data is distributed evenly across the classes, while two of them are skewed, and the text of the articles is or can be split into title and body text. In some cases additional information are also available (e.g. authors, user engagements, etc.), however, in the experiments we reduced the set of attributes to the following three: title, body text and label.

4.1   FakeNewsAMT and Celebrity Datasets

The FakeNewsAMT (FNAMT) and Celebrity (Celeb) datasets[1] were introduced in [42]. The paper describes the construction process of these together with some fake news classification experiments using different linguistic features. FakeNewsAMT was built by collecting legitimate news from trusted news portals (e.g. ABCNews, CNN, USAToday, etc.), and then using crowdsourcing via AMT fake versions of these were produced. The articles in this dataset form six categories (sports, business, entertainment, politics, technology, education), each containing 40 texts, thus totaling 2×240 stories.

Celebrity was compiled by collecting news about public figures from online magazines (e.g. Entertainment Weekly, People Magazine, RadarOnline, etc.), whereas the truth values of the articles were determined using fact-checking sites. The dataset consists of 2×250 articles.

---

[1] http://web.eecs.umich.edu/~mihalcea/downloads.html#FakeNews.

Table 1. Datasets used in the experiment, showing the sizes, i.e. number of documents in the datasets.

Dataset name	Abbreviation	Size
FakeNews AMT	FNAMT	480
Celebrity	Celeb	500
BuzzFeed	BF	101
Random	Rand	150
McIntire's dataset	MD	6 335
UTK-MLC	UTK	20 800
PolitiFact	PF	917
GossipCop	GC	20 880

4.2   Horne and Adali's Datasets

Horne and Adali [15] compiled the following two sets: an election dataset using BuzzFeed articles and a random political news dataset. The BuzzFeed Political News Data (BF) is based on the dataset described in [55], gathered by searching for real and fake news, and analyzing their Facebook engagements using BuzzSumo.[2] The articles were categorized based on known fake and credible sources. Although both [55] and [15] specify $60 + 60$ news stories,[3] the dataset made available for download contains 48 fake and 53 legitimate news.[4]

The Random Political News Data (Rand) consists of 3×75 stories, each one belonging to one of the following three categories: fake, real and satire. The fake news were randomly collected using Zimdars' list of fake news sources [69], while the legitimate ones come from 2014's most trusted news outlets list published by the Pew Research Center [34]. In our experiments only the fake and real categories were used.

4.3   McIntire's Dataset

McIntire's dataset (MD) [30] is based on a Kaggle fake news dataset,[5] extended to incorporate real news in order to be used for fake news classification. The author used AllSides[6] to scrape 5279 articles (from sources like New York Times, WSJ, Bloomberg, etc.) and randomly selected the same number of news from the above-mentioned Kaggle dataset. While [30] reports the above-mentioned counts, the final dataset received from the author contains 3171 real and 3164 fake news articles.

---

[2] https://buzzsumo.com/.

[3] The original dataset can be found at https://docs.google.com/spreadsheets/d/1ysnzawW6pDGBEqbXqeYuzWa7Rx2mQUip6CXUUUk4jIk/edit.

[4] https://github.com/BenjaminDHorne/fakenewsdata1.

[5] *Getting Real about Fake News*, https://www.kaggle.com/mrisdal/fake-news/home.

[6] https://www.allsides.com/unbiased-balanced-news.

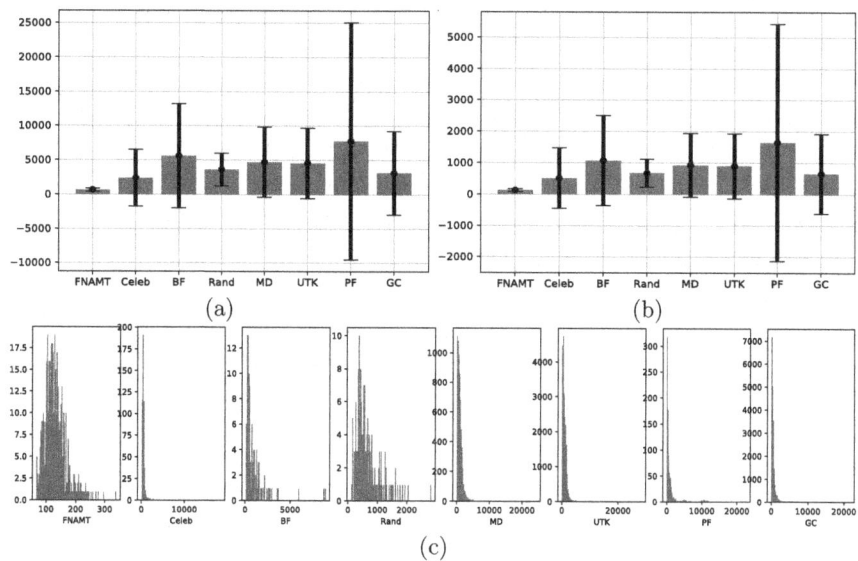

Fig. 1. Document length statistics (mean and standard deviation) in the corpora used in the experiments in terms of (a) characters and (b) words, considering the body text of the documents. The histograms in (c) show the word-based length distributions of the datasets.

4.4   UTK-MLC Dataset

The UTK-MLC (University of Tennessee, Knoxville, Machine Learning Club) dataset[7] (UTK) is taken from a Kaggle InClass Prediction Competition run by the UTK Machine Learning Club. It comprises a training set of 20 800 articles distributed approximately equally between the two classes, and a test set with 5200 news articles. Since no information could be obtained about the compilation procedure, nor the test labels could be accessed, only the training set was used in the experiments.

4.5   FakeNewsNet: The PolitiFact and GossipCop Datasets

The FakeNewsNet[8] data repository has recently been introduced in [53]. It contains two datasets, PolitiFact (PF) and GossipCop (GC), and—unlike the previous ones—beside news content, social and spatiotemporal information (i.e. user engagement records with location information and timestamps) are also available. PolitiFact[9] and GossipCop[10] are fact-checking websites for determining the veracity of American political and worldwide celebrity news, respectively.

---

[7] https://www.kaggle.com/c/fake-news.
[8] https://github.com/KaiDMML/FakeNewsNet.
[9] https://www.politifact.com/.
[10] https://www.gossipcop.com/.

Table 2. 5-fold cross-validation F_1 scores ($\times 100$) obtained using GOFTeC models for the eight datasets.

Features	FNAMT	Celeb	BF	Rand	MD	UTK	PF	GC
Baseline binary	56.32	73.15	79.03	84.19	92.73	97.60	88.76	89.56
Baseline freq	53.90	68.96	73.95	80.11	91.91	96.69	86.13	88.53
Baseline tf-idf	44.46	64.02	78.38	76.52	92.42	96.47	86.31	88.56
Uni+bigrams binary	50.86	74.49	73.68	83.89	93.18	97.98	88.58	**90.88**
Punctuation tokenizer	64.02	78.24	77.84	90.13	95.08	99.72	89.85	**90.97**
χ^2 5000	82.81	83.15	80.53	94.33	94.98	99.78	89.08	90.34
Stat. features	84.32	83.15	80.53	94.33	95.23	99.79	89.10	90.38

The authors used all real and fake news data from PolitiFact, while available GossipCop stories having a rating less than 5—considered to be fake articles—were extended by real news from E! Online.[11]

After excluding empty records and news having no title and body text, the PolitiFact and GossipCop datasets are left with a total of 917 (533 real, 384 fake) and 20 880 (16 035 real, 4845 fake) articles, respectively.

Table 1 enumerates the datasets used in our experiments, displaying their sizes, while Fig. 1 shows character and word-based average document lengths (and standard deviations) in the corpora, as well as word-based length distributions.

5   Experimental Results

For conducting the classification experiments the eight datasets presented in Sect. 4 are used. The datasets were not explicitly split into training and test sets, therefore the performance is measured using stratified 5-fold cross-validation F_1 scores [29]. In the case of the neural networks the training data resulting from the cross-validation split was further divided into training (75%) and validation data (25%). The validation data was used for early stopping.

For every dataset we highlighted the three best results across the tables (typeset with bold numbers).

The experiments were implemented and run in Python 3.6.9, using *scikit-learn* [40] (v. 0.21.3) and *keras* [4] (v. 2.3.1) (*tensorflow* v. 2.3.0).[12]

The first three baseline models, *binary*, *freq* and *tf-idf* apply binary weighting, plain word occurrences and tf-idf weights; all of these use unigrams only.

In *uni+bigram binary* unigrams are augmented by bigrams while using binary weighting. The fifth model, *punctuation tokenizer*, modifies the base tokenizer of *scikit-learn* by generating also punctuation marks (or sequences of these) as valid tokens. The χ^2 model employs χ^2 feature selection [61] using a fixed

---

[11] https://www.eonline.com/.

[12] The code is available on GitHub: https://github.com/miafranc/fakenews2020.

Table 3. 5-fold cross-validation F_1 scores ($\times 100$) obtained filtering out the features that appear in more than half of the data as well, and using χ^2 with tf-idf weights.

Classifier	FNAMT	Celeb	BF	Rand	MD	UTK	PF	GC
LR	86.16	92.10	89.26	97.28	**95.90**	99.81	92.24	90.61
Linear SVM	84.69	91.89	**96.72**	99.35	94.63	**99.80**	90.70	89.23
Multinomial NB	**94.20**	94.75	91.84	99.35	95.06	99.39	**93.75**	77.97

Table 4. 5-fold cross-validation F_1 scores ($\times 100$) using a final L_1-normalization over the dimensions.

Classifier	FNAMT	Celeb	BF	Rand	MD	UTK	PF	GC
LR	**96.73**	100	**99.13**	100	92.29	97.68	78.43	86.90
Linear SVM	94.61	100	**99.13**	100	93.17	98.04	82.89	89.26
Multinomial NB	92.61	**99.60**	94.11	100	**95.83**	98.39	89.41	86.90

number of features, taking the 5000 highest-scoring n-grams. Finally, statistical text features (*stat. features*) are concatenated to the document vectors: these are intended to represent the complexity, level of understandability of the text [42]. We generated word and sentence length histograms of equal bin widths on range $[1, 30]$, with an extra bin summing up the cases when exceeding the maximum value. The histograms obtained are normalized by the number of words and number of sentences in the document, respectively. The remaining four regular expression-based features count the number of all caps words, words starting with capital letters, numbers (sequences of digits), and possible abbreviations— normalized by the number of words. The results—using a logistic regression classifier—are shown in Table 2.

In the final GOFTeC experiments we tried to slightly tweak the classifiers' performance using simple such techniques: (a) filtering out features that appear in more than half of the data, (b) using tf-idf scores instead of raw counts in χ^2 feature selection, (c) L_1-normalizing the final features. Here, beside logistic regression (LR) we also tested linear support vector machine (SVM) and multinomial naive Bayes (NB) classifiers. The evaluation scores obtained are shown in Tables 3 (a, b) and 4 (c).

In the second set of experiments we compared 5 neural network models for text categorization: character-based CNNs (C-CNN) [65], the word-based CNN of [19] (W-CNN), a basic LSTM whose architecture is shown in Fig. 2, attention-based BLSTM (A-BLSTM) [66], and GCNs [62]. Instead of using one-hot encoding and pre-trained word vectors, in all the models except GCN an embedding layer was inserted after the input layer. Since document lengths varied across the datasets used (see Fig. 1), we used different character and word-based lengths in the input layers, approximately equal to the average lengths. For the first four models the following settings were used: a batch size of 50, 1000 number of epochs with early stopping, setting the patience to 50 epochs and using cyclical

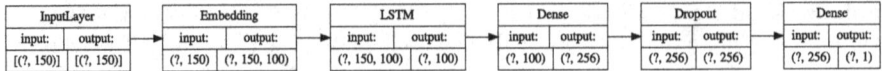

InputLayer		Embedding		LSTM		Dense		Dropout		Dense	
input:	output:	input:	output:	input:	output:	input:	output:	input:	output:	input:	output:
[(?, 150)]	[(?, 150)]	(?, 150)	(?, 150, 100)	(?, 150, 100)	(?, 100)	(?, 100)	(?, 256)	(?, 256)	(?, 256)	(?, 256)	(?, 1)

Fig. 2. Architecture of the LSTM-based classifier: the input dimensionality is 150 (for FNAMT), the activation of the inner dense layer is ReLU, and the dropout rate is set to 0.5.

Table 5. 5-fold cross-validation F_1 scores ($\times 100$) of the five neural network models used in the experiments.

Classifier	FNAMT	Celeb	BF	Rand	MD	UTK	PF	GC
C-CNN	67.95	52.43	80.30	70.49	**98.86**	**99.87**	89.87	87.32
W-CNN	60.53	65.76	80.62	71.35	95.13	99.51	87.62	90.83
LSTM	60.53	39.21	69.80	61.04	54.87	95.86	73.34	86.79
A-BLSTM	55.72	67.89	83.57	86.65	88.96	98.74	84.40	89.84
GCN	41.09	56.01	77.67	88.00	94.08	95.80	89.93	**90.97**

learning rates (triangular, base learning rate $= 10^{-5}$, maximum learning rate $= 10^{-3}$, step size $= 8$) [56]. For the GCN 200 training epochs with a patience of 10 epochs, and a fixed learning rate of 0.02 were applied; in all cases, the Adam optimizer was used [21]. The results are shown in Table 5.

6   Discussion and Analysis of the Results

In the GOFTeC experiments the binary representation, as shown in Table 2, proved to be superior to the other baselines, which is in line with the finding that taking into account just the presence or absence of terms can sometimes yield better performance than more sophisticated models [38]. While the advantage of considering bigrams as well is not so obvious from the results, it indeed resulted in better scores in the subsequent experiments; using higher-order n-grams, in general, tend to produce more accurate predictions [8], up to a given n. One of the best improvements in the entire process was brought by the new tokenizer:[13] it introduces tokens made up of non-word characters, usually punctuation marks, which can be useful in predicting the veracity of news articles, since, by [42, 59], the overuse of punctuation marks (e.g. !!, !!!) can be a good indicator of deception. While some machine learning algorithms also determine the relevant features during learning, using a smaller subset of relevant features may provide better results [13]. χ^2 feature selection produced the best scores among the methods tested—we reported only these results here. Statistical text features are intended to represent the complexity, level of understandability of the text, which, by [42], can help in detecting fake news. Some of the existing readability

---

[13] This is also true for the neural networks, nevertheless, here we do not report the results obtained using different tokenizers.

metrics (e.g. Flesch–Kincaid [20]) use simple attributes as the number of words, number of sentences and number of syllables to determine text complexity, thus we concatenated histogram features to the document vectors, together with a few regular expression-based features, as described in the previous section.

The rationale behind the first tweak applied was the following: terms occurring with approximately the same frequency in every class in the training data will have a near zero contribution in determining the class label, therefore it can be omitted. An application of this idea can be found, for example, in spam filtering systems [12]. Filtering out words that appear in more than 50% of the training data clearly eliminates those occurring in all or almost all documents, discarding at the same time other terms as well. Though not equivalent to the initial intent, we experimented with this simplified setting. Second, we combined tf-idf and χ^2 feature selection by using tf-idf scores when building the contingency table. While co-occurrences are intrinsically natural numbers, χ^2 can be used on arbitrary positive reals too, the assumption being that using tf-idf instead of raw frequencies better reflects term importance. The tf-idf scores were used only in feature selection; the weighting for n-grams representation remained the previously chosen binary scheme. Last, we experimented with data normalization. Document length normalization is a common step in text categorization, in order to omit the length of the document from the equation [29, 49]. Feature scaling can also have beneficial effects if feature ranges differ [14]. While L_2-normalizing along the second dimension did not yield any better results, for the smaller datasets, L_1-normalization of the features resulted in better evaluation scores.

The results obtained with neural networks clearly show that—as it was expected—small datasets cannot provide sufficiently good performance [11], while the best GOFTeC models were outperformed by more than 3% on McIntire's dataset. Convolutional networks provided somewhat better results than the recurrent models, graph convolutional methods—being the most resource intensive approaches among those tested—holding many opportunities and challenges for transductive learning. Since using character n-grams for categorizing documents has already been proved to be very effective [3,17,27], the good performance of character-based CNNs was not surprising.

After performing the experiments, we analyzed the results focusing on the following two main issues: (i) Are there any articles that are consistently misclassified by the different machine learning models, that is are there some patterns for which the methods tested systematically fail? (ii) Which are the most relevant linguistic features for the legitimate and fake classes? Could the most important features be used to explain some of the incorrect predictions and to prove some of the assumptions made when building the classifiers?

To answer the first one, we used the first split from the 5-fold cross-validation partitions, and for every classifier the false positive and false negative predictions were recorded (positive and negative meaning legitimate and fake news, respectively) and summed up for every dataset. Since the second tweak—in the case of the GOFTeC experiments—did not improve the results consistently, the first tweak was applied here only (additional term filtering and χ^2 with tf-idf). Thus,

Table 6. Misclassified articles: (a) a false positive example from FakeNewsAMT and (b) a false negative from McIntire's dataset. Both texts were classified wrongly 7 times out of 8.

German Power Spat With Denmark Gets Fixed	'Britain's Schindler' Dies at 106
A long-running squabble between Denmark and Germany over power cables linking the countries has improved the amount of electricity Nordic producers can supply to Europe's biggest market to the greatest level in 17 years. Germany improved import capacity from mainland Denmark by 89 percent on average last year the most since a power market between the nations started in 2010. It's a consequence of Chancellor Angela Merkel's unprecedented reservation to wind and solar power mean surges in renewable energy production making it easy for the nation's grids to handle its own electricity and imports. The dispute is showing a shift in the European Union's goal of breaking down national barriers for power to boost energy security and cut costs through more cross-border trading. Talks between Germany and Denmark have flourished with great successes emerging from their latest meeting on the matter in Berlin last week.	A Czech stockbroker who saved more than 650 Jewish children from Nazi Germany has died at the age of 106. Dubbed "Britain's Schindler," Nicholas Winton arranged to transport Jewish youngsters from Prague after Germany annexed Czechoslovakia in March 1939. Though the children were originally set to arrive in Britain by plane, the German invasion forced Winton to transport them by train through Germany before they eventually reached England by boat. Winton arranged eight trains, known as the Kindertransports (children's transports), to evacuate the children, and died on the anniversary of the 1939 departure of the one carrying the largest number of children: 241. Winton was knighted by Queen Elizabeth II in 2003 for his efforts, despite keeping it secret for nearly 50 years.
(a)	(b)

the incorrect predictions of 8 classifiers (3 GOFTeC and 5 neural network models) were summarized. There were no articles being misclassified by all methods, i.e. there are no eight-vote documents. The situation is different, however, for the seven-vote articles: in this case we found 161 false positives and 3 false negatives. Table 6 shows two such examples, a false positive from the FakeNewsAMT and a false negative from McIntire's dataset. Examining the texts, we found no obvious cues for the misclassification. Nevertheless, the above numbers suggested that there could be a problem in the GossipCop dataset: 160 of the 161 false positives misclassified by 7 classifiers are coming from this dataset. Examining these results more closely, we found the following issues: 63 of these texts are about privacy policy and personalized ads on a given website (usually starting as "About Your Privacy on this Site..."), 6 texts are about enabling HTTP cookies, there are 4 documents with missing title or body text and also having very short content (between 3–21 words), and mislabeled data can be found as well in the dataset.[14] Hence, the results obtained for this dataset should be treated considering these points. Moreover, we recommend cleansing the dataset before using it in further experiments.

For the second problem we used the first cross-validation split as above, and being aware that data differ across datasets (e.g. celebrity vs. political news) wanted to find important features valid for all datasets. To achieve this, we considered logistic regression (first tweak) and determined the most important

---

[14] In the case of the articles "Grammy winners 2018: the complete list" and "Carrie Fisher", for example, we were unable to find any indication that there are false information in these.

Table 7. Most 20 important positive and negative LR features. The number on the right shows how many times the n-gram appeared among the most important 100 features.

Positive (real)				Negative (fake)				
–	6	,"	3	!	5	am	2
. "	5	. "	3	sources	4	caused	2
'	5	"	3	stated	3	future .	2
. but	4	' s	3	wikileaks	3	taken	2
two	3	republican	3	breaking	3	several	2
earlier	3	this year	2	police	3	claimed	2
. and	3	three	2	. this	3	source	2
stat. feature 33	3	she said	2	these	2	–	2
;	3	he said	2	recent	2	' s	2
war	3	but the	2	bought	2	website	2

features by appropriately ranking the weights of the model—in this way for every dataset the 100 most important dimensions were identified for both classes. The results were summed up by a voting scheme, producing the scores shown in Table 7.[15] The numbers on the right show how many votes a feature has obtained, where the maximum value would be 8, because of the 8 datasets. Unfortunately, based on the features with the most votes, we could not explain beyond any doubt the wrong predictions from Table 6. Both articles have long sentences, and the only feature multiply present in both is the possessive 's: 5 and 3 times, respectively. What the reader cannot see, because the font used for typesetting the present text is not capable to differentiate between some characters, is that the apostrophes used in the possessive 's-es in the two texts of Table 6 and respectively the two parts of Table 7 do differ: the first one is the unicode character U+0027, and the second one is U+2019. While this could cause misclassification, we are not sure whether it is not the result of some more complex feature combinations. Similarly, we cannot explain why the possessive 's-es appeared among the most important features. However, as it was previously assumed— and in fact already proved by the improvements brought by the introduction of the punctuation tokenizer—the punctuation marks play an important role in the prediction, the exclamation mark being the most important feature for the fake class—at least following the present methodology. One of the fifth most important features for legitimate articles is a statistical text feature, *stat. feature 33*, corresponding to the extra bin in the sentence length histogram (see Sect. 5). Thus, text complexity metrics are indeed of central importance in automatic fake news detection [42]. However, the above analysis also pointed out some possible weaknesses of our methods, as terms usually considered stopwords (*but, and,*

---

[15] We could have used other approaches here as well, for example χ^2 to rank the bag-of-n-grams features, Grad-CAM [50] for the convolutional neural networks, etc.

Table 8. Previous classification results reported in the literature. The first part shows the results obtained within the works that introduced the datasets. Since the methodology may differ, the results cannot be compared directly.

	FNAMT	Celeb	BF	Rand	MD	UTK	PF	GC
Introduction	78	76	71–78[†§]	77	91.7[†]	98.59[‡]	70.6	72.5
[47][†]	83.3	79						
[9][†]	95	78						
[2]	81	83					77	
[1][†]		87						
[18]					90			
[10]							92.5	82.9
[52]							92	
[67]							89.6	89.5
Our best scores	96.73	100	99.13	100	98.86	99.87	93.75	90.97

[†]Accuracy/[‡]Accuracy on test set/[§]4 top features only

the) have also been included and got a good ranking too. Therefore, although the results obtained using no external data are decent, supposedly using a simple stopword list could already improve on these. We consider this good news, as it means that more accurate predictions are possible.

7   Conclusion

The results of the experiments performed reveal that text categorization methods, without using any external knowledge base, can learn to distinguish fake news from legitimate ones. The goal of this study was not to find the best deception detection method, but to demonstrate that the text of the news articles in itself is sufficient to determine veracity—not meaning at all that these linguistic feature-based models cannot be attacked or tricked [68]. However, in Table 8 we enumerated some classification results reported in the literature using the same datasets as in our experiments—only as a reference, as the results obtained are not always directly comparable.

Good old-fashioned text categorization methods often outperform end-to-end neural network models, or produce only slightly worse results. Of course, the systematic feature engineering and selection, and the tweaks applied play an important role in achieving this.

In order to be used effectively in practice, however, instead of just reporting an article as being fake, one might consider assigning confidence scores to the predictions, as well as providing supporting evidence [35]. Future work includes explaining the decisions made by the classifiers, cleaning the datasets, analyzing the important features more thoroughly if possible, and also experimenting with other approaches.

Acknowledgements. We would like to thank George McIntire for making available his fake news classification dataset [30].

Appendix: List of URLs

This appendix contains the URLs from the paper in order of appearance, all retrieved on 2 Dec. 2020.

1. FakeNewsAMT and Celebrity datasets [42]: http://web.eecs.umich.edu/~mihalcea/downloads.html#FakeNews
2. BuzzSumo: https://buzzsumo.com/
3. BuzzFeed News: Election content engagement [55]: https://docs.google.com/spreadsheets/d/1ysnzawW6pDGBEqbXqeYuzWa7Rx2mQUip6CXUUUk4jIk
4. Horne and Adali's datasets: https://github.com/BenjaminDHorne/fakenewsdata1
5. *Getting Real about Fake News* Kaggle dataset: https://www.kaggle.com/mrisdal/fake-news/home
6. AllSlides: https://www.allsides.com/unbiased-balanced-news
7. UTK-MLC Kaggle dataset: https://www.kaggle.com/c/fake-news
8. FakeNewsNet data repository: https://github.com/KaiDMML/FakeNewsNet
9. PolitiFact: https://www.politifact.com/
10. GossipCop: https://www.gossipcop.com/
11. E! Online: https://www.eonline.com/
12. Python code of the experiments: https://github.com/miafranc/fakenews2020

References

1. Ahmed, H., Traore, I., Saad, S.: Detecting opinion spams and fake news using text classification. Secur. Priv. **1**(1), e9 (2018)
2. Barua, R., Maity, R., Minj, D., Barua, T., Layek, A.K.: F-NAD: an application for fake news article detection using machine learning techniques. In: IEEE Bombay Section Signature Conference (IBSSC), pp. 1–6. IEEE (2019)
3. Bojanowski, P., Grave, E., Joulin, A., Mikolov, T.: Enriching word vectors with subword information. Trans. Assoc. Comput. Linguist. **5**, 135–146 (2017)
4. Chollet, F., et al.: Keras (2015). https://github.com/fchollet/keras, Accessed 2 Dec 2020
5. Conroy, N.J., Rubin, V.L., Chen, Y.: Automatic deception detection: methods for finding fake news. In: Proceedings of the 78th ASIS&T Annual Meeting: Information Science with Impact: Research in and for the Community, p. 82. American Society for Information Science (2015)
6. Cook, J., et al.: Quantifying the consensus on anthropogenic global warming in the scientific literature. Environ. Res. Lett. **8**(2), 024024 (2013)
7. Deerwester, S., Dumais, S.T., Furnas, G.W., Landauer, T.K., Harshman, R.: Indexing by latent semantic analysis. J. Am. Soc. Inf. Sci. **41**(6), 391–407 (1990)

8. Fürnkranz, J.: A study using n-gram features for text categorization. Technical Report, OEFAI-TR-98-30, Austrian Research Institute for Artifical Intelligence (1998)
9. Gautam, A., Jerripothula, K.R.: SGG: spinbot, grammarly and GloVe based fake news detection. In: IEEE Sixth International Conference on Multimedia Big Data (BigMM), pp. 174–182. IEEE (2020)
10. Giachanou, A., Zhang, G., Rosso, P.: Multimodal fake news detection with textual, visual and semantic information. In: Sojka, P., Kopeček, I., Pala, K., Horák, A. (eds.) TSD 2020. LNCS (LNAI), vol. 12284, pp. 30–38. Springer, Cham (2020). https://doi.org/10.1007/978-3-030-58323-1_3
11. Goodfellow, I., Bengio, Y., Courville, A.: Deep Learning. MIT Press, Cambridge (2016). http://www.deeplearningbook.org, Accessed 2 Dec 2020
12. Graham, P.: Better Bayesian filtering (2003). http://www.paulgraham.com/better.html, Accessed 2 Dec 2020
13. Guyon, I., Elisseeff, A.: An introduction to variable and feature selection. J. Mach. Learn. Res. **3**(Mar), 1157–1182 (2003)
14. Han, J., Pei, J., Kamber, M.: Data Mining: Concepts and Techniques. Elsevier, Amsterdam (2011)
15. Horne, B., Adali, S.: This just. in: fake news packs a lot in title, uses simpler, repetitive content in text body, more similar to satire than real news. In: The Workshops of the 11th International AAAI Conference on Web and Social Media, pp. 759–766 (2017)
16. Joachims, T.: Text Categorization with Support Vector Machines: Learning with Many Relevant Features. Technical Report, LS VIII-Report, Universität Dortmund, Dortmund, Germany (1997)
17. Joulin, A., Grave, É., Bojanowski, P., Mikolov, T.: Bag of tricks for efficient text classification. In: Proceedings of the 15th Conference of the European Chapter of the Association for Computational Linguistics, vol. 2, Short Papers, pp. 427–431 (2017)
18. Khan, J.Y., Khondaker, M.T.I., Iqbal, A., Afroz, S.: A benchmark study on machine learning methods for fake news detection. arXiv preprint arXiv:1905.04749 (2019)
19. Kim, Y.: Convolutional neural networks for sentence classification. In: Proceedings of the 2014 Conference on Empirical Methods in Natural Language Processing (EMNLP), pp. 1746–1751 (2014)
20. Kincaid, J.P., Fishburne Jr., R.P., Rogers, R.L., Chissom, B.S.: Derivation of New Readability Formulas (Automated Readability Index, Fog Count and Flesch Reading Ease Formula) for Navy Enlisted Personnel. Technical Report, 8–75, Naval Technical Training Command, Millington, TN (1975)
21. Kingma, D.P., Ba, J.: Adam: A method for stochastic optimization. arXiv preprint arXiv:1412.6980 (2014)
22. Kipf, T.N., Welling, M.: Semi-supervised classification with graph convolutional networks. arXiv preprint arXiv:1609.02907 (2016)
23. Kumar, S., Shah, N.: False information on web and social media: A survey. arXiv preprint arXiv:1804.08559 (2018)
24. LeCun, Y., Bottou, L., Bengio, Y., Haffner, P.: Gradient-based learning applied to document recognition. Proc. IEEE **86**(11), 2278–2324 (1998)
25. Leiserowitz, A., Maibach, E., Rosenthal, S., Kotcher, J., Ballew, M., Goldberg, M., Gustafson, A.: Climate change in the American mind: December 2018. Yale Project on Climate Change Communication (Yale University and George Mason University, New Haven, CT) (2018)

26. Leiserowitz, A., Maibach, E., Roser-Renouf, C., Feinberg, G., Rosenthal, S.: Climate change in the American mind: March, 2015. Yale Project on Climate Change Communication (Yale University and George Mason University, New Haven, CT) (2015)
27. Lodhi, H., Saunders, C., Shawe-Taylor, J., Cristianini, N., Watkins, C.: Text classification using string kernels. J. Mach. Learn. Res. **2**(Feb), 419–444 (2002)
28. Long, Y., Lu, Q., Xiang, R., Li, M., Huang, C.R.: Fake news detection through multi-perspective speaker profiles. In: Proceedings of the Eighth International Joint Conference on Natural Language Processing, vol. 2: Short Papers, pp. 252–256 (2017)
29. Manning, C.D., Raghavan, P., Schütze, H.: Introduction to Information Retrieval. Cambridge University Press, Cambridge (2008)
30. McIntire, G.: How to Build a "Fake News" Classification Model. ODSC (2017), https://opendatascience.com/how-to-build-a-fake-news-classification-model, Accessed 2 Dec 2020
31. McIntyre, L.: Post-truth. MIT Press, Cambridge (2018)
32. Melford, C., Fagan, C.: Cutting the funding of disinformation: The ad-tech solution. Technical report, The Global Disinformation Index (2019)
33. Mikolov, T., Chen, K., Corrado, G., Dean, J.: Efficient estimation of word representations in vector space. arXiv preprint arXiv:1301.3781 (2013)
34. Mitchell, A., Gottfried, J., Kiley, J., Matsa, K.E.: Political Polarization & Media Habits. How Liberals and Conservatives Keep Up with Politics. Pew Research Center, From Fox News to Facebook (2014)
35. Mohseni, S., Ragan, E.D., Hu, X.: Open issues in combating fake news: Interpretability as an opportunity. arXiv preprint arXiv:1904.03016v1 (2019)
36. Moses, L.: 43 percent of social media users don't know where the stories they read originally appeared. Digiday (2016). https://digiday.com/media/57-percent-readers-aware-brands-theyre-reading-social/, Accessed 2 Dec 2020
37. Nowak, J., Taspinar, A., Scherer, R.: LSTM recurrent neural networks for short text and sentiment classification. In: Rutkowski, L., Korytkowski, M., Scherer, R., Tadeusiewicz, R., Zadeh, L.A., Zurada, J.M. (eds.) ICAISC 2017. LNCS (LNAI), vol. 10246, pp. 553–562. Springer, Cham (2017). https://doi.org/10.1007/978-3-319-59060-8_50
38. Paltoglou, G., Thelwall, M.: A study of information retrieval weighting schemes for sentiment analysis. In: Proceedings of the 48th Annual Meeting of the Association for Computational Linguistics, pp. 1386–1395. Association for Computational Linguistics (2010)
39. Pariser, E.: The filter bubble: How the new personalized web is changing what we read and how we think. Penguin (2011)
40. Pedregosa, F., et al.: Scikit-learn: machine learning in Python. J. Mach. Learn. Res. **12**, 2825–2830 (2011)
41. Pennington, J., Socher, R., Manning, C.: GloVe: global vectors for word representation. In: Proceedings of the 2014 Conference on Empirical Methods in Natural Language Processing (EMNLP), pp. 1532–1543 (2014)
42. Pérez-Rosas, V., Kleinberg, B., Lefevre, A., Mihalcea, R.: Automatic detection of fake news. In: COLING, pp. 3391–3401. Santa Fe, New Mexico (2018)
43. Pierri, F., Ceri, S.: False news on social media: A data-driven survey. arXiv preprint arXiv:1902.07539 (2019)
44. Rennie, J.D., Shih, L., Teevan, J., Karger, D.R.: Tackling the poor assumptions of Naive Bayes text classifiers. In: Proceedings of the 20th International Conference on Machine Learning (ICML-03), pp. 616–623 (2003)

45. Rubin, V.L., Chen, Y., Conroy, N.J.: Deception detection for news: three types of fakes. In: Proceedings of the 78th ASIS&T Annual Meeting: Information Science with Impact: Research in and for the Community. American Society for Information Science (2015)
46. Ruchansky, N., Seo, S., Liu, Y.: CSI: a hybrid deep model for fake news detection. In: Proceedings of the 2017 ACM on Conference on Information and Knowledge Management, pp. 797–806. ACM (2017)
47. Saikh, T., De, A., Ekbal, A., Bhattacharyya, P.: A deep learning approach for automatic detection of fake news. arXiv preprint arXiv:2005.04938 (2020)
48. Salton, G., Wong, A., Yang, C.S.: A vector space model for automatic indexing. Commun. ACM **18**(11), 613–620 (1975)
49. Sebastiani, F.: Machine learning in automated text categorization. ACM Comput. Surv. **34**(1), 1–47 (2002)
50. Selvaraju, R.R., Cogswell, M., Das, A., Vedantam, R., Parikh, D., Batra, D.: Grad-CAM: visual explanations from deep networks via gradient-based localization. In: Proceedings of the IEEE International Conference on Computer Vision, pp. 618–626 (2017)
51. Shearer, E., Matsa, K.E.: News use across social medial platforms 2018. Pew Research Center (2018). https://www.journalism.org/2018/09/10/news-use-across-social-media-platforms-2018/, Accessed 2 Dec 2020
52. Shu, K., Awadallah, A.H., Dumais, S., Liu, H.: Detecting fake news with weak social supervision. IEEE Intell. Syst. (2020)
53. Shu, K., Mahudeswaran, D., Wang, S., Lee, D., Liu, H.: FakeNewsNet: a data repository with news content, social context and dynamic information for studying fake news on social media. arXiv preprint arXiv:1809.01286 (2018)
54. Shu, K., Sliva, A., Wang, S., Tang, J., Liu, H.: Fake news detection on social media: a data mining perspective. ACM SIGKDD Explor. Newsl. **19**(1), 22–36 (2017)
55. Silverman, C.: This Analysis Shows How Viral Fake Election News Stories Outperformed Real News On Facebook. BuzzFeedNews (2016). https://www.buzzfeednews.com/article/craigsilverman/viral-fake-election-news-outperformed-real-news-on-facebook, Accessed 2 Dec 2020
56. Smith, L.N.: Cyclical learning rates for training neural networks. In: 2017 IEEE Winter Conference on Applications of Computer Vision (WACV), pp. 464–472. IEEE (2017)
57. Tacchini, E., Ballarin, G., Della Vedova, M.L., Moret, S., de Alfaro, L.: Some like it hoax: automated fake news detection in social networks. In: Proceedings of the Second Workshop on Data Science for Social Good (2017)
58. Tajfel, H.C., Turner, J.: The social identity theory of intergroup behavior. In: Jost, J.T., Sidanius, J. (eds.) Political Psychology: Key Readings, pp. 276–293. Psychology Press, New York (2004)
59. Veszelszki, A.: Linguistic and non-linguistic elements in detecting (Hungarian) fake news. Acta Universitatis Sapientiae Communicatio **4**, 7–36 (2017)
60. Yang, Y.: An evaluation of statistical approaches to text categorization. Inf. Retrieval **1**(1–2), 69–90 (1999)
61. Yang, Y., Pedersen, J.O.: A comparative study on feature selection in text categorization. In: International Conference on Machine Learning, pp. 412–420 (1997)
62. Yao, L., Mao, C., Luo, Y.: Graph convolutional networks for text classification. In: Proceedings of the AAAI Conference on Artificial Intelligence, vol. 33, pp. 7370–7377 (2019)
63. Zellers, R., et al.: Defending against neural fake news. arXiv preprint arXiv:1905.12616 (2019)

64. Zhang, T., Oles, F.J.: Text categorization based on regularized linear classification methods. Inf. Retrieval **4**(1), 5–31 (2001)
65. Zhang, X., Zhao, J., LeCun, Y.: Character-level convolutional networks for text classification. In: Advances in Neural Information Processing Systems, pp. 649–657 (2015)
66. Zhou, P., et al.: Attention-based bidirectional long short-term memory networks for relation classification. In: Proceedings of the 54th Annual Meeting of the Association for Computational Linguistics, vol. 2: Short Papers, pp. 207–212 (2016)
67. Zhou, X., Wu, J., Zafarani, R.: Safe: Similarity-aware multi-modal fake news detection. arXiv preprint arXiv:2003.04981 (2020)
68. Zhou, Z., Guan, H., Bhat, M.M., Hsu, J.: Fake news detection via NLP is vulnerable to adversarial attacks. In: Proceedings of the 11th International Conference on Agents and Artificial Intelligence (ICAART), vol. 2, pp. 794–800 (2019)
69. Zimdars, M.: False, Misleading, Clickbait-y, and/or Satirical "News" Sources. Google Docs (2016). https://docs.google.com/document/d/10eA5-mCZLSS4MQY5QGb5ewC3VAL6pLkT53V_81ZyitM, Accessed 2 Dec 2020

Context for API Calls in Malware
vs Benign Programs

Monika Chandrasekaran[1], Anca Ralescu[1(✉)], David Kapp[2],
and Temesguen M. Kebede[2]

[1] EECS Department, University of Cincinnati, Cincinnati, OH, USA
`chandrmk@mail.uc.edu`, `Anca.Ralescu@uc.edu`
[2] AFRL, WPAFB, Dayton, USA
{`david.kapp,temesgen.kebede.1`}`@us.af.mil`

Abstract. The current progress in computer technology is matched by
the increase in the malware and cyber-attacks, resulting in a nearly constant battle between establishing a complete malware detection technique and newly evolving smart malicious code. The analysis of malware is made difficult by the fact that, to a large extent, malware and
benign code use the same instructions. This suggests that the difference
in behavior might be due not to the instructions used, but in *how* they
are used. In particular, the *context* in which instructions are used seems
to play an important role in deciding between malicious and benign code.
This work describes progress towards defining and extracting the context
of API from Portable Execution files of the Windows operating system.
It is suggested that the context can be used as a feature in a machine
learning algorithm towards identifying attempts to corrupt the system
and to elude the antivirus scanners through code obfuscation.

Keywords: API calls · Context · Malware · Skip-gram model

1 Introduction

Malware refers to malicious code aimed at disrupting the normal functionality of a computer system, damage and steal data, obtain unapproved access
into the system for any illegitimate economic benefits. They are executed in the
background causing a great deal of loss. The prime characteristic of malware is
to disguise the intended actions and make their way into system without the
knowledge of the user. They are broadly classified based on their characteristics as *virus, trojan, worm, backdoor, adware, rootkit, spyware* [17]. Ideally, a
malware detector looks for traces of malware in order to prevent the system
malfunction. Signature based methods are widely used for malware detection
[3,13]. However, such methods are highly dependent on a database of malware
signatures and fail to detect evolving malware or code obfuscation. The function of an anti-virus software, that to safeguard the computer, consists of three
main tasks: Scan the system, Detect the malware, and Remove the malware. The
malware detector scans the program and generate a signature which is stored in

© Springer Nature Switzerland AG 2021
D. Simian and L. F. Stoica (Eds.): MDIS 2020, CCIS 1341, pp. 222–234, 2021.
https://doi.org/10.1007/978-3-030-68527-0_14

the malware signature database. The signature is a unique byte sequence representation for each malware. When new malicious code is injected a match is performed against the signature database for identification. If a match is found, the malware is identified and system access is denied. If a match is not found, the code is considered benign and allowed to further execute. Most of anti-virus software have incorporated signature based techniques. The major drawback of signature based method lies in their failure to detect the zero-day malware, i.e., malware whose signatures are not in the database. Such malware passes the scanning process and gain access to the system. By the time its nefarious effect is observed and a signature is collected a considerable amount of damage has been incurred to the system under attack. Furthermore, this process requires the database to be updated daily leading to a high cost. The newly evolving smart malware start from an existing malware structure with a different byte sequence using obfuscation methods [13,19], allowing the malware once again bypass the scanning as the signature cannot be matched. Machine learning techniques have emerged as a possible tool to overcome these shortcomings. However, the issues of feature extraction and selection have been particularly challenging. By defining the notion of context of an API, this paper puts forward an approach to detect those APIs whose contexts are different according to the type (malware of benign) code in which they appear. From this point on, the paper is organized as follows: Subsect. 1.1 describes the general framework for a machine learning approach to the problem in hand; Sect. 2 describes the dataset on which the study is carried on; Sect. 3 describes the skip-gram model used in this work to extract context representations of various APIs; Sect. 4 describes the experimental setup, and illustrates the approach with concrete examples. The paper ends with a short section on limitations of the approach and future directions, and a Conclusions section.

1.1   A Framework for Malware Detection Based on API Calls

The general mechanism of malware detection using machine learning involves the following stages:

- *Feature Extraction*
- *Feature Selection*
- Classification ML Algorithms
- Performance Metrics Evaluation

The paper is concerned with the first two stages. Initially, a dataset needs to be generated that comprises an examples of malware and benign executable files. The complexity (dimensionality) of this dataset is reduced by *feature extraction*, which is followed by *feature selection* where those features which are best for classification are selected. A number of different approaches have been adopted for feature selection. Usually, this largely reduces the input data to a representation which requires lower computational overhead. Some of the features [16,19] for selection relevant to the current work include Byte-n-gram, Opcode-n-gram,

API calls [1,14], Portable Executables, String features, Function features. The right feature selection is essential to achieve a better accuracy. Based on examples of the benign and malware code, a classifier is trained using the selected features. Possible machine learning approaches to malware detection include Decision trees, KNN algorithm, Support Vector Machine, Bayesian Network etc [4,13] An ideal classifier is expected to provide a clear demarcation between malicious and benign executables and reduces the rate of misclassification. The trained classifier is then tested on a test dataset – examples of each malware and benign code not used in model training, and the performance of the classifier is evaluated with respect to various performance metrics, which include Accuracy, Precision, Recall, and F-measure. A robust system displays high accuracy, high true positive rate and low false positive rate (positive refers to the class of interest, in this case, the malware class)[[7,9,18]

2   The Dataset [2]

The approach taken in this work uses the EMBER dataset [2]. This is an open dataset for training machine learning models for malware detection. The dataset contains the features extracted from 1.1M binary files divided among malicious, benign, and unlabeled sets (300K each) for training, and 100K each malicious and benign for testing. It is a collection of JSON lines files with each line holding a single JSON object. Each object is composed of the following data:

- A unique hash identifier for each file
- The time of first occurrence of the file
- A label 0-Benign,1-Malware,-1 Unlabelled
- Raw features grouped under eight categories:
  1. general file information,
  2. header information,
  3. imported function,
  4. exported function,
  5. section information,
  6. byte histogram,
  7. byte entropy histogram,
  8. string information.

The imported function feature used in this work is obtained by parsing the import address table and recording the library imported API calls. The API calls were selected based on the access to registers, system file access, process category and system information. The major dynamic linked libraries used in our experiment consist of `advapi32.dll`, `kernel32.dll`, `ntdll.dll`, `user32.dll` with the following functionalities [12]:

- `advapi32.dll` is associated with Advanced API Services Library which is the core of Windows system process designed to support several API's including security and registry calls.It also manages the creation and management of windows services and system shutdown process.When this file gets compromised the intruder gains access to the sensitive information from the system.

- `kernel32.dll` is located in the kernel of the Windows operating system and handles memory management, thread creation, input/output operations, interrupts and synchronization functions. They manifest the base API's to the application .When they are infected the system will be unable to boot and function.
- `ntdll.dll` exports the Windows Native API, which is the interface used by user-mode components of the operating system that must run without support from Win32 or other API subsystems.It holds the NT kernel functions.This file is highly trivial for an operating system since many applications call them.When this gets tampered OS will become unstable and no operations can be performed.
- `user32.dll` helps to create and change the Windows user interface such as menus, views and options. It enables the implementation of graphical user interface similar to Windows default format. It holds functions for message handling and plays a major role in establishing communication between operating system and various user applications. Any application that has a graphical user interface makes use of user32.dll.When this file gets corrupted one can no longer use or open an application.

3   Word to Vector

The skip gram model [10, 15] is implemented for word-to-vector conversion, thus making possible the use of machine learning algorithms to perform computations on numerical vectors instead of words. A window size is fixed and for each API call a *target vector* and a *context vector* are generated. The target vector represents the central API around which the window spans. The context vector consists of the API calls that are in closer proximity to the API in target vector. The well-known example shown in Fig. 1, taken from [10] Some API Share Contexts in Benign and Malware.[1] The window spans from two words before to two words after the centered word and generate the pairings. The vector encoding consists of 1 if the word exist and 0 if the word is not in the window proximity. For example, the vector for '(the, quick)' will be $[1, 1, 0, 0, 0, 0, 0, 0, 0]$.

For the problem considered here, the 'words'are the API calls. The dimension of the input (hot) vectors is equal to the size of the collection of APIs. As an informal example, if we assume the set of API to be $\{api1, api2, api3, api4\}$, and a window size of 1, then input hot vectors could be

$$[\text{target}[1, 0, 0, 0] \ \text{context}[0, 1, 0, 0]],$$
$$[\text{target}[0, 1, 0, 0] \ \text{context}[1, 0, 0, 0], \ [0, 0, 1, 0]]$$
$$[\text{target}[0, 0, 1, 0] \ \text{context}[0, 1, 0, 0], \ [0, 0, 0, 1]]$$
$$[\text{target}[0, 0, 0, 1] \ \text{context}[0, 0, 1, 0]]$$

---

[1] This example is used here because most people have an intuitive understanding of the words used in it. By contrast, in the domain of API calls, such intuition may be present only in some very experienced domain experts.

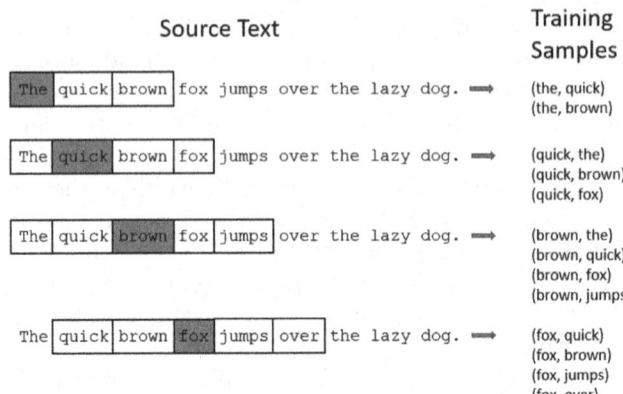

Fig. 1. Illustration of the skip gram model [10]

The network trains for frequency of associations of APIs with the target API. However, once the network is trained, of interest are the values accumulated in the hidden layer, which, taken as a vector (usually of much lower dimension than the input), form the actual vector representation of the target API (given the skipgrams/pairs of APIs used for training). Eventually, as API names are converted to vectors, for further analysis, a proximity/similarity measure is needed. In our experiments we performed the well known `cosine` similarity. For two vectors, of the same dimension A, and B, the well known `cosine` similarity is defined as the cos of the angle made by these vectors, shown below:

$$cosSim(A, B) = \cos(\angle(A, B)) = \frac{A \cdot B}{\|A\| \|B\|},$$

where $A \cdot B$ is the dot product of the vector A and B, and $\| \cdot \|$ denotes the vector norm.

4   Experimental Setup

For the EMBER dataset used in these experiments, the API calls from the import feature of the data were selected. The API calls from the crucial DLL files are extracted. The experiments used 1000 malware and benign files, with the total of 8000 API calls in benign files and 10500 API calls in the malware files.

4.1   Data Preparation

The relevant API calls are extracted from the dataset into separate files of malware and benign data collection (using Python 3 programming) [5,11]. The raw data was parsed, the import features were identified, using the Regex (regular expression) package in Python to parse specific text. Ideally, the obtained regular expressions incorporate the delimiters and parse the special characters. Depending upon the label for each file the extracted data are written into malware or benign files. Snippets of the raw and processed data are shown in Figs. 2 and 3 respectively.

{"sha256": "c31950077f3dc66b0decad88326773cb74102c0da3007db0c29177c5064b1269", "md5":
"ba2d2b60b02cf229acd85a66284b3e43", "appeared": "2018-09", "label": 1, "avclass": "high",
"histogram": [39415, 6597, 5938, 5563, 5578, 6479, 5484, 4771, 6193, 7648, 4453, 5514,
5596, 6070, 5714, 5376, 6418, 5962, 5274, 5166, 4937, 5179, 4572, 4702, 4877, 6410, 4169,
3793, 5041, 4747, 5294, 3974, 4944, 4802, 4476, 4686, 5129, 4216, 5138, 4014, 4858, 6641,
3638, 5214, 5757, 6149, 4819, 4391, 5230, 6064, 5135, 5395, 4784, 4119, 3955, 4618],
"byteentropy": [14336, 0, 0, 0, 0, 0, 0, 0, 0, 0, 0, 0, 0, 0, 0, 0, 0, 0, 0, 0, 0, 0, 0,
0, 0, 0, 0, 0, 0, 0, 0, 0, 5747, 29, 34, 13, 42, 23, 47, 24, 28, 13, 33, 17, 17, 25, 18,
34, 1823, 12, 16, 15, 16, 13, 21, 12, 10, 15, 10, 13, 26, 16, 15, 15, 0, 0, 0, 0, 0, 0, 0,
0, 0, 0, 0, 0, 0, 0, 0, 0, 0, 0, 0, 0, 0, 0, 0, 0, 0, 0, 0, 0, 0, 0, 0, 0, 0, 2762, 79,
63, 67, 63, 57, 80, 62, 86, 69, 73, 95, 111, 118, 124, 187, 16207, 568, 450, 418, 396,
400, 552, 438, 435, 484, 452, 643, 762, 769, 1195, 2455, 5740, 283, 189, 187, 185, 191,
160, 270, 201, 242, 215, 370, 366, 347, 474, 820, 7265, 448, 372, 316, 330, 336, 291, 471,
352, 442, 385, 580, 558, 541, 661, 988, 10497, 775, 608, 563, 566, 559, 526, 1049, 660,
770, 727, 1318, 1149, 1150, 1463, 2196, 6243, 675, 513, 482, 464, 468, 543, 892, 556, 645,
593, 1189, 1026, 1030, 1332, 1781, 0, 0, 0, 0, 0, 0, 0, 0, 0, 0, 0, 0, 0, 0, 0, 0, 181139,
166230, 165836, 151521, 189583, 167558, 161942, 146570, 155091, 163849, 152163, 146027,
170957, 168198, 142931, 152933], "strings": {"numstrings": 6134, "avlength":
5.668568633844147, "printabledist": [379, 319, 325, 318, 370, 274, 310, 288, 353, 356,
304, 381, 425, 360, 337, 329, 374, 340, 288, 346, 301, 363, 332, 322, 343, 356, 289, 327,
347, 334, 314, 381, 394, 476, 345, 421, 520, 446, 356, 400, 380, 410, 385, 528, 448, 437,
364, 410, 402, 409, 331, 402, 468, 385, 339, 381, 367, 429, 336, 430, 399, 409, 353, 366,
375, 386, 277, 345, 404, 390, 317, 316, 352, 383, 359, 396, 386, 397, 381, 396, 340, 329,
274, 284, 398, 341, 290, 351, 320, 379, 252, 335, 344, 331, 294, 338], "printables":
34771, "entropy": 6.57100772857666, "paths": 0, "urls": 0, "registry": 0, "MZ": 21},
"general": {"size": 1353344, "vsize": 3784704, "has_debug": 0, "exports": 0, "imports": 8,
"has_relocations": 0, "has_resources": 1, "has_signature": 0, "has_tls": 0, "symbols": 0},
"header": {"coff": {"timestamp": 1308078076, "machine": "I386", "characteristics":
["CHARA_32BIT_MACHINE", "RELOCS_STRIPPED", "EXECUTABLE_IMAGE", "LINE_NUMS_STRIPPED",
"LOCAL_SYMS_STRIPPED"]}, "optional": {"subsystem": "WINDOWS_GUI", "dll_characteristics":
[], "magic": "PE32", "major_image_version": 1, "minor_image_version": 0,
"major_linker_version": 2, "minor_linker_version": 25, "major_operating_system_version":
4, "minor_operating_system_version": 0, "major_subsystem_version": 4,
"minor_subsystem_version": 0, "sizeof_code": 176128, "sizeof_headers": 4096,
"sizeof_heap_commit": 4096}}, "section": {"entry": "", "sections": [{"name": "", "size":
61440, "entropy": 7.907082960866848, "vsize": 176128, "props": ["CNT_INITIALIZED_DATA",
"MEM_EXECUTE", "MEM_READ", "MEM_WRITE"]}, {"name": "", "size": 0, "entropy": -0.0,
"vsize": 8192, "props": ["CNT_INITIALIZED_DATA", "MEM_EXECUTE", "MEM_READ", "MEM_WRITE"]},
{"name": "", "size": 40960, "entropy": 7.6697093044308575, "vsize": 65536, "props":
["CNT_INITIALIZED_DATA", "MEM_EXECUTE", "MEM_READ", "MEM_WRITE"]}, {"name": ".rsrc",
"size": 40960, "entropy": 5.153040813942048, "vsize": 40960, "props":
["CNT_INITIALIZED_DATA", "MEM_EXECUTE", "MEM_READ", "MEM_WRITE"]}, {"name": "", "size":
180224, "entropy": 7.986451938502913, "vsize": 2613248, "props": ["CNT_INITIALIZED_DATA",
"MEM_EXECUTE", "MEM_READ", "MEM_WRITE"]}, {"name": ".data", "size": 876544, "entropy":
7.980184351257665, "vsize": 876544, "props": ["CNT_INITIALIZED_DATA", "MEM_EXECUTE",
"MEM_READ", "MEM_WRITE"]}]}, "imports": {"kernel32.dll": ["GetModuleHandleA"],
"user32.dll": ["MessageBoxA"], "advapi32.dll": ["RegCloseKey"], "oleaut32.dll":
["SysFreeString"], "gdi32.dll": ["CreateFontA"], "shell32.dll": ["ShellExecuteA"],
"version.dll": ["GetFileVersionInfoA"], "MSVBVM60.DLL": ["EVENT_SINK_GetIDsOfNames"]},
"exports": [], "datadirectories": [{"name": "EXPORT_TABLE", "size": 0, "virtual_address":
0}, {"name": "IMPORT_TABLE", "size": 544, "virtual_address": 2908160}, {"name":
"RESOURCE_TABLE", "size": 40796, "virtual_address": 253952}, {"name": "EXCEPTION_TABLE",
"size": 0, "virtual_address": 0}, {"name": "CERTIFICATE_TABLE", "size": 0,

Fig. 2. Raw data

4.2 Vector Generation

The skip-gram model, implemented in Python3, has the following parameters:

– Window size = 2
– Dimension of word embedding per hidden layer: $N = 100$
– Epoch = 50
– Learning rate: $\eta = 0.01$

```
 1  setfiletime
 2  comparefiletime
 3  searchpatha
 4  getshortpathnamea
 5  getfullpathnamea
 6  movefilea
 7  lstrcata
 8  setcurrentdirectorya
 9  getfileattributesa
10  getlasterror
11  createdirectorya
12  setfileattributesa
13  sleep
14  gettickcount
15  getfilesize
16  getmodulefilenamea
17  exitprocess
18  getcurrentprocess
19  copyfilea
20  lstrcpyna
21  getcommandlinea
22  getwindowsdirectorya
23  closehandle
24  getuserdefaultlangid
25  getdiskfreespacea
26  globalunlock
27  globallock
28  globalalloc
29  createthread
30  createprocessa
```

Fig. 3. Processed data

The API-to-vector network is trained using back error propagation. Softmax is applied to the output layer in order to obtain the association probabilities. Figure 4 shows the vector generated for `regrestoreall` API.

```
regrestoreall [ 0.13580457   0.06362823   0.2583795   -0.1188621    0.54005928 -0.29375591
 -0.28132525   0.01268397   0.29925385   0.47511831   0.31913659 -0.51390703
  0.72180565   0.40876341  -0.39304414  -0.37538265  -0.31027718   0.38775487
  0.56876678   0.80744709   0.51436789   0.00343007   0.18991917   0.77530234
  0.29581286  -0.86196544  -0.50541325   0.284479     0.79870754   0.45419901
  0.02336165  -0.28942552   0.77389533   0.9489149   -0.53248182   0.037239
 -0.2592598    0.9559395    0.54913655  -0.21480771  -0.92991117   0.44822009
  0.73628647   0.8281983   -0.10935859   0.58543093   0.34939087  -0.02156114
  0.48836279   1.11697599   0.69965552   0.45772039  -0.27250248   0.46249601
 -0.45993566   0.01057832  -0.00626786   0.16972193   0.40125325   0.76761992
  0.40243569   0.74489708   0.75570514   0.06111903   0.08337334  -0.00772947
  1.22186018   0.14440962  -0.21528143   0.35003189   0.81918887  -0.15006589
 -0.87973833   0.14918573   0.68012185  -0.47262579  -0.12026342   0.38795692
 -0.75706634  -0.97552023  -0.3607557    0.27384491  -0.44772097   1.15488558
  0.61973482   0.56128824  -0.1931362   -0.36324852  -0.7674405    0.61326638
  0.89906314  -0.63702231   0.70644213  -0.79209158  -0.52928087  -0.10907524
 -0.15724559  -0.42681088   0.7510478   -0.43222179]
```

Fig. 4. Word-to-vector conversion for *regrestoreall*

4.3 Measuring Similarity

The vector similarity, measured by the cos of the angle between two vectors [6,8], is used to identify the API calls that are closer in proximity to the target API call. The associations with the target API, output by the network, are interpreted in this study as *context* for the target API. In this interpretation, the results indicate that for some APIs the contexts in which they appear are different according to whether the corresponding program is malicious or benign. Figures 6 and 5 show that the contexts for `regqueryinfokeyw` in a malware and a benign program are different: in the malware, the context is {`reqqueryvaluexw`, cryptreleasecontext,openservicew, dataservicew} and for the benign program the context is reqqueryvaluexw, cryptreleasecontext,regdeletekeyw, regnumkeyw}. Many such API's whose contexts differ in malware and benign files were obtained as illustrated in Table 1.

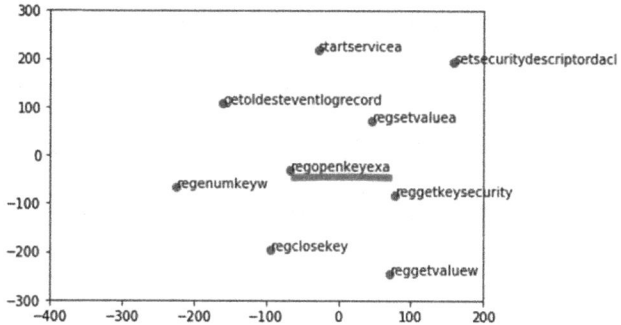

Fig. 5. The context for *regopenkeyexa* in benign programs

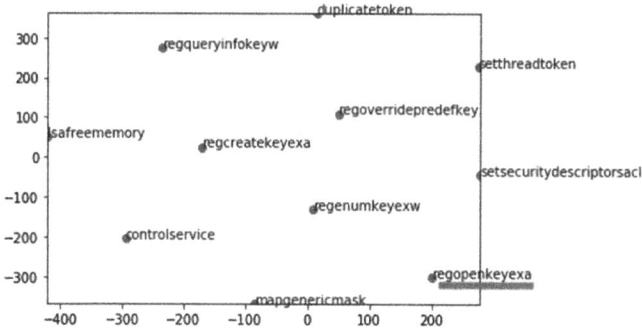

Fig. 6. The context for *regopenkeyexa* in malware programs

4.4   Some API Share Contexts in Benign and Malware

Complementing the identification of APIs whose contexts are different according to whether the APIs are in benign or malicious programs, some APIs contexts are the same (or at least have a nonempty intersection) regardless the type of program in which they appear. Figures 7 and 8 illustrate the contexts for the API eventunregister, which has similar contexts in malware and benign programs.

In the malware file the context is {allocateuniqueid, getlengthrequired, regloadkeya, createservicew, closeservicea, gettokeninformation, openservicew, openservicea} (Fig. 7).

Table 1. The context for API in benign and malware files

API	Context in benign	Context in malware
crypthashdata	{cryptreleasecontext, mapgenericmask, initiatesystemshutdowna addace}	{cryptgethashparam, cryptdestroyhash, regenumkeyw, regdeletevaluew}
cryptacquirecontextw	{getaclinformation, cryptverifysignature setsecuritydecriptoracl regdeletekeya}	{crytpacquirecontexta, cryptdestroykey cryptdecrypt setfilesecurityw}
crhptgethashparam	copysid,openprocesstoken regenumkeya cryptacquirecontexta}	{crypthashdata cryptdestroyhash regdeletevaluew regquetinfokeyw}
cryptreleasecontext	{ashaupdate regopenkeyexw eventregister, crypthashdata}	{reenumkeyw, regqueryinforkeyw, regdeletvaluew regsetvalueexw}
openprocesstoken	{regstorekeyw, gettokeninformation, accesscheck, mapgenericmask}	{initilaizeacl, allocateandinitializesid, addaccessallowance, duplicatetokenex}
adjusttokenprivileges	{lsaclose regqueryinforkeya openservicew}	{deregistereventsource, reverttoself, setservicestatus, deleteservice}
regclosekey	{openeventlogw, regsavekeyw, regqueryinfokeyw lsastoreprivayedate}	{regqueryvalueexw regopenkeyexw, cryptcreatehash convertstringsecuritydescr-iptortosecuritydescriptora}

In the benign files the context is {getlengthrequired, gettoken information, openservicea, regloadkeya, createservicew, readeventloga, eventregister, traceevent} (Fig. 8).

Therefore, the APIs whose contexts are identical in malicious and benign programs, are not indicative of the nature of the file in which they appear. Moreover, the extent to which such APIs fail to reflect the nature of the file in which they appear, can be quantified, by using the well known Jaccard index. The similarity of two sets, A, and B, measured by the Jaccard index is defined by equation:

$$J(A, B) = \frac{|A \cap B|}{|A \cup B|},$$

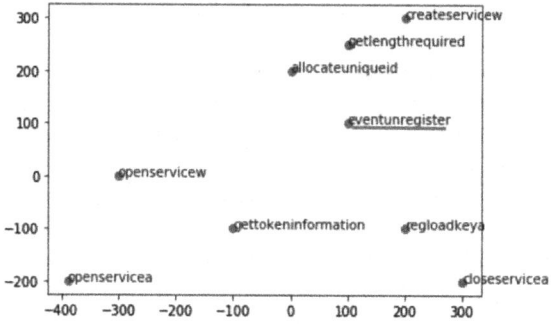

Fig. 7. The context for *eventunregister* in malware files

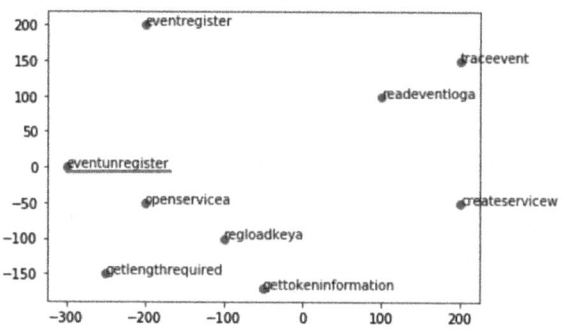

Fig. 8. The context for *eventunregister* in benign files

where $|A|$ denotes the cardinality (number of elements) of the set A.

Taking as an example the API eventunregister and denoting by $C_B(\cdot), C_M(\cdot)$, the context of an API in the benign and malicious files respectively, obtains:

$$
\begin{aligned}
C_B(\text{eventunregister}) = \{&\text{getlengthrequired, gettokeninformation,} \\
&\text{openservicea, regloadkeya, createservicew,} \\
&\text{readeventloga, eventregister, traceevent}\}
\end{aligned} \tag{1}
$$

and

$$
\begin{aligned}
C_M(\text{eventunregister}) = \{&\text{allocateuniqueid, getlengthrequired,} \\
&\text{regloadkeya, createservicew, closeservicea,} \\
&\text{gettokeninformation, openservicew, openservicea}\},
\end{aligned} \tag{2}
$$

Therefore, the intersection and union of these contexts are shown in Eqs. (3) and (4) respectively.

$$
\begin{aligned}
C_B(\text{eventunregister}) \cap C_M(\text{eventunregister}) = \\
\{&\text{getlengthrequired, regloadkeya, createservicew,} \\
&\text{gettokeninformation, openservicea}\},
\end{aligned} \tag{3}
$$

and

$C_B(\texttt{eventunregister}) \cup C_M(\texttt{eventunregister}) =$
$\{\texttt{getlengthrequired}, \texttt{gettokeninformation}, \texttt{openservicea},$
$\texttt{regloadkeya}, \texttt{createservicew}, \texttt{readeventloga},$
$\texttt{eventregister}, \texttt{traceevent}, \texttt{allocateuniqueid},$
$\texttt{closeservicea}, \texttt{openservicew}\},$

$$(4)$$

Therefore, the Jaccard index for these contexts is

$$J(C_M(\texttt{eventunregister}), C_B(\texttt{eventunregister})) = \frac{5}{11}.$$

The higher the Jaccard index for two contexts, the less relevant is the corresponding API to the type of file in which it appears. Obviously, the Jaccard index for contexts which do not overlap, i.e., exactly those contexts which can be used to classify benign and malicious files, is equal to 0. Thus the Jaccard index can be used to extract the features which discriminate between benign and malicious code. Furthermore, to increase the robustness of the approach, the Jaccard index may be used as a graded measure. For that purpose, it may be convenient to explicitly convert it into a *degree of non-overlap*. More specifically, given two contexts, C_1, and C_2, the degree of non-overlap $\mu(C_1, C_2)$ is defined as shown in Eq. (5):

$$\mu(C_1, C_2) = 1 - J(C_1, C_2) \tag{5}$$

With this definition, higher μ values convey higher non-overlap, and therefore better discrimination between malicious and benign code. Although not explored here, this measure of non-overlap can help differentiate between pairs of contexts according to how close they are to complete non-overlap. For example, when for the pairs of contexts (C_1, C_2) and (C_3, C_4), $0 << \mu(C_1, C_2) < \mu(C_3, C_4) \approx 1$, features C_3 and C_4 are preferable. However, C_1 and C_2 may also be useful.

5   Limitations and Future Work

The communication between the operating system and user applications is made through system calls interface. The UNIX operating system has a well defined set of system calls. The Windows OS calls its system calls as native API. The choice in this study of the Windows OS was motivated by the fact that this operating system is widely used, and therefore, any results obtained for this system will have a large impact. While this can be viewed as a limitation of the current study, it can also serve as an inspiration for developing a similar approach for other operating systems.

For the Windows OS considered in this study, different directions could be investigated. For example, one can experiment with various measures of similarity such as Hamming-Cosine Similarity, Distance Weighted Cosine Similarity where the cosine similarity is combined with the Hamming distance, or with a weighted distance, the similarity measure for text classification and clustering in conjunction with different classification methods (e.g., Decision Trees, Support Vector Machines, the KNN algorithm) to differentiate between malware and benign files.

6   Conclusion

We have proposed a novel method for API call sequence analysis. The results shown here suggest that those APIs which appear in different contexts for benign and malware are good candidates as features to differentiate malware from benign files. This approach can be used to generate a dynamic signature database of API contexts that helps in detecting a new malware which would otherwise easily escape the static malware signature database. Moreover, as a program executes detection of the malware specific context can be used to flag a program as possible malicious before the actual API that the context points to, is actually executed. This feature can be incorporated in a cybersecurity system that needs to identify and analyze the new evolving malware.

Acknowledgment. This research was partially supported by the AFRL Award #FA8650 to the University of Cincinnati.

References

1. Alazab, M., Venkataraman, S., Watters, P.: Towards understanding malware behaviour by the extraction of API calls. In: 2010 Second Cybercrime and Trustworthy Computing Workshop, pp. 52–59 (2010)
2. Anderson, H.S., Roth, P.: Ember: an open dataset for training static PE malware machine learning models. arXiv preprint arXiv:1804.04637 (2018)
3. Bazrafshan, Z., Hashemi, H., Fard, S.M.H., Hamzeh, A.: A survey on heuristic malware detection techniques. In: The 5th Conference on Information and Knowledge Technology, pp. 113–120 (2013)
4. Gavriluţ, D., Cimpoeşu, M., Anton, D., Ciortuz, L.: Malware detection using machine learning. In: 2009 International Multiconference on Computer Science and Information Technology, pp. 735–741. IEEE (2009)
5. Guido, A.C.M.: Introduction to machine learning with Python: A guide for data scientists (2016)
6. Li, B., Han, L.: Distance weighted cosine similarity measure for text classification. In: Yin, H., et al. (eds.) IDEAL 2013. LNCS, vol. 8206, pp. 611–618. Springer, Heidelberg (2013). https://doi.org/10.1007/978-3-642-41278-3_74
7. Liu, L., Wang, Bs., Yu, B., et al.: Automatic malware classification and new malware detection using machine learning. Front. Inf. Technol. Electron. Eng. **18**, 1336–1347 (2017)
8. Lin, Y., Jiang, J., Lee, S.: A similarity measure for text classification and clustering. IEEE Trans. Knowl. Data Eng. **26**(7), 1575–1590 (2014)
9. Souri, A., Hosseini, R.: A state-of-the-art survey of malware detection approaches using data mining techniques. Hum.-Centric Comput. Inf. Sci. **8**(1), 1–22 (2018). https://doi.org/10.1186/s13673-018-0125-x
10. McCormick, C.: Word2vec tutorial-the skip-gram model (2016)

11. McKinney, W.: Python for data analysis :data wrangling with pandas, numpy, ipython (2017)
12. Microsoft: Server core functions by dll (windows) (2019)
13. Moser, A., Kruegel, C., Kirda, E.: Limits of static analysis for malware detection. In: Twenty-Third Annual Computer Security Applications Conference (ACSAC 2007), pp. 421–430. IEEE (2007)
14. Ranveer, S., Hiray, S.: Comparative analysis of feature extraction methods of malware detection. Int. J. Comput. Appl. **120**(5), 1–7 (2015)
15. Rong, X.: word2vec parameter learning explained. arXiv preprint arXiv:1411.2738 (2014)
16. Santos, I., Penya, Y.K., Devesa, J., Bringas, P.G.: N-grams-based file signatures for malware detection. ICEIS **2**(9), 317–320 (2009)
17. Vinod, P., Jaipur, R., Laxmi, V., Gaur, M.: Survey on malware detection methods. In: Proceedings of the 3rd Hackers' Workshop on computer and internet security (IITKHACK 2009), pp. 74–79 (2009)
18. Ye, Y., Li, T., Adjeroh, D., Iyengar, S.S.: A survey on malware detection using data mining techniques. ACM Comput. Surv. **50**, 1–40 (2017)
19. Ye, Y., Wang, D., Li, T., Ye, D.: IMDS: Intelligent malware detection system. In: Proceedings of the 13th ACM SIGKDD International Conference on Knowledge Discovery and Data Mining, pp. 1043–1047. Association for Computing Machinery (2007)

Machine Learning Based Query Exploration

Diana-Georgiana Mocanu$^{(\boxtimes)}$ [ID]

Babeș-Bolyai University, Cluj-Napoca, Romania
dianagmocanu@gmail.com

Abstract. With the continuous advances in technology, we can observe an increase in the rate at which we acquire new information. Compared with the previous decades, in today's world, we have access to a volume of data on the order of zettabytes. Therefore, scientists find it more challenging to interact with structured databases and implicitly to write relevant and concise SQL queries that will uncover relevant results.

Starting from an initial query, given as input by the user, we will further generate the so-called negated query. The tuples resulted from the execution of this query will be the ones that are undoubtedly unwanted by the user. There are multiple possible ways of building the negation relative to the number of initial conditions. With the help of a machine learning algorithm, that uses the labeled dataset obtained from the positive and negative sets, we can rewrite a different and improved query. This will contain all the initial positive tuples but also some new, very similar ones. The initial search could be expanded from an intuitively written query to a more inclusive one that will select similar tuples with less intuitive feature values. Through this method the user could potentially highlight less important feature correlations and even uncover hidden patterns in the dataset. Most importantly this could be achieved without any knowledge of the logic behind the actual features.

Keywords: Query exploration · Query rewritten · Pattern mining

1 Introduction

With the continuous advances in technology, we have reached an exponential growth rate of acquiring new data. Did humans reach their limit of being able to process the information that is accessible? According to different studies the rate of generating data from both machine and human is 10 times bigger than the one of traditional (business) data [1]. Moreover, it is estimated that by the year 2025 the volume of data will exceed the hardware capacity to store it [2]. Therefore, the methods that are used today have become almost useless, humans having to spend more time making sense of the data than ever before.

The motivation behind this paper is to improve the user's experience and simplify the process of query exploration by developing an application that would be able to generate new queries based on an original one given as input by the user.

One of the principal motives is the overall time spent by users when they are trying to compose a correct interrogation that will produce a complete result. They need to put in a lot of time and effort to understand the schemas of different databases and the

D. Simian and L. F. Stoica (Eds.): MDIS 2020, CCIS 1341, pp. 235–250, 2021.
https://doi.org/10.1007/978-3-030-68527-0_15

existing relationships between tables, especially when they interact for the first time with that specific database. Taking all this into consideration, it can be more time consuming for a person to compose a query than the overall execution time of this query and the intermediate ones.

Lately, the process of querying scientific data has changed. It is no longer direct, a query and a result but more of a try and error process, an incremental one. It is possible that after each execution of what would become the intermediate queries, we will need to make some modifications based on what we discovered new about the data in that previous step. Moreover, it is possible that after we obtain some unexpected results, we will be obliged to do some additional research regarding the data. Therefore, more tries may be needed for obtaining the wanted results, as, frequently, data can be difficult to classify in such a way that it would be easy to find the most relevant results [3].

Another obstacle that can appear is the size of the data and the number of existing attributes. In most of the cases, abstract abbreviations are used for the names of the attributes, the relationships present inside the data are quite complex, and in some cases even unknown. Moreover, the data stored in these datasets come from different measurements, which can be difficult to understand. In this case, being a human is one of the greatest minuses, as our memory is limited and cannot be upgraded or changed entirely. Making sense of the data and being able to find connections between different attributes in order to find the optimal query is a challenge. However, the solution to this problem can be found, with the help of machine learning techniques, which have become one of the most popular and universal answers nowadays.

Taking everything into consideration, we can see why the domain of Data Exploration has become so popular over the last few years [3–6]. It implies an easier way of understanding big datasets by enabling the user to explore the data or even by offering live assistance during the process of query creation. In other words, the role of data exploration is to rethink and enhance the classical way of creating SQL queries along with their predefined role of retrieving concise and exact data.

In this paper we propose a simpler and more interactive way of writing queries. Starting from an initial query we will obtain a new alternative query following three simple steps. First, we will construct the learning dataset, based on which we will generate a decision tree, and using this decision tree we will create the new query. This query will be based on some pattern that was discovered through the help of the decision tree and will give the user the wanted result that can be the desired extra tuples, new the data exploration information or even an optimized query, without the need of putting extra effort in understanding the specific characteristics of that dataset. The constructed learning dataset will be composed from two labels a positive and a negative one, and therefore the decision tree will try to split the received data into these two possible classes. When we have the tree it will be easy to calculate conditions using the decision rules that were applied at each node level.

1.1 Problem Statement

The statement of the chosen problem can easily be put into words as follows: Having a database and a query written using the SQL syntax, generate a new, non-equivalent, query based on the pattern that can be found using the results of the initial one.

Therefore, the result of the new query will contain all the tuples produced by the old one, but also new ones uncovered by the discovered relations inside the given dataset.

1.2   Contributions

Out main purpose is creating an algorithm that will help us in solving the problem described above. Although we are building on the logic of a previous paper (insert ref), we are still going to implement everything from scratch. The most important component of our study is the algorithm itself.

We are determined to overcome the difficulties encountered in the past by finding a better solution for the manner in which the negated query is obtained. Undoubtedly this will be one of the most challenging parts of our paper, as there are multiple ways of finding the complementary of a query. We propose two main approaches to this particular issue: greedy negation and random negation. The first method will partially negate the query multiple times and then choose which predicates should be negated in order to obtain the best results. The second method will select a list of random tuples from the entire dataset excluding the positive data. Provided it has a significant size, will create a population that correctly describe the entire dataset.

To provide the visualization of the results we have chosen a client-server architecture, the client, can then be easily changed without affecting the logic of the algorithm.

2   The Negation of SQL Queries

2.1   SQL Queries

A query is a way of obtaining information from a database that corresponds to some logical conditions. In this paper, we will use the notation of Q for a query and $\ulcorner$ Q for its negation. Furthermore, we will base our algorithm on simple queries that are composed of the SELECT, FROM, and WHERE clause. The predicates from the conditional clause will be composed with the help of disjunctions or conjunctions. Therefore, the queries will look like this:

$$\text{SELECT * FROM table 1 WHERE } a > b \text{ AND } c < d \text{ OR } e > f \qquad (1)$$

2.2   Query Negation

In the first place, I would like to explain the term of negation. If we take for example the Stanford Encyclopedia of Philosophy negation is the phenomenon of semantic opposition. It denotes the change in the meaning of expression such that the new connotation is in some way opposed to the original expression. We can affirm that negation is unique to humans, as it is part of every human language, whereas it seems to be absent in the animal communication system. Some research that has at base apes and non-primates show that the stages that children go through for the acquisition of negation like understanding the function of rejection, refusal, and non-existence, can be learned by apes but it cannot be said the same about the stages of truth-conditional negation and denial [7]. In terms of syntax, we can say that negation is profoundly complex as it can be expressed in different categories and parts of speech [8] for instance in particles (not, never), adjectives (absent), adverbs (hardly, scarcely), verbs (fail, lack), nouns (absence, failure), prepositions (off, out) and so on [9].

A **negation query** is a construct of an initially received query with some small changes, one of which is the alteration of the original condition. For resolving this problem, we choose to follow two different paths in achieving the wanted results. The first one is to negate the whole condition choosing to follow the logical approach by applying De Morgan's laws.

Let us take into consideration the above query (1), if we denote the condition with C then $\vdash C$ will be equivalent with.

$$\vdash ((a > b) \wedge (c < d) \vee (e > f)) => \vdash (a > b) \vee \vdash (c < d) \wedge \vdash (e > f) =>$$

$$(a <= b) \vee (b >= d) \wedge (e <= f) \tag{2}$$

The second option is to negate just parts of the condition, which will generate an exponential number of possible new conditions.

The **first negation algorithm** has the main purpose to offer a more diverse learning dataset. To achieve this, the entire condition from the WHERE clause is negated, and from the resulting set, we will choose a number n of random tuples. This application is meant for large databases; therefore, this algorithm is based on the law of large numbers.

The law of large numbers comes from the domain of probability and statistics and it implies that by repeating a specific experiment multiple times and then, computing the average result will offer a better approximation of the real result [10].

This means that with the increase in the number of tuples, the rate of a randomly selected set to be uniformly distributed around the initial space increases, and, therefore, it will better represent the entire population. Moreover, this fulfills one of the key criteria of a training data set: the data used must be representative of the entire domain, and it offers enough information for the machine learning algorithm to be able to generalize the population [10]. Consequently, we will be able to obtain more accurate results with an increase in the size of the data set.

Let us take as example the following query written using the famous iris dataset.

$$\begin{gathered} \text{SELECT} * \text{FROM iris\_data} \\ \text{WHERE petal\_width} > 1.3 \text{ AND petal\_length} < 6 \\ \text{OR sepal\_length} > 6 \end{gathered} \tag{3}$$

If we take the query described above, then the new negated query will be constructed under the form of:

SELECT * FROM iris_data
WHERE NOT (petal_width > 1.3 AND petal_length < 6 OR sepal_length > 6)
ORDER BY RAND() LIMIT n

$$(4)$$

Where the initial condition is negated entirely. The RAND() function will generate for each row in the resulted table a random number, after which they will be ordered (by the ORDER BY clause) and we will select only the first n tuples.

The number n of tuples that will construct the negative data set can be calculated based on the following formula:

$$n = \frac{i}{100}(t - p) \text{ and } i \in [1, 100] \tag{5}$$

Where n represents the number of negative tuples, t is the total number of entries in the table on which the query is written, and p represents the size of the positive data. Here 'i' represents the percentual relative coefficient that is in an inverse proportional relationship with the number of randomly selected tuples, which can be given as input by the user.

To give a clearer explanation for the 'i' coefficient's role we can rely on a couple of examples. The smaller the coefficient will be, the fewer tuples will be selected as negatives, and more tuples will be able to be added to the improved query. The opposite happens if the coefficient is close to 100; more tuples will be selected as negative and fewer new tuples will be added to the improved query.

Although theoretically for the value for 'i' of 100 there would be no newly selected tuples, we could potentially compute a different query giving the same results. This could either simplify our current interrogation by using fewer conditions or smaller intervals, or it could even uncover hidden data correlations by choosing completely new attributes.

Another way to choose n is to be equal with the number of positive tuples, but if the size of p is small compared with the size of the entire dataset then the pattern will be found only in a small percentage of the dataset, and therefore the population will not be generalized well enough. In these conditions the first formula is better as the size of the resulted set can be easily controlled and influenced.

The **second negation algorithm** takes on a greedier approach. Here we do not negate the whole condition anymore but only parts of it. Therefore, the problem can be easier simplified in a combinatory problem, we want to find a combination of sub-conditions where one or more are negated fulfilling the condition that its execution will offer a dataset close in size with the positive one.

Maybe the main challenge here was the creation of a set containing all possible combinations of the sub-conditions where at-least one was negated at each time. For resolving this problem, the combination is created according to a binary number that will belong to the interval $[1, 2^n - 1]$, where n is the number of sub-conditions. For instance, if we take the previous query (3). If we take the number 2, in binary it is 010.

If for '0' keep the condition as it is and for '1' we will take the complementary condition, consequently the new condition will be equal with:

$$\text{SELECT} * \text{FROM iris\_data}$$
$$\text{WHERE petal\_width} > 1.3 \text{ AND NOT peta\_length} < 6 \text{ OR sepal\_length} > 6 \quad (6)$$

Compared with the first negation algorithm, this one offers a more fixed result, because as long the input query remains the same the generated one will also remain unchanged. The explanation is a simple one, at each iteration, the program will choose the same combination for generating the minimum difference between the positive tuples and the negative ones. In contrast, the first algorithm response is more dynamic, since the selection domain is random, and it can be easily observed that the smaller the dataset the bigger the variation between runs will be.

3 Generating a New Query

In this chapter, we will describe in more detail the process of generating a new query based on the initial one (Q) from the user. Based on Q we will calculate ⌐ Q based one of the methods explained above. After that, we will create two learning sets a positive and negative one which will be further used in the CART algorithm that is used for creating the decision tree. After this, we will parse the obtained tree and construct the new predicates.

3.1 Dataset Creation

The next step, after negating a query, is the construction of the learning dataset. It will be composed of two labels: positive and negative. This dataset plays an important role in the creation of the decision tree and therefore on the result. This is because, the patterns that are going to be found are dependent on the overall size of the learning dataset and therefore on the distribution of the two categories.

The positive set consists of those tuples obtained through the execution of the query given as input by the user. These are taken further as an example of what the user will want to see in the result obtained through the execution of the resulted interrogation.

The negative category represents the part of the data that was obtained through the execution of the negated query from the previous step. Moreover, this category represents the counterexamples, what probably the user does not want to see in the result.

For the creation of the datasets, the initial query should be a correct one, containing the SELECT, FROM, and WHERE clauses. The predicates included by the conditional clause need to be valid, including columns that do exist in the table. Moreover, the execution of the written query should produce a result set that is not empty. These criteria should also be fulfilled by the negative set.

If we take the query (3), then we will obtain the following snapshot of the data (Fig. 1). As it can be easily observed, the first three tuples are labeled as positive while the last two negatives are acquired from the execution of the negated query.

Sepal length	Sepal width	Petal length	Petal width	id	Class
6.3	3.3	6	2.5	101	+
7.1	3	5.9	2.1	103	+
7.3	2.9	6.3	1.8	108	+
4.6	3.4	1.4	0.3	7	-
5.5	3.5	1.3	0.2	37	-

Fig. 1. Snapshot of the learning dataset

3.2   Tree Creation, Parsing, and New Conditions

The decision tree is built with the help of the machine learning algorithm: CART. For classification, we use the dataset that was constructed at the previous step. The machine learning algorithm is used from the scikit-learn, which is a Python library. The algorithm takes as parameters a set of data that hold the training samples and the set of corresponding labels [11]. Moreover, the decision tree class can be easily customizable as it offers the possibility to modify some parameters such as the total number of leaf nodes that the final tree will contain. It can be observed that the smaller the value the greater the power of generalization will be. This is very useful if we want to offer the user the ability to choose the overall number of similar conditions that he would like to obtain or to impose an upper limit. A big number of leaf nodes imply multiple conditions, and therefore more effort from the user part to be able to comprehend all of them.

Once the decision tree has been fit, we can start the construction of the new conditions. This process consists of parsing the tree from the root to the leaves' computing at each node the equivalent conditions. Each such path gives us a possible condition, which is composed of the conjunction of the rules that were used for splitting the data at each node. If they reached a positive leaf, then we will memorize it. This process is going to be repeated until all leaf nodes are checked, therefore, until we obtain a complete set of all sequences of rules for reaching a specific positive leaf node. If we have the previous example, then we will obtain a tree similar with the one depicted in Fig. 2.

It can be easily observed that at each node, the data is further split until we reach the leaves. Here it is clear that the subset belongs to a specific label, or the imposed limit of leaves nodes has been reached.

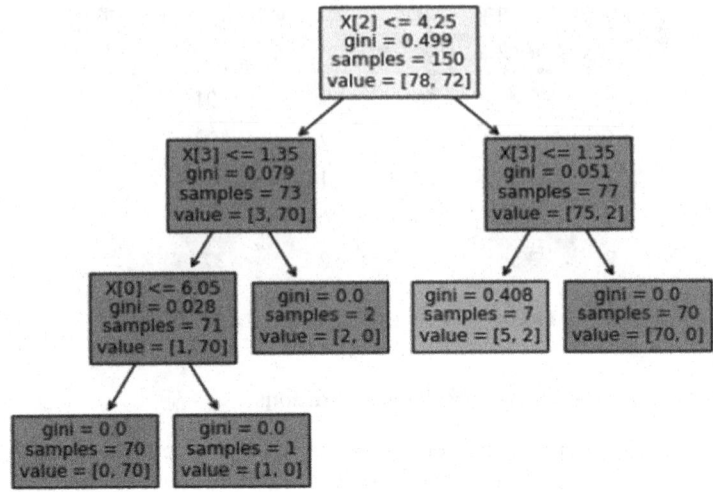

Fig. 2. Decision tree

3.3 Query Construction

The new query is formed from three parts, the first and second fragment are represented by the SELECT and FROM clause that remains unchanged compared with the initial query and the WHERE clause.

For this last fragment, we can use any of the conditions obtained from parsing the tree. That specific interrogation will uncover the tuples that would have belonged to the corresponding leaf node. If we want to obtain an even more general condition, we can construct the disjunction of all the conjunctions already constructed. Therefore, the rewritten query will be a simple one in terms of SQL syntax.

If we continue with the previous example, then we can obtain an alternative query similar with the one presented in (7).

$$\text{SELECT * FROM iris\_data}$$
$$\text{WHERE (petal\_length} > 4.25 \text{ AND petal\_width} > 1.35) \text{ OR (petal\_length} <= 4.25$$
$$\text{AND petal\_width} > 1.35 \text{ AND sepal\_length} > 6.05)$$

$$(7)$$

This new interrogation is constructed of the disjunction of two possible positive paths.

4 Usage and Experimental Results

All the concepts theorized in the previous chapters were concretized in a client-server application, developed with the help of two different programming languages. In practice, two different applications communicate with the help of some protocols.

The first one written in Python oversees the main logic of our application while the second one, in React Js, is offering a basic, functional interface for enriched user experience (Fig. 3).

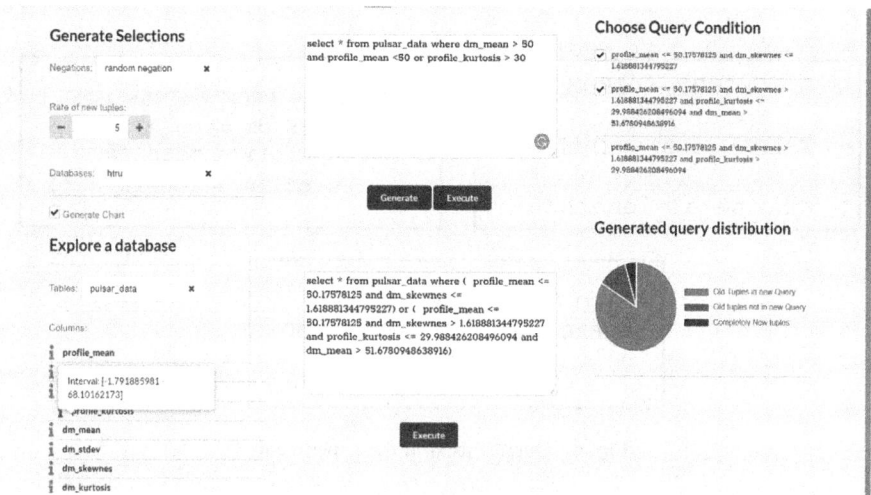

Fig. 3. User interface prototype

4.1   Experimental Results

The focus of this application is to generate new queries based on the patterns that were discovered among different rows between specific values. In addition, the first negation algorithm that we have described is based on a random selection of the negative data set, which implies a variation between different executions of the same initial query. Because of this dynamic nature of the generated queries calculating their overall precision can be a real challenge.

As a result, we should first put down some unbiassed objectives that should be fulfilled by the rewritten query. In the first place, the new interrogation should be similar, to some extent, to the initial one. Since the main purpose of this process is to facilitate the activity of query exploration, the execution of the modified query should reveal some new tuples. Therefore, the resulted size should not hold too big of a difference compared with the length of the initial query result's or measured up against the entire dataset. One of the reasons, why the generated query should not be too large, is the complexity implied for the user to be able to interpret and use the information further.

For the following experiments we have used three databases, the first one is the popular iris database with a total size of 150 tuples and five attributes, the second one is the HTRU2 data set [12] which contains data about pulsar stars, which radio emissions are even traceable on Earth. The number of tuples stored by this dataset is 17898 and it contains nine attributes. The third one is the Beijing Multi-Site Air-Quality Data Set

[13] and it includes information about air pollution that was measured hourly, containing 33176 tuples composed of eighteen attributes. Furthermore, we will define queries for each table that will be used for all the experiments. The condition for each query will be composed of one formed by two conjunctions and one disjunction. The predicates were chosen aleatory (Fig. 4).

```
SELECT * FROM
    iris_data
WHERE petal_width >
      1.3
and petal_length<6
```

```
SELECT * FROM pulsar_data
    WHERE dm_mean > 50
    and profile_mean <50
    or profile_kurtosis > 30
```

```
SELECT * FROM air_data
WHERE co < 150 and O3<260
       or pm2 > 200
```

Fig. 4. Queries used in the experiments

We can split our experiments into two big categories based on the negation algorithm used. The number of the resulted tuples obtained from the execution of the iris query is equal to 78, the pulsar: 573, while for the air interrogation is 3527.

The system on which we have conducted the experiment was a HP notebook with an i7 intel core and 8 GB RAM without using any additional boosting software or hardware.

Experiment 1
The greedy experiment can further be split into two different sections based on the database that was used.

For the iris data set the execution of the query written on the iris data set took around 0.3400 s. The new rewritten query result will be composed entirely of the tuples belonging to the initial query. If the disjunction part from the initial query will be neglected, then the new generated query will reveal eleven new tuples as it can be observed in Fig. 5.

For the HTRU dataset, the time for the generation of the new query took around 1.1409 s and it found five new alternative conditions. The minimum number of new tuples that could be obtained is 73 and the maximum of 1000. Even if the difference between different alternatives is this substantial the user has the possibility to choose the best suitable condition for its wishes. In the same style as for the iris dataset if the query is executed without the disjunction, then we will only have one predicate and 896 new tuples will be obtained. The same observation can be made about the air data set. Here the algorithm took 4.091 s for the generation of the new query. The variation between the alternative condition obtained is between 4 and 2900.

We can observe that the greedy algorithm performs better on smaller databases. This is because, on datasets that contain a bigger number of values, it has a harder time finding relevant patterns, and exists a high chance of returning numerous new tuples almost the entire size of the database used.

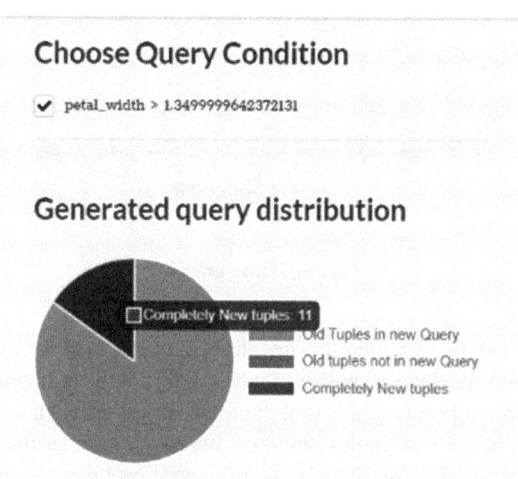

Fig. 5. Example 1 conjunction result.

Experiment 2

This experiment was performed using the same queries as in the previous example. Compared with experiment one, this time we have used the negation algorithm based on the random selection. Because of this, we have run the algorithm multiple times for both queries, changing each time the i, the percentual relative coefficient defined in the previous sub-chapter.

Because of the overall number of executions, we believe that it is easier to show the results in the form of different charts. The following Fig. 6 was constructed based on the values from both the pulsar and air tables, but as it can be seen there exists a link between the size of the obtained negative data set and the overall time needed for execution. The sizes that were used in the graph vary from 173 to 29649 tuples. We should also mention that for obtaining those values the:

$$i \in \{1, 20, 30, 50, 70, 90, 100\} \tag{8}$$

The random negation offers the user more flexibility in the context of altering the result since it offers the possibility of changing the percentual coefficient. As can be observed in the following diagrams there exist an inverse proportionate association between i and the numbers of new tuples acquired as well as the chance of obtaining different results in the rewritten query from one execution to another. With the increase in the values of i the probability of obtaining new tuples diminishes. This is because the

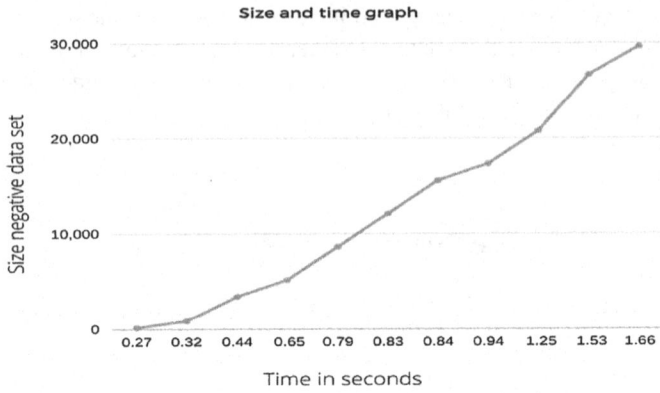

Fig. 6. Time graph

smaller the coefficient the smaller the negation data set will be and the bigger the chance that the random selection will be more random and uniformly distributed.

Another interesting observation is that the total size of the data set can influence the variation of the number of new tuples obtained, hence the smaller the dataset the bigger the variation. For the iris data set there was a big variation for the same coefficient from one execution to another, in the generated query but also the newly obtained tuples. This can easily be observed if we compare the two graphs one for the iris data set and one for the HTRU one (Fig. 7).

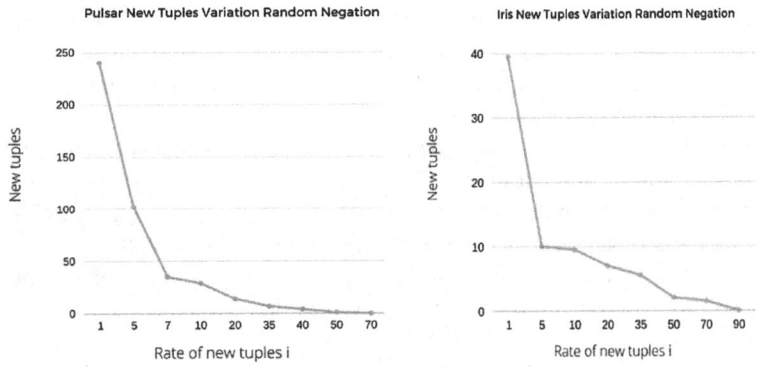

Fig. 7. Pulsar and iris results

In addition, the generated queries obtained in this experiment offered quite interesting results as can be observed in Fig. 8. Here not only we obtained new tuples but also some interesting conditions, which implies predicates that were not even used in the original one. Moreover, in multiple identic runs when talking about the parameters used, we obtained varied but similar results, since some of the used values for the same columns are very similar.

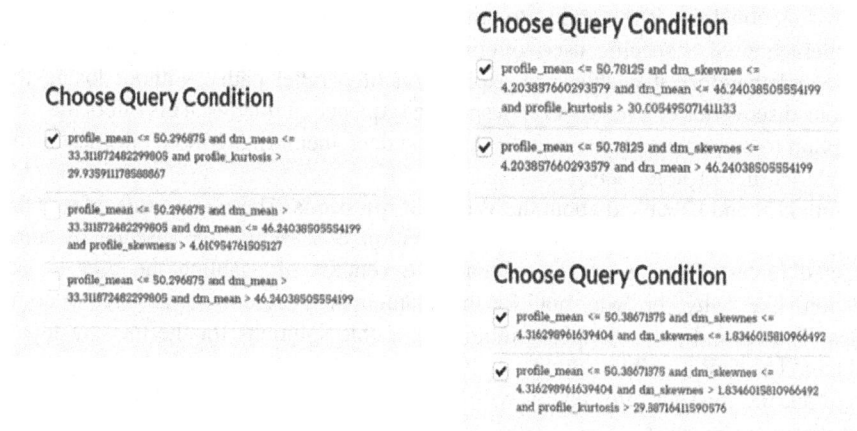

Fig. 8. Alternative conditions

5   Related Work

Query exploration has become a very popular research field in the last years. I am going to talk in more detail about other related work that was done in the domain of query exploration.

One such solution is presented in [3], where the authors propose a more **"guided interaction"** in the process of formulating queries. The authors describe three complementary principles for guided interactions: enumeration, insight, and responsiveness. Through enumeration, they refer to listing all the possible query autocompletion, in this way they intend on eliminating the inconvenience represented by the knowledge of the schema and data needed. The need for unsolicited insights is augmented through overcoming the dependencies among data that may be unknown to the user. It accentuates the need of giving multiple insights with useful information for the user that will better attend him in writing a query that will translate its initial intents. Furthermore, through responsiveness, they emphasize the need for an instantaneous response even if it can affect the accuracy. This is augmented by the need of the user to get those insights when he is writing the query and to be relevant at that specific moment, even if this implies that those pieces of information can be at a specific extend inaccurately.

Another interesting solution is detailed in [14] that is named DBNav. In this paper, the authors outline a new terminology the one of **query steering**. They articulate the need for a "tour guide" that will assist the user along with the query sessions, and that will be able to provide the user with noteworthy information about the data. Through query sessions, they refer to a sequence of queries that have in common the same user intent, and that every single one of them will represent a starting point for the next formulated query. DBNav is based on a more personalized interaction with the database as it acquires information about its users based on their profiles but also on their interaction with the application constructing the so-called "application profile". One of its key functionalities is the ability to "time travel". The application will memorize all the interaction of a specific user, offering the ability to visualize precedent written query, and therefore it enables the exploration of parallel paths, without losing the previous discoveries or even to rerun a specific sequence of queries. This offers the user the option to better visualize and understand the data, memorizing more effortlessly the possible worthy of note facts.

Both [15] and [5] talked about the **Why-not questions**. These questions refer to the ability of users to request additional information concerning the absence of some expected tuples. These papers accentuate the benefits of enabling the user to ask questions like "why" or "why not" for the unanticipated presence or absence of some tuples. The second paper proposes multiple possible solutions for the missing tuples problem. The first possible solution is the creation of a refined query that through its execution, the missing information will be available. Therefore, enabling the user to even discover the cause.

This paper is based on [4]. We extend the concept presented by adding two different negations algorithm, where the second one is a reinterpretation of the original negation that was used in the initial paper. If in the initial paper the C4.5 algorithm was used, we experimented with a different decision tree algorithm, but that is similar to the previously mentioned one. We extend the initial idea to a more complex and complete application by offering the user the possibility to execute queries, explore the schema of a database but also to observe the distribution of the values of a specific column as an addition to the main functionality.

6   Conclusion

In this paper, we have researched and developed a more interactive and time-efficient method for the process of query exploration. Our solution represents a symbiosis between machine learning techniques and SQL queries. Starting from an initial interrogation we generate a set of examples and counterexamples, which are used in the creation of the decision tree. From the obtained model we further extract new predicates that are used in the alternative query. The execution of this obtained interrogation will reveal some new similar tuples, besides the initial ones.

Through our application, a data analyst that works every day with new data has the possibility to investigate in a simpler and user interactive way a chosen database. He also has access to smaller functionalities that will help him better comprehend a specific data set. After a query has been rewritten the user can now assess its 'quality', based on criteria made accessible through a pie chart. This chart offers information about the number of new tuples obtained, the initial ones still present, and ones that are not missing. Therefore, the user can form a more precise idea about the direction in which the exploratory session is heading, without the need for performing an exhausting analysis of the obtained results. Moreover, our analyst also has the option to execute different queries, to explore the schema of a database, or choose different data sets without being obliged to change between different applications.

We have conducted some experiments to test capabilities of our solution, and we obtained promising results. We observed that the random negation precision increases directly proportional to the size of the dataset. However, as it happens with every software application, we observed some limitations. One such shortcoming is imposed by the front-end interface that will halt for a few seconds when the volume of data to be displayed and rendered is too big, and even two seconds can be a considerable amount of time. Another one is represented by the performance of the library used on the server-side that maintains the communication with the database. Its performance depends on the size of the data that needs to be retrieved, and for more than two million tuples it needs a lot of time for retrieving.

We can always find further improvements that can be done to an application, it is in our human nature to seek perfection. One such possible addition to our application will be the implementation of more diverse negation algorithms. Another one is to solve this problem using different algorithms for finding a pattern inside a dataset. Hence offering even more freedom of choice to a data analyst who searches for unknown patterns. Furthermore, a future improvement will be the addition of a history-based query retrieval. Therefore, the user will be able to access all the queries used in our application in that specific session, having the possibility to go back in time to explore multiple parallel paths.

References

1. Ffoulkes, P.: InsideBIGDATA Guide to the Intelligent Use of Big Data on an Industrial Scale. InsideBIGDATA, Massachusetts (2017)
2. Reinsel, D., Gantz, J., Rydning, J.: Data age 2025: the evolution of data to life-critical. In: IDC (2018)
3. Nandi, A., Jagadish, H.V.: Guided interaction: rethinking the query-result paradigm. In: PVLDB (2011)
4. Cumin, J., Petit, J.-M., Scuturici, V.-M., Surdu S.: Data exploration with SQL using machine learning techniques.In: EDBT (2017)
5. Tran, Q.T., Chan, C.-Y.: How to ConQueR why-not questions. In: SIGMOD, pp. 15–26 (2010)
6. Tran, Q.T., Chan, C.-Y., Parthasarathy, S.: Query by output. In: SIGMOD, pp. 535–548 (2009)
7. Heine, B., Kuteva, T.: The Genesis of Grammar. Oxford University Press, Oxford (2007)
8. Negation. https://plato.stanford.edu/entries/negation/#Int, Accessed 3 Sept 2020
9. Clark, H.H.: Semantics and Comprehension. Mouton (1976)
10. Brownle, J.: A Gentle Introduction to the Law of Large Numbers in Machine Learning. https://machinelearningmastery.com/a-gentle-introduction-to-the-law-of-large-numbers-in-machine-learning, Accessed 3 Sept 2020
11. Buitinck, L., et al.: API design for machine learning software: experiences from the scikit-learn project. In: European Conference on Machine Learning and Principles and Practices of Knowledge Discovery in Databases (2013)

12. Lyon, R.J., Stappers, B.W., Cooper, S., Brooke, J.M., Knowles, J.: D: Fifty years of pulsar candidate selection: from simple filters to a new principled real-time classification approach. Mon. Not. R. Astron. Soc. **459**(1), 1104–1123 (2016)
13. Chen, S.X., Xu, Z., He, J., Dong, A., Guo, B., Zhang, S.: Cautionary tales on air-quality improvement in Beijing. Proc. Roy. Soc. A **473**(2205), 20170457 (2017)
14. Cetintemel, U., et al.: Query steering for interactive data. In: CIDR (2013)
15. Tran, Q.T., Chan, C.-Y., Parthasarathy, S.: Query reverse engineering. VLDB **23**(5), 721–746 (2014)

Data-Driven Insights on Secondary Education: A Case Study on Teachers' Demography and Qualification

Dessislava Petrova-Antonova[1] and Olga Georgieva[2]

[1] GATE Institute, Sofia University, 125 Tsarigradsko shose Blvd.,
1113 Sofia, Bulgaria
dessislava.petrova@gate-ai.eu
[2] FMI, Sofia University, 5 James Bourchier Blvd., 1164 Sofia, Bulgaria
o.georgieva@fmi.uni-sofia.bg

Abstract. The paper presents an application approach based on Big Data Value Chain concept to data collected for teachers' demography and qualification. The proposed approach discovers and further enables to account for the teacher aging as a sensitive factor of the education process. The first step of the study ensures a reliable and holistic dataset by careful preprocessing of the raw data. Different types of data analysis have been applied in the analytical step. As the statistical analysis was not able to discover all existing relations a non-trivial approach was proposed to discover models and connections of the three main teachers' age groups. The analysis of linked data enriches the retrieved information by getting more insight at the relation between the teachers' groups and the municipality of the school they teach. By describing education tendencies and by modeling the significant dependencies of the teachers' age groups the proposed approach enables to reveal information useful for policies making.

Keywords: Big data value chain · Secondary education · Data analytics

1 Introduction

Teachers are one of the most important factors, which influences the student performance and forces their achievements [1–3]. Keeping well-qualified teachers in classrooms is a primary goal of any educational reform. The Teaching and Learning International Survey (TALIS) provides an insight into the level of professionalism among teachers and school leaders and challenges emerging in 21st century education. The first volume of the survey shows that teachers and school leaders still feel like need training to meet today's requirements of new, modern kind of professionalism, including cross-curricular skills, ICT adoption and multilingual and/or multicultural teaching [4]. The second volume of TALIS propose to "turn schools into intellectually stimulating places to work" taking into account that the quality of schools' systems depend on the quality of their teachers and school leaders. The new message from this volume is that the quality of teachers and school leaders is in correlation with the quality of their qualification, working environment and their intention for collaboration and continuous development. In addition to TALIS, several research works are focused

© Springer Nature Switzerland AG 2021
D. Simian and L. F. Stoica (Eds.): MDIS 2020, CCIS 1341, pp. 251–263, 2021.
https://doi.org/10.1007/978-3-030-68527-0_16

on exploration of correlation between teachers' effectiveness and students' performance [1, 5, 6]. One of the reasons for lower teaching productivity is turnover of effective teachers, especially when schools do not replace teachers who leave with others who have equal effectiveness [7]. The replacement of experienced teachers with new inexperienced teachers is pointed as another reason [6]. At the same time, the schools find more difficult fully qualified teachers in some fields than in others, like science, technology, engineering and mathematics (STEM) [8]. The lack of STEM skills causes staffing shortages and demographic imbalance.

Several studies give answers to the question why teachers leave the educational system. This could be personal reasons such as family circumstances or career changes, but also lack of administrative support and teaching autonomy, students' misbehavior, low salary, missing professional development opportunities [4, 9–11]. The stability of several educational systems across the world is affected by the attrition among the teachers [12]. The second volume of TALIS provides and analysis of pressing concerns for the renewal of the teaching workforce taking into account the percentages of teachers who want to leave within the next five years [4]. According to the analysis countries such Lithuania, Bulgaria and Estonia with high average age of teachers, have high percentage (equal to or above 40). A moderate positive country-level correlation between teachers wanting to leave the educational system and the proportion of teachers' age equal to or above 50 is found.

This paper is focused on some aspects of the above concerns and problems by exploring the nature of the Bulgarian teacher workforce based. It is important to understand the existing processes, explain dependencies and reasoning tendencies to support decision making related to teachers' quality policies and practices. The following research questions (RQ) are addressed:

- **RQ1:** What are demographic characteristics of the teachers in Bulgaria, including age, administrative region and populated location?
- **RQ2:** What differences exists in teachers' educational degree, qualification and to what extent such differences exist in between beginning teachers, more experienced teachers and teachers and retirement teachers?

In order to answer the above questions, a Big Data Value Chain approach is applied, which provides a powerful tool for acquiring data, combining data from different sources and providing access to it with low latency while ensuring data integrity and preserving privacy [14]. The primary goal of the Big Data Value Chain is to discover models, correlations, deviations and other facts and events that are hidden in the data itself. In this paper the concept of Big Data Value Chain is adopted to perform a study in the phases as follows:

- **Data acquisition and preprocessing** – on this phase, a dataset for analytics is prepared, including activities such as data collection, cleaning, aggregation, consistency checking, etc.
- **Data analysis** – this phase applies data mining techniques such as cross correlation analysis and clustering in order to get new knowledge from data;
- **Data insight** – this phase includes interpretation of the obtained results during the analysis phase.

The rest of the paper is organized as follows. Section 2 outlines data acquisition and preprocessing. Section 3 describes data analysis within the performed study, whereas Sect. 4 discusses the obtained results. Section 5 concludes the paper and gives directions for future work.

2 Data Acquisition and Preprocessing

The data is obtained from a software platform for pedagogical specialists in Bulgaria, which serves the Project BG05M2OP001–2.010–0001 "Qualification for professional development of pedagogical specialists" of the Ministry of Education and Science. The whole number of records in the data set is 67 218. It is anonymized for the purpose of the case study presented in this paper.

2.1 Data Acquisition

The platform collects data for schools and teachers in a relational database, with two purposes: (1) offering and searching for pedagogical job positions and (2) organizing qualification courses for pedagogical specialists. The data needed for the purpose of the current work is extracted from 11 tables of the relational database (Fig. 1).

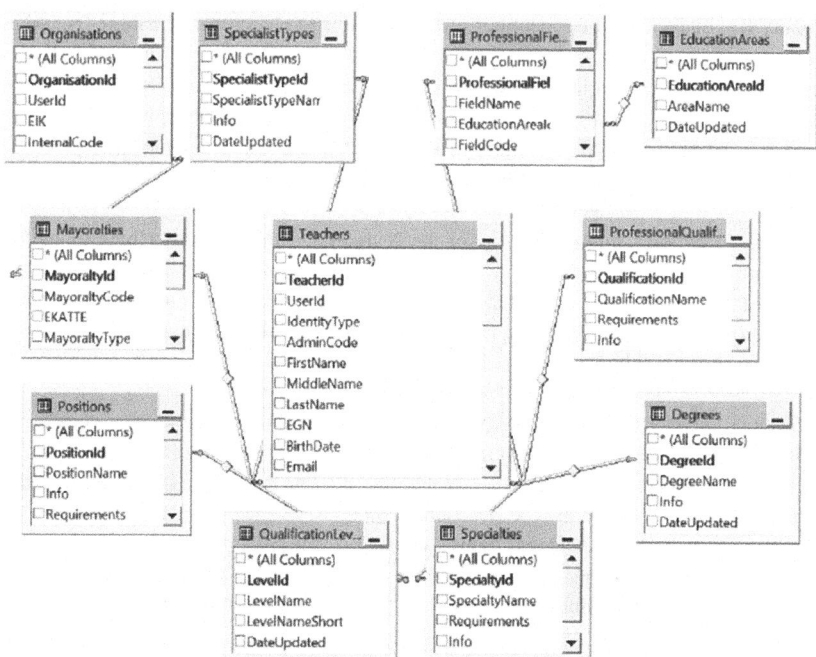

Fig. 1. Relational database diagram.

The main data is stored in table "Organizations", holding schools' data, and table "Teachers", holding teachers' data. The data extracted from table "Teachers" is anonymized, so personal data such as data of birth, name, email, etc. is removed. During the extraction process the age of each teacher is calculated, based on the date of birth. The obtained dataset is enriched with data for municipality and district of the schools and stored in.csv format. Table 1 shows the structure of the final dataset used for the analysis.

Table 1. Structure of the case study dataset.

Field name	Data type	Description	Acquisition method
Teacher ID	Integer	Unique identifier of the teacher	Direct from database
Admin code	String	Administrative code of the educational organization, where the teacher work	Direct from database
Age	Integer	Aggregated based on date of birth	Indirect through calculation
Degree year	Integer	Year of obtained educational degree	Direct from database
Professional experience	Integer	Years of Total working experience	Direct from database
Teacher experience	Integer	Years of Working experience in the educational system	Direct from database
Mayoralty ID	Integer	Unique identifier of the mayoralty	Direct from database (predefined values from 1 to 5260)
Degree ID	Integer	Unique identifier of the type of degree (bachelor, master, PhD)	Direct from database (predefined values from 1 to 7)
Professional field ID	Integer	Unique identifier of the professional field according to the Bulgarian educational system	Direct from database (predefined values from 1 to 51)
Specialty ID	Integer	Unique identifier of the degree's specialty	Direct from database (predefined values from 1 to 40)
Professional qualification ID	Integer	Unique identifier of the field of professional qualification (Biology, Math, History, Engineer, Chemistry, Physics, etc.)	Direct from database (predefined values from 1 to 103)
Qualification level ID	Integer	Unique identifier of the professional qualification level	Direct from database (predefined values from 1 to 6)
Specialist type ID	Integer	Unique identifier of the job position type (teacher, director, psychologist, trainer, etc.)	Direct from database (predefined values from 1 to 7)
Position ID	Integer	Unique identifier of the job position	Direct from database (predefined values from 1 to 48)

(continued)

Table 1. (*continued*)

Field name	Data type	Description	Acquisition method
Organization subcategory ID	Integer	Unique identifier of the type of educational organization (kindergarten, primary school, secondary school, etc.)	Direct from database (predefined values from 1 to 26)
Education	String	Type of education (primary, secondary, higher, etc.)	Indirect through replacement of string with value from 1 to 5
Municipality ID	Integer	Unique identifier of school's municipality	Indirect through aggregation based on Mayoralty ID (predefined values from 1 to 265)
District ID	Integer	Unique identifier of school's district	Indirect through aggregation based on Mayoralty ID (predefined values from 1 to 28)

2.2 Data Preprocessing

The preprocessing is needed to prepare a reliable, informative and holistic dataset. The data standardization requires the structure of a dataset to be converted into a uniform format. The continuous data as *Age*, *Experience* were directly adopted in the data set, whereas the original values of nominal data as *Degree*, *Professional field*, *Specialty*, *Qualification* and ext. Were appropriately valued by integers to allow further mathematical transformations and analysis (Table 1). The categories for these qualitative variables are nomenclature values, which are used in the Bulgarian education system and related regulations of Ministry of Education and Science. Also *Education* values given as strings in the data are not appropriate for analytic processing and thus, they were appropriately replaced with integers.

The activities related to data preprocessing are based on the data anomaly patterns, defined in [13] as follows:

- **Illegal Values** – values in a given column are outside of the domain range;
- **Inconsistent values** – syntactically correct values but not compliant with other attribute values;
- **Missing values** – values are not presented in a given column;
- **Column headers containing attribute values** – column headers are not titles of the attributes.
- **Incorrect column headers** – column headers do not represent the attributes stored in them;
- **Columns not related to data model** – dataset hold data columns, which are not relevant to the data model;
- **Rows not related to data model** – data acquisition returns records that are not relevant to the data model;

- **Multiple values stored in one column** – single column stores values of several attributes;
- **Single value split across multiple columns** – value of a single attribute exists in more than one column;
- **Duplicate rows** – dataset contains duplicate records for the same entity.

Table 2 shows data inconsistencies handled according to data anomality patterns.

Table 2. Cleaning of data inconsistencies.

No	Data inconsistency	Cleaning activity	Anomality pattern
1	Check if professional experience < Teacher experience	Set professional experience = Teacher experience	Missing values
2	Check if professional experience is negative value	Set professional Experience to absolute value	Illegal values
3	Check if teacher experience is negative value	Set teacher experience to absolute value	Illegal values
4	Outliers removed	3 removed	Duplicate rows
5	Check if professional experience > Age-20	Correct professional experience value	Inconsistent values
6	Check if teacher experience > Age-20	Correct professional teacher value	Inconsistent values
7	Data standardization	Transform 'Education' column from string to int codes	N/A
8	"0" value of TeacherID	Record deleted	Illegal Values
9	Mayorality field misssing values	30 values set based on the admin code	Missing values
10	ProfessionalField field missing values	17 values set base on SpecialtyId and ProfessionalQualificationId	Missing values
11	OrganisationSubCategory field missing value	1 value set based on the admin code	Missing values

The anomality patterns covered are as follows: Incorrect column headers, Columns not related to data model, Rows not related to data model, Multiple values stored in one column and Single value split across multiple columns are avoided through appropriate SQL queries to the relational database and resulted in a well-structured.csv file.

The municipality and district data were not explicitly presented in the dataset as only the mayoralty identifier was available. That is why data enrichment was performed based on the mayoralties' EKATTE codes (Unified Classifier of Administrative-Territorial and Territorial Units).

3   Data Analysis

Data features have different impact to the data mining and consequent knowledge extracted. Features as *Teacher Id* and *Admin Code* just give more information about the items. They are useful in linking the Teachers data to other data sets but not proper for information retrieve inside the data set. Other features – *Age, Degree year, Professional experience* and *Teacher experience* are measurable in years. They could give us information about achievements of teachers in time. Their possible dependency could be investigated by correlation analysis. The nominal categorical data as *Degree, Education, Qualification* and ext. May have dependencies that are not easily seen and need specially organized data treatment procedures.

3.1   Statistical Analysis for Preliminary Information Acquisition

Cross-correlation analysis was applied by Pearson's linear correlation coefficient calculating pairwise manner of the data features. Only meaningful correlations above 0,35 are presented in Table 3 and further discussed.

The factor *Age* is in strong linear dependence with factors *Professional experience* and *Teaching experience*. *Professional experience* and *Teaching experience* are in very strong dependence having correlation coefficient of 0,9167. These results prove that most of the teachers relay on their education as a profession chosen for their entire life. In this context the achieved relatively strong inverse linear dependence of factors *Age* and *Degree year* with a correlation coefficient −0,7776 could be explained as a

Table 3. Results of cross-correlation analysis.

	Age	Degree year	Obtained professional qualification	Current professional qualification	Professional qualification	Qualification level	Professional experience	Teaching experience	Current position
Age	1	−0,7776	−0,0074	0,0019	0,0305	−0,3235	0,8947	0,8210	0,0496
Degree year	−0,7776	1	−0,0597	−0,0610	−0,0562	0,2886	−0,7495	−0,7296	−0,0856
Obtained Professional qualification	−0,0074	−0,0597	1	0,5104	0,2903	0,0545	−0,0122	−0,0124	0,3538
Current Professional qualification	0,0019	−0,0610	0,5104	1	0,3700	0,0503	0,0044	−0,0043	0,2610
Professional qualification	0,0305	−0,0562	0,2903	0,3700	1	0,0216	0,0292	0,0188	0,2015
Qualification level	−0,3235	0,2886	0,0545	0,0503	0,0216	1	−0,3729	−0,4111	0,0111
Professional experience	0,8947	−0,7495	−0,0122	0,0044	0,0292	−0,3729	1	0,9167	0,0416
Teaching experience	0,8210	−0,7296	−0,0124	−0,0043	0,0188	−0,4111	0,9167	1	0,0170
Current position	0,0496	−0,0856	0,3538	0,2610	0,2015	0,0111	0,0416	0,0170	1,0000

comparatively large part of teachers graduated without delay of their study. The presented relatively strong linear dependence between *Degree year* and the two factors *Professional experience* and *Teaching experience* with levels above 0,72 supports the above conclusions.

The dependency *Obtained Professional qualification* and *Current Professional qualification* is moderately strong as correlation coefficient is 0,5104. It should be underlined that *Current Professional qualification* and *Professional qualification* as well as *Current Professional qualification* and *Current Position* are not strongly dependent. These correlations mean that the obtained qualification do not coincides with the current qualification and/or position for relatively large number of teachers. The result shows that large part of the teachers is motivated to change their professional qualification. The connections of *Qualification level* with *Professional experience* as well as with *Teaching experience*, respectively, are at the same magnitude ($-0,3759$ and $-0,411$) however with a negative sign. The two dependencies have a similar level strength as the above discussed but in negative direction meaning that the qualification level is obtained during the years of teaching. Further interpretation and expert understanding is need for more deep process description.

The lack of linear relation between continuous and nominal data is partly explainable with the difference in their natural origin. This imposes further investigation of their possible dependences. Another preliminary analysis we did shows that the multidimensional data space exploration does not present any valuable grouping as no meaning clusters could be separated.

3.2 Information Retrieval Methodology of Teacher's Data

The correlation analysis results conclude that the teachers' age is the most defining factor affecting the other characteristics of the guild. On another hand, the aging is a sensitive factor for policy making as it gives information about the tendencies in the education process.

Bearing in mind all above results and conclusions, the further investigation has been proposed to follow several consequent steps to ensure exact, full and efficient data knowledge retrieve.

1. Determine main teachers' age groups by defining natural data grouping according to cluster analysis.
2. Between age groups analysis.
3. Search for informative dependencies between distinct age groups and the others factors presented in the data set.

Age Data Clustering to Define Meaningful Teachers Groups. Objective function clustering methods are mostly applicable as they find data groups naturally spread over the whole data space. Optimization procedure minimizes the clustering criterion in order to determine clusters' centers. Each defined cluster is represented by its cluster center. here, we apply the well-known K-means algorithm initialized by three number of clusters, $c = 3$ as we are interested in three main age groups of young, middle aged and elderly teachers.

Between Age Groups Analysis Results. The age of the teachers presented in the data set varies from 20 to 84 years. The cluster centers as most representative of the obtained groups (Table 4) show that middle and elderly teachers are closer each other of about 11 years; the difference of young teachers and middle aged is of 13 years. The middle-aged group is the largest one, whereas young teachers' group is significantly smaller comparing to the rest two. Young teachers are in double less than middle aged.

Table 4. Teachers' age groups.

Cluster of	Cluster center [years]	Number of teachers	Age interval
1 (Young teachers)	32,85	14181	[20 39]
2 (Middle aged teachers)	46,6	29378	[40 52]
3 (Elderly teachers)	57,74	23659	[53 84]

Search for Informative Dependencies. The results are obtained by cylindrical extension of the obtained teachers' age groups to other features presented in the data set.

A significand distinction of the degree levels (Degree ID) through different groups is marked (see Fig. 2, where the levels are numbered from 1 to 7 as it is described in the figure's legend).

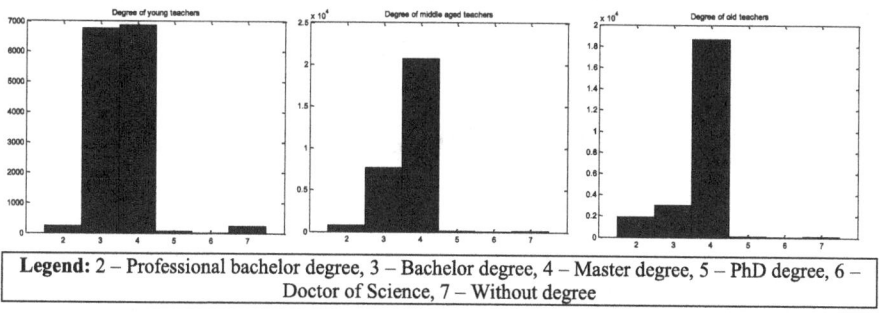

Legend: 2 – Professional bachelor degree, 3 – Bachelor degree, 4 – Master degree, 5 – PhD degree, 6 – Doctor of Science, 7 – Without degree

Fig. 2 Degree ID of a) young b) middle aged and c) old teachers.

It could be seen that young teachers are almost equally spread in two large degree groups – bachelor degree group (level 3) and master degree group (level 4) with over 6800 teachers in each. Small portions are with professional bachelors (level 2), without degree level (level 7) and very few have PhD degree (level 6) (Fig. 2a). The middle aged teachers are dominantly master degree teachers but there is quite large group of bachelors (Fig. 2b). Most of the old teachers are masters, whereas relatively small portions are bachelors (Fig. 2c). It is a clearly seen tendency for young teachers to be bachelors without pursuing a master's degree.

Histogram analysis of each age group could grasp some tendencies according to the characteristics of *"Current Professional Qualification"(cPQ)* (see Fig. 3 with numbered qualifications presented in the figure's legend). The comparison among the presented qualifications shows that for young and middle aged teachers the qualification of *"Early childhood and primary education teacher"* is preferred than qualification *"Early childhood education teacher"*, which is the case for the old teachers. The qualification *"Primary education teacher"* is highly presented with over 3000 of middle aged and old teachers, whereas the group of young teachers with this qualification is quite less presented - less than 800. The parts of middle and old aged declared that they have *"Other professional qualification"* is larger than the *"Without professional qualification"* but this comparison is in opposite for the young teachers. The middle aged teachers possess the largest group of teachers with qualification in *"Informatics"* as well as *"Physiology"*.

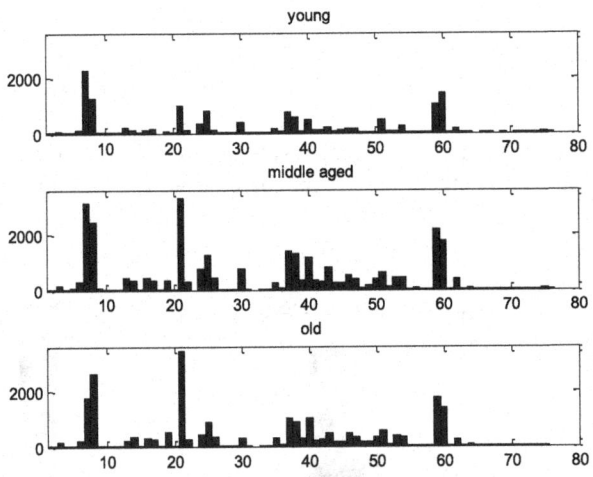

Legend: 7 – *Early childhood and primary education teacher*, 8 - *Early childhood education teacher*, 21 – *Primary education teacher*, 30 – *Physiology*, 43 – *Informatics*, 59 – *Other professional qualification*, 60 – *Without professional qualification*

Fig. 3. *Current Professional Qualification* histogram distribution of the three groups.

4   Data Insight

The obtained results through correlation analysis, cluster analysis and further histogram analysis could be summarized as follows. Presently there is some an unbalance as the young teachers are significantly less than the largest group of middle-aged teachers. In addition, the elderly teachers are quite large group commensurate with the middle aged one.

It could be underlined that young teachers prefer to be bachelors without pursuing a master's degree, whereas the two older groups have master degree. This particularly could be explained by the objectivity resulted of the adopted two-level university degree education (bachelor and master degree) in EU countries. The young teachers currently prefer *Professional Qualification 7*, whereas the elderly have *cPQ* 8 and *cPQ* 21. This information could be partially explained with the changes in the country education system.

The raw data contain the administrative code of the school city place of the teacher. However, this is not enough in order to get insight to the connection between the teachers and the municipality area of the school where they teach. That is way the teacher's data sets are linked with the municipality data set. The same approach for data analysis and interpretation is applied for the linked data information retrieve.

The results could be summarized as follows. Due to demographic characteristics the teachers in the villages are quite less than the teachers in the cities, however with different proportion among the distinct age groups (Fig. 4). The relations of teachers in cities over teachers in villages are 4,71 for young; 6,28 for middle aged and 8,27 for elderly teachers, respectively. It is seen that there is a tendency of cities to attract young teachers.

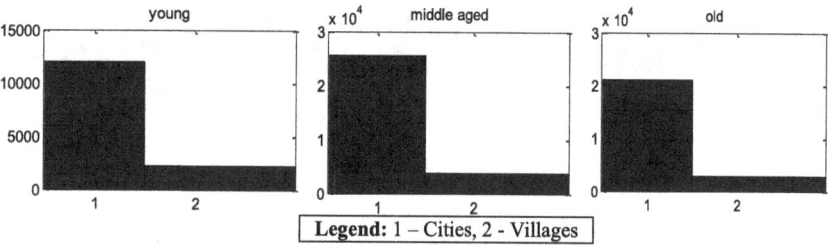

Fig. 4. Number of teachers in cities and villages according the age groups.

More information is visible for the distribution of the teachers in the distinct municipalities (Fig. 5). The young teachers in *Sofia*, the capital of Bulgaria, are relatively much more than the middle aged in comparing to the established relation of about 1:2 for the whole country. Other differences are visible according to the elderly teachers. For instance, the old teachers in *Gabrovo* are relatively large group than the respective group in the two commensurate districts *Vidin* and *Pernik*. For the two largest towns after the capital city the retrieved information shows that the young teachers in *Plovdiv* are quite large amount than their number in *Varna*. However, the difference of middle aged and elderly groups is quite commensurable.

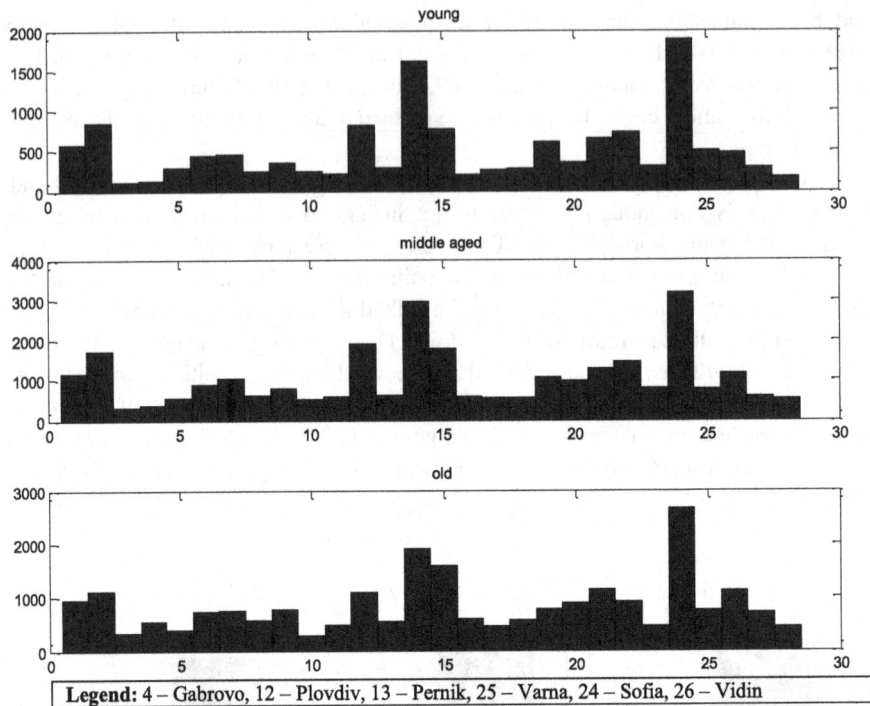

Legend: 4 – Gabrovo, 12 – Plovdiv, 13 – Pernik, 25 – Varna, 24 – Sofia, 26 – Vidin

Fig. 5. Number of teachers spread over the districts.

5   Conclusion and Future Work

The paper presents an application approach based on Big Data Value Chain concept to data collected for teachers and the schools of their teach. It is considered as a case study covering data for pedagogical specialists in Bulgaria. It performs a research of the main dependencies in the provided factors hidden in the data. Under detailed, deep and careful processing of the raw data the first step of the study ensures a reliable and holistic dataset. Different types of data analysis have been applied in the analytical step. As the statistical analysis was not able to discover all existing relations a non-trivial approach was proposed to discover models and connections among three main teachers' age groups. The analysis of linked data enriches the retrieved information by getting more insight at the connection between the teachers and the district of the school they teach.

The proposed approach discovers and further enables to account for the teacher aging as a sensitive factor of the education process. By describing education tendencies and by modeling the significant relations of the teachers' age groups the obtained information reveals important dependencies useful for policies making. For instance, it could be used for assessment and possible improvement of the qualification structure. By following the changes in the qualifications of teachers from the obtained to the current one within the qualification field (natural science, philological subjects and etc.) and out of it, the strength of their cross relation could be estimated.

The approach could be deepened with whole economic and social analysis of the regions according to the existing education tendencies. Further it could be generalized through data of other education systems.

Acknowledgement. This research work has been supported by GATE project, funded by the Horizon 2020 WIDESPREAD-2018–2020 TEAMING Phase 2 programme under grant agreement no. 857155, by the Bulgarian National Science fund under project ITDGate agreement no. DN 02/11 and by the Science Fund of Sofia University under project no. 80–10-61/13.04.2020.

References

1. Hanushek, E., Rivkin, S.: Generalizations about using value-added measures of teacher quality. Am. Econ. **100**, 267–271 (2010)
2. Rockoff, J.: The impact of individual teachers on student achievement: Evidence from panel data. Am. Econ. **94**, 247–252 (2004)
3. Ronfeldt, M., Loeb, S., Wyckoff, J.: How teacher turnover harms student achievement. Am. Educ. Res. J. **50**(1), 4–36 (2013)
4. OECD: TALIS 2018 Results (Volume II): Teachers and Schools Leaders as Valued Professionals, TALIS. OECD Publishing, Paris (2018). https://doi.org/10.1787/19cf08df-en
5. Hirsch, E., Emerick, S., Church, K., Fuller, E.: Teacher working conditions are student learning conditions: A report on the 2006 North Carolina Teacher Working Conditions: The Center for Teaching Quality, Hillsborough, NC (2007)
6. Rivkin, S.G., Hanushek, E.A., Kain, J.F.: Teachers, schools, and academic achievement. Econometrica **73**, 417–458 (2005)
7. Finster M.: Identifying, Monitoring, and Benchmarking Teacher Retention and Turnover: Guidelines for TIF Grantees, Department of Education for Teacher Incentive Fund (TIF) (2015)
8. Cowan, J., Goldhaber, D., Hayes, K., Theobald, R.: Missing elements in the discussion of teacher shortages. Educ. Res. **45**(8), 460–462 (2016)
9. Boyd, D.J., Grossman, P.L., Ing, M., Lankford, H., Loeb, S., Wyckoff, J.: The influence of school administrators on teacher retention decisions. Am. Educ. Res. J. **48**, 303–333 (2011)
10. Johnson, S.M., Kraft, M.A., Papay, J.P.: How context matters in high-need schools: the effects of teachers' working conditions on their professional satisfaction and their students' achievement. Teach. Coll. Rec. **14**(10), 1–39 (2012)
11. Burkhauser, S.: How Much Do School Principals Matter When It Comes to Teacher Working Conditions? Educ. Eval. Policy Anal. **39**(1), 126–145 (2016). https://doi.org/10.3102/0162373716668028
12. Viac, C., Fraster, P.: Teachers' well-being: a framework for data collection and analysis. OECD Education Working Press, No. 213. OECD Publishing, Paris (2020)
13. Sukhobok, D., Nikolov, N., Roman, D.: Tabular data anomaly patterns. In: 2017 International Conference on Big Data Innovations and Applications (Innovate-Data), pp. 25–34 (2017)
14. Curry, E.: The big data value chain: definitions, concepts, and theoretical approaches. In: Cavanillas, J.M., Curry, E., Wahlster, W. (eds.) New Horizons for a Data-Driven Economy, pp. 29–37. Springer, Cham (2016). https://doi.org/10.1007/978-3-319-21569-3_3

Automatic Identification of Watermarks and Watermarking Robustness Using Machine Learning Techniques

Dana Simian[1,2]([⊠]) [iD], Ralf D. Fabian[1,2] [iD], and Mihai D. Stancu[1] [iD]

[1] Research Center in Informatics and Information Technology,
Lucian Blaga University of Sibiu, Sibiu, Romania
{dana.simian,ralf.fabian}@ulbsibiu.ro, stancu.mihai.dorian@gmail.com
[2] Department of Mathematics and Informatics,
Lucian Blaga University of Sibiu, Sibiu, Romania

Abstract. The goal of this article is to propose a framework for automatic identification of watermarks from modified host images. The framework can be used with any watermark embedding/extraction system and is based on models built using machine learning (ML) techniques. Any supervised ML approach can be theoretically chosen. An important part of our framework consists in building a stand-alone module, independent of the watermarking system, for generating two types of watermarks datasets. The first type of datasets, that we will name artificially datasets, is generated from the original images by adding noise with an imposed maximum level of noise. The second type contains altered watermarked images obtained from the original ones by using different transformations. The module also performs an automatic labeling process of these data, building watermarks' containers. Then, many models can be built using the watermarks containers and different ML techniques. Comparing the performances of all the obtained models allows the choice of the best model, or provides details for building ensemble learning. To validate the proposed framework, we conducted experiments using a particular watermarking system, built by us and many models based on artificial neural networks (ANN) and support vector machines (SVM). As a side result we identified a possible methodology for evaluating the robustness of a watermarking system, by using ANN and the two types of datasets generated in our proposed methodology.

Keywords: Artificial neural networks · Support vector machines · Watermarking · Robustness

1 Introduction

The rapid increase in multimedia content generation opportunities and ease of access has led to the development of digital embedding techniques that add supplementary information which can be later retrieved, for multiple purposes, e.g., author identification, content classification, profile identification and tracking of multimedia documents in order to prevent illegal usage, copying and distribution. Intended or unintended alterations of digital media may occur while

© Springer Nature Switzerland AG 2021
D. Simian and L. F. Stoica (Eds.): MDIS 2020, CCIS 1341, pp. 264–284, 2021.
https://doi.org/10.1007/978-3-030-68527-0_17

processing and transmitting documents or in illegal use and distribution, obscuring any later information extraction. Therefore it is important that the additional information embedded in media documents be identified despite of any changes to the initial state of the media contents. Watermarking is one of the embedding techniques successfully applied for digital images, video and sound files. In this article we will focus on watermarking for digital images, but the results could be extended to other multimedia contents.

A watermark is, in essence, a binary image pattern embedded in the host data. Depending on the visibility of the watermark for the human end-user we distinguish between visible and invisible watermarking techniques [15]. An invisible watermark is stored in imperceptible data. Spatial and frequency domains can be used for embedding the watermark. The watermarking techniques using frequency domains embed information in the coefficients of the transformed image obtained using an invertible transformation such as discrete cosine transform (DCT), discrete Fourier transform (DFT) and discrete wavelet transform (DWT). In spatial domain, changing the grey levels of pixels is used to insert supplementary data [3,15]. Watermarking systems can be characterized by several properties that are highly application dependent. Among others, we will refer to security and robustness. The watermark is called robust if it could be detectable after embedding even if the image has been altered. A watermark is secure if it is not possible to change its role. Hence, robustness is a necessary property for a watermark to be secure. Robustness is required when the main goal of the watermark is ownership identification even if the original data is altered [15]. Fragile watermarking is used for detecting changes in the host data [15]. The watermark extraction and recovery process could need both the original and the embedded image, or only the embedded image, corresponding to a non-blind, respectively blind watermarking technique [3,15]. More details about watermarking techniques' features can be found in the comprehensive survey [8].

Numerous studies are devoted to the behaviour of different watermarking algorithms against geometric and processing attacks [11,16]. Recovery and identification of the watermark is highly content dependent. Different adaptive techniques are used to compensate this drawback. Machine learning (ML) techniques like artificial neural networks (ANN) and support vector machines (SVM) were used to enhance the robustness and transparency of different image watermarking techniques. In [20] a blind wavelet-based image watermarking technique is proposed, that uses an ANN for storing the relationships between the wavelet coefficients before and after the watermark embedding process. In [6] a robust hybrid watermarking method is proposed, using spatial domain, particle swarm optimization (PSO) and k-nearest neighbor algorithm (kNN). The extraction algorithm is trained on features from the blocks in which the watermark is embedded and control blocks, and builds using PSO two centroids belonging to each of the watermark binary values. These centroids are used by the kNN classifier to extract the watermark. A Least Square–Support Vector Machine (LS-SVM) is used in [22] for increasing the robustness of a colour image watermarking scheme against geometric distortions, including rotation, scaling, and translation. An LS-SVM training model is applied in the decoding step for correcting the watermarked color image before extraction. The SVM classifier is used against geometrical attacks for image authentication in a watermark scheme with geometrical invariants [19]. The SVM is trained to learn the relationships between the set of characteristics of rotation – scale – translation (RST) invariants and the secret key generated by the watermarking method. In the mean value modulation-based image watermarking method in multiwavelet domain proposed in [12], a SVM is used to learn the mean value relationship. In [7], the output coefficients for watermark embedding are generated using an SVM. In [2] a full counter-propagation neural network (FCNN) is used for storing the watermark. Moreover, the neural network integrates both the watermark embedding and extracting procedure. A Back Propagation Neural Network (BPNN) is used in the quantization step of the embedding process, in order to increase the robustness of the watermarking algorithm to different attacks [13]. Recently, several solutions for improving the digital watermarking process using Deep Neural Networks (DNN) were proposed. In [4] is presented a DNN based framework for embedding the watermarks in the images, in the training process. The same DNN model serves later for images ownership identification, based on the embedded watermarks. A very new direction of study, is oriented to protect ML generated models by using watermarks embedded in the models during training [5,18].

In all references we have studied, adaptive models have been used for improving the performances of embedding – extraction of a specific watermarking algorithm and there are no independent adaptive models used for automatic recognition of the extracted watermark.

The aim of this paper is to study the possibility of designing a module based on ML techniques for automatic recognition of the extracted watermark from altered host images, that can be added to any watermaking system and to study its independence from this system. Any ML technique could be adopted. Using this module we will treat watermark classification separated from implanting watermarks into images and their subsequent retrieval. From the practical point of view this is an important separation of tasks as it can be realized by independent and not co-interested parties. To the best of our knowledge, there are not reported results in this direction. We propose a methodology (framework) which supposes three stages. In the first stage, is performed a theoretical performances analysis using a system for generating artificially altered watermarks. The main aims of this stage are to provide a comparative report of the performances of the considered ML models and to study the models performances' evolution according to the level of noise. In the second phase, other ML models are built using a training set of altered watermarks. We designed and implemented a module for generating this set of altered watermarks from the original set of watermarks, by using real attacks. This module can be used with any watermarking system. In the last stage, we performed a comparative study of the results obtained in first two stages. To compare the model performances we used as metrics the accuracy, precision and recall. Comparing the performances of all the models generated in the first and second stage allows the choice of the best model, or/and provides details for building ensemble learning.

In order to validate the proposed framework we conducted experiments using ANN and SVM models for automatic watermark recognition after extraction. The proposed module and the resulted models were tested on an improved version of the watermarking system presented in [16]. However the results do not depend on the embedding watermarking system.

As an interesting side result we identified a possible methodology for evaluating the robustness of a watermarking system, by using ANN and the two types of datasets generated in our proposed methodology.

The next of the article is organized as follows. In Sect. 2 we introduce the problem and propose the framework for watermarks identification. Sect. 3 describes the design of the independent module used to generate the sets of altered watermarks. In order to test our framework we chose two machine learning techniques that are described in Sect. 4. Initial selection of the framework components and the generation of the training sets can be found in Sect. 5. Our framework works with any watermarking system. Anyway, to prove the functionality of our proposed methodology we used a particular watermarking system, which is briefly presented in Sect. 6. The experimental setup and the models trained on datasets of type 1 are given in Section 7. The characteristics of the models trained on the datasets of type 2 are presented in Sect. 8. Here are also reported the results of a new evaluation of the models from Sect. 5 on the datasets of type 2. Discussions of the results, conclusions and further directions of study are given in Sect. 9.

2   Problem Formulation. The Proposed Framework

Let us consider a set of n watermarks that will be encoded in digital images. An important problem to be solved is the prediction of the class that an extracted watermark belongs to, assuming that the original image suffered different transformations before watermark extraction. This problem lies in the watermark recognition process.

The watermark recognition task is a special case of pattern recognition [15]. Without loss of generality we consider only black and white image watermarks.

After a watermark is embedded into an image and the image is then altered (e.g.: rotated, cropped, contrast changed, etc.) the embedded watermark is altered as well, different levels of noise being added. Many results in designing robust watermarking systems against different transformations [8,11,13], aiming to increase detectability and recognition of the embedded watermark were reported over time. Watermark recognition after extraction is usually left to human users.

Our study is oriented in a different direction. In this article we aim to answer two questions:

Question 1. Can we design an independent system that can be used together with any watermarking system for automatic recognition of an extracted altered watermark and if not, what are its limits?

Question 2. Can we obtain more information by comparing the performances of ML models used for watermarks identification and trained on the two types of datasets defined in this article (artificial datasets and altered datasets)?

It is proved that a No Free Lunch Theorem holds in Machine Learning [1]. This theorem states that averaged on all problems in a class, the computation cost of all ML techniques is about the same. It is necessary to use, compare and combine many ML techniques and models in order to find the most suitable one for a particular problem from a generic class. We can not find a ML technique to be the best for automatic identification of extracted, altered watermarks, for any set of watermarks, any set of host images and any watermarking system. In order to find the best ML model for a particular case of watermarks recognition and classification we propose a general frame, based on a module that can be associated with any embedding-extracting watermarking system. The aim of this module is to build containers with altered watermarks to be used in the training process of the ML models. More details about this module are given in Sect. 3.

On the other hand, to ensure the independence of our system from the watermarking system and its degree of robustness, we build a set of artificially altered watermaks, introducing different levels of noise in the original watermarks.

The proposed methlogy is presented next.

General framework for automatic identification of altered watermarks

1. Initial selection
   (a) Selecting the set of original watermarks to be embedded and the set of images to be watermarked.
   (b) Choosing the ML algorithms to be used for models generation.
2. Generating, from the initial set of watermarks, two types of modified watermarks.
   (a) Generating a set marked with type 1, namely artificial altered watermarks, by adding different levels of random noise.
   (b) Generating a set marked with type 2, more precisely modified watermark containers. This set is generated using the proposed independent module (see Sect. 3) and a real watermarking system. The module is able to build training sets using any watermarking system.
3. Main steps
   (a) Step 1. The first evaluation of the models - results for artificially generated training sets (type 1 datasets). This step involves training and evaluating the ML approaches chosen on these training sets, analyzing the results for each type of ML model and different noise levels, and comparing the results for all ML approaches considered.
   (b) Step 2. Training new models on type 2 dataset and evaluating the models performances.
   (c) Step 3. Practical evaluation. This step involves evaluation of the ML models built in the Step 1 on the datasets of type 2 and analysis and comparison of the results reported in all the three steps.

3   The Module for Building Training Datasets

The aim of this section is to propose a module that can be added to any water-marking system for generating training datasets with altered watermarks. The module should perform the following operations:

1. Choose the images where the original watermarks will be embedded in (one, more or all from a specified folder).
2. Select the watermarks to be embedded (one, many or all from a specified folder).
3. Choose the transformations to be applied to the selected images.
4. Perform *loop generation algorithm*:
   > *For all selected watermarks*
   >> Initialize the watermark's container.
   >
   > *For all selected images*
   >> *For all selected watermarks*
   >>> Input the image and the watermark in the watermarking system and command the embedding watermark process;
   >>> Take the watermarked image returned as output by the watermarking system;
   >>>
   >>> *For all selected transforms*
   >>>> Applied the transformation to the image;
   >>>> Input the image in the watermarking system and command the extraction process;
   >>>> Save the extracted watermark in the original watermark's container.
   >
   > *Return the training set: pairs {extracted watermark, label}*
5. Generation of a final report (number of images, number of watermarks, number of transformations applied for image alteration)

Watermarks' containers could be implemented in different ways (i.e. they could be folders, different kinds of data structures or different strategies for labeling the extracted watermarks).

In order to enable the comparison of the performances of the proposed module with different watermarking techniques, we considered a subset of the transformations that are usually used for evaluation of watermarking robustness [2,7,8,23,24]: image color saturation, brightness, contrast, sharpness, rotation, scaling, cropping, noise, compression. These transformations, when applied on images in which watermarks have been embedded, add noise to the watermarks after the watermarks have been extracted. The level and distribution of noise may depend on the host image, the transformation applied and the transformation parameters.

The module is independent of the internal watermark embedding mechanism. The only requirements are that the watermarking system provide two explicit methods for embedding, respectively extracting the watermark.

4 Models

As we stated in the previous section, any ML models could be adopted in our proposed framework. Based on the reported results in pattern recognition [14] we consider for testing our framework two models, artificial neural networks (ANN) and support vector machines (SVM). The aim of the next subsections is to provide details about the considered models. Both techniques taken into account are supervised learning techniques, so they require a training set of examples for learning.

4.1 ANN Model

We consider in our study a 3 layer ANN model, optimizing the log-loss function using the stochastic mini-batch gradient descent-based algorithm ADAM (Adaptive Moment Estimation) [9]. The error is summed over minibatches. It was proved experimentally that ADAM algorithm outperforms other stochastic optimizers even in the case of non-convex objective functions [9].

Let denote by g_t the gradients of the stochastic objective function f with parameters w at timestep t:

$$g_t = \Delta_w f_t(w_{t-1}) \tag{1}$$

The algorithm computes adaptive learning rates for each parameter using the updated exponential moving averages of the gradient, m_t, and square gradient, v_t. The moving averages are estimates of the first moment (the mean) and the second raw moment (the uncentered variance) of the gradient. The updating rules for m_t and v_t are:

$$m_t = \beta_1 \cdot m_{t-1} + (1 - \beta_1) \cdot g_t \tag{2}$$
$$v_t = \beta_2 \cdot v_{t-1} + (1 - \beta_2) \cdot (g_t)^2 \tag{3}$$

β_1, $\beta_2 \in [0,1)$ represent exponential decay rates for the moment estimates.

The exponentially decaying average of m_t is similar to momentum.

Bias-corrected moving averages are computed:

$$\hat{m}_t = \frac{m_t}{1 - (\beta_1)^t} \tag{4}$$

$$\hat{v}_t = \frac{v_t}{1 - (\beta_2)^t} \tag{5}$$

These corrections are necessary to compensate the bias through 0 of the moving averages, that occurs especially during the initial timesteps, and especially when the decay rates are small [9].

The update rule for the weights is:

$$w_t = w_{t-1} - \alpha \cdot \frac{\hat{m}_t}{\sqrt{\hat{v}_t} + \varepsilon} \tag{6}$$

The hyperparameter α represent the stepsize and ε represents a value used for numerical stability. The updating rule is iterated over mini-batches.

In order to avoid overfitting and increasing the prediction accuracy of our ANN we used k-fold cross validation for model selection. The log-loss function is used to compute the probability given by any perceptron from the output layer. The predicted label of an unknown instance is given by the output perceptron with the highest probability.

We chose ADAM due to the reported results regarding its behaviour on terms on training time and validation score for large enough datasets (thousands or more instances), as it is reported on Scikit learn documentation.

4.2   SVM Model

SVMs were introduced by Vapnik in [21] and were used first for solving binary classification. SVM builds a classification hyperplan, f, with maximal margin. In the case of linear classification the classifier is defined by:

$$f(X) = \langle W, X \rangle + b; \quad X \in \chi, \ b \in R \tag{7}$$

where W represents the normal to the classification hyperplan, $(-b)$ is its distance to the origin and χ is the space of instances.

This leads to the quadratic optimization problem:

$$\min_{W,b} \left(\frac{1}{2} \|W\|^2 \right) \tag{8}$$

subject to

$$y_i(\langle W, X_i \rangle + b) \geq 1, \ \forall (X_i, y_i) \in D \tag{9}$$

with X_i the input vectors (instances), $y_i \in \{-1, 1\}$ the class labels and D the training set. The nice property of the SVM classifier is that only the instances situated on the margins contribute to the solution. These vectors are named *support vectors*. To handle noise that could make data non-linearly separable, slack variables ζ_i are introduced in the problem formulation:

$$\min_{W,b} \left(\frac{1}{2} \|W\|^2 \right) + C \sum_{i=1}^{|D|} \zeta_i \tag{10}$$

subject to

$$y_i(\langle W, X_i \rangle + b) \geq 1 - \zeta_i, \ \forall \zeta_i \geq 0, \ (X_i, y_i) \in D \tag{11}$$

The slack parameters control the acceptable distance from the hyperplane for a miss classified instance. The parameter C is very important for the model. It controls the trade-off between the size of the margin and the training errors. For data that are non linearly separable, the kernel trick is used for mapping the data in a hyperplane with higher dimension where they become linearly separable, without knowing the explicit expression of the mapping function. Common kernels are:

- polynomial: $K(X,Y) = (\gamma X^T Y + c_0)^d$
- sigmoid: $K(X,Y) = tanh(\gamma X^T Y + c_0)$
- RBF: (Radial Basis Function) $K(X,Y) = exp(-\gamma \|X - Y\|^2)$

Other functions satisfying Mercer's theorem can be used as kernels. The choise of kernel and of its parameters is strongly dependent on data and is an important problem in the design of a SVM model. Different methods are used for this purpose: empirical, grid search, genetic algorithms, hybrid techniques for kernel optimization [17]. For our model we opted for a grid search between the kernels defined in (9)–(11), together with a grid search of the kernels' parameters.

The problem we face in this article requires classification into multiple classes. There are many strategies for building a multiple-class classifier. Our model for SVM multi-class classification uses the *one against one* strategy. For a classification problem with n classes there are generated n(n − 1)/2 binary SVM classifiers. Each classifier is trained on two out of all classes. The output is obtained using a voting system. Each classifier is tested against the test dataset and the winning class receives one vote. The output is the class with the most votes [10].

5   Initial Selection and Generating the Modified Watermarks

We chose a dataset of watermark patterns consisting of binary patterns representing silhouettes of different objects. For the considered classification methods, particular shapes are not relevant, since training samples show degradation due to transformations accrued over the host image. Training instances per class are obtained by adding different levels of random noise. Different noise distributions could be selected. Even at a sample size of 16x16 pixels, many classes may still exhibit distinctive class-specific features.

Binary shape datasets used for testing are collected from LEMS Vision Group, and others:

- CalTech 101 Silhouettes Data Set,
  https://people.cs.umass.edu/~marlin/data.shtm
- Binary Shape Databases (closed binary shapes collected by the Brown University),
  https://vision.lems.brown.edu/content/available-software-and-databases
- Two-dimensional articulated shapes (silhouettes) for partial similarity experiments,
  http://tosca.cs.technion.ac.il/book/resources_data.html
- PhyloPic, free silhouette images of animals, plants, and other life forms,
  http://phylopic.org
- Silhouettesfree, free silhouette images,
  http://silhouettesfree.com

We generated many datasets with different levels of maximum noise, uniformly distributed. For the experiments reported in this article we chose the datasets with maximum noise levels: 20%, 40%, 60% and 80%. In Fig. 1 are presented examples of artificially altered watermarks with different levels of random noise with uniform distribution.

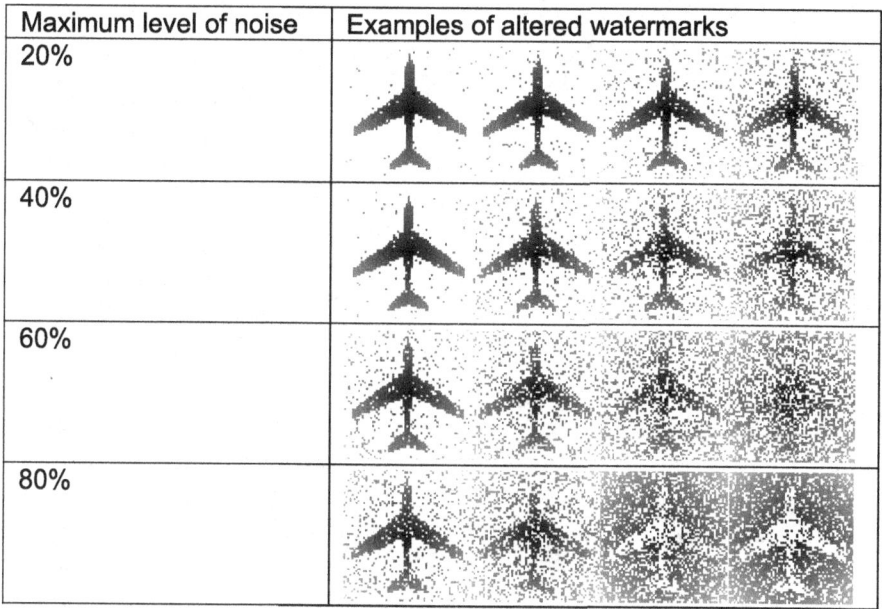

Maximum level of noise	Examples of altered watermarks
20%	
40%	
60%	
80%	

Fig. 1. Examples of altered watermarks of type 1 (artificially altered)

The characteristics of each dataset used for training and validation are:

- Initial (non-altered) images from the training set: 93 non-altered images
- Total number of the examples in the training set after adding noise: 9300. For each non-altered watermark there were generated 100 noisy variants, with noise levels varying between 0 and the maximum noise value.

For generating training sets of type 2 we have to use the module describe in Sect. 3 applied on a particular watermarking system. More details about this system are provided in Sect. 6.

The host images have been selected from the USC-SIPI Image Database. We chose 9 benchmark images from a broad range of categories regarding content, size, brightness, color palette, noise and amount of details illustrated. These images are shown in Fig. 2.

Fig. 2. Benchmark host images selected from USC-SIPI Image Database

We generated the dataset by applying the following transformations to the host images after watermarks have been embedded:

- Color saturation: 0, 0.1, 0.3, 1.7, 1.8, 1.9, 2.1, 2.2
- Brightness: 0.6, 0.7, 1.2, 1.3, 1.4
- Contrast: 0.6, 0.7, 1.2, 1.3, 1.4
- Sharpness: −0.5, −0.3, −0.1, 1.7, 1.9, 2.1, 2.2, 2.5

The values for the transformations are according to the ImageEnhance module from Python PIL library. The upper and lower limits have been empirically chosen such that watermark extraction would not give pure noise as a result. The increments between values have also been chosen empirically in order to provide a wider range of noise levels of the extracted watermarks.

In Fig. 3 are shown a few examples of altered watermarks of type 2.

Transformation	Examples of altered watermarks
Brightness	
Contrast	
Sharpness	

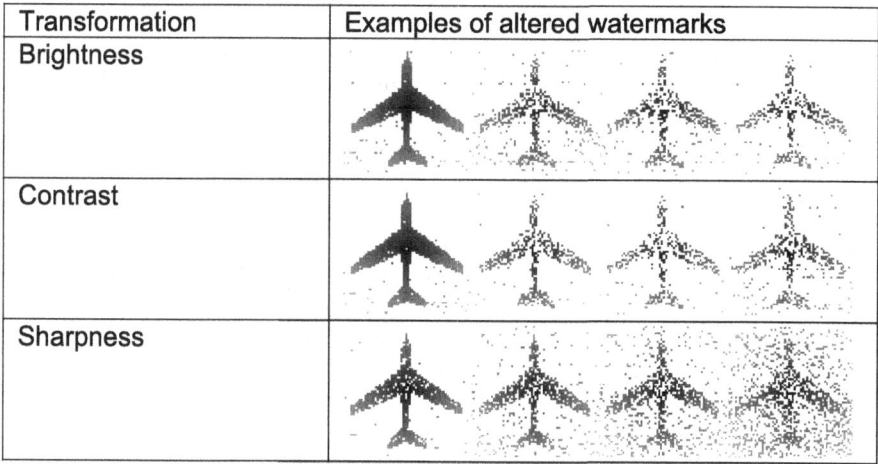

Fig. 3. Examples of altered watermarks of type 2

6   Watermarking System

The watermarking system used for validation of our approach, was built on peculiarities of several visual content-based techniques designed for minimal image quality loss. We used an improved version of the watermarking system presented in [16].

For the watermarking process we use as input a host image and a watermark pattern to be embedded. The host image can be any monochrome or colour image and the watermark pattern itself consists of a binary image. The watermarking technique works with intensity/luminance data and requires a colour space transformation performed preliminary such that only the luminance channel is used for further processing.

To address watermark invisibility and robustness, the watermarking system works in the frequency domain and as such exploits peculiarities of the human visual system. A transformation of a host image from spatial domain is done by applying a DCT (Discrete Cosine Transform).

The method performs blind watermarking by using quantisation index modulation to insert each bit of the pattern in one of the DCT coefficients, choosing carefully from low, middle or high frequency range. Lower frequency coefficient modifications affect significantly the visual quality but increase robustness, whereas modifications in the high frequency range are less perceptible, but decrease robustness since most compression image techniques rely on eliminating data from this range. After embedding, an IDCT (Inverse Discrete Cosine Transform) is applied to bring the image back into spatial domain.

Among the reasons for choosing this watermarking method, there are mainly: robustness and tuning settings. In the context of this paper, by robustness we understand the capability to resist against certain image processing operations.

As for the parameters of the embedding process there are: watermark pattern size, block size, data per block, quantisation step size, frequency offset and redundancy.

The extraction process is similar to the embedding process. Since the watermarking technique is blind, recovering does not require the original unmarked data (e.g., digital image) and uses as input only an image with embedded watermark and the embedding settings.

To measure the similarity between the extracted watermark and the original one, different objective metrics are applicable. Experimental results in [16] illustrate how correlation coefficient (CC) and bit error rate (BER) can decide if the watermark is a match by comparing the result against a threshold value.

The effectiveness of the whole embedding and extraction process depends very much on the actual visual content of the host image. Both quality and robustness vary from one image to another. To be more conclusive, [16] presents a magnified view of a BER evaluation for a set of test images. Embedding the same watermark pattern with equal settings in different host images, produce different results. Degradation during embedding is not only caused by quantization steps, it is also influenced by actual pixel data too. Since the watermarking process has to deal with a variety of visual media, a watermark classification and author identification strategy applied after extraction has to deal with a high amount of imprecision.

7   Experimental Setup

We implemented the models presented in Sects. 4 on a system with the following specifications: Intel(R) Core(TM) i7 6700K CPU @ 4.00 GHz, 64 GB RAM, 500GB SSD. For both ANN and SVM some hyperparameters were chosen using grid search and others in empirical way (see Subsects. 7.1 and 7.2). The implementation was made in Python using the machine learning library Scikit-learn. The datasets were splitted into training (80%) and testing (20%) datasets. K-fold cross validation (CV), with k = 3, was used to evaluate the accuracy in the training process. The higher the value for k, the longer the training time is.

7.1   Experimental Results with Datasets of Type 1 - ANN

Characteristics of the neural network:

- $64 \times 64 = 4096$ neurons in the input layer (equal to the number of pixels in the watermark).
- 93 neurons in the output layer (equal to the number of watermarks). Each neuron in the output layer corresponds to a specific watermark. The neural network makes the prediction based on the neuron from the output layer with the strongest activation.
- 100 neurons in the hidden layer (empirical choice).

Hyperparameters in ADAM:

- alpha = 0.0001
- beta1 = 0.9, beta2 = 0.999
- batch size = min (200, number of examples)
- learning rate = (constant)
  For chosing the learning rate value we used a grid search between 0.1, 0.01, 0.001
- epsilon = 1e−08

Activation function: we made a grid search between two activation functions:

- Logistic sigmoid function $f(x) = 1/(1 + e^{-x})$
- Rectified linear unit function $f(x) = max(0, x)$

For the evaluation of the model's prediction performance we use cross validation accuracy and prediction (classification) accuracy on the test sets.

The parameters of the best models after grid search and their performances' evaluation are given in Table 1. We denoted by ANN x the ANN model trained on a dataset with maximum noise level x%.

Table 1. Characteristics and results for best ANN models trained on datasets with different noise levels.

	ANN 20	ANN 40	ANN 60	ANN 80
Activation function	Linear	Sigmoid	Sigmoid	Linear
Learning rate	0.01	0.001	0.001	0.001
Mean train time (sec)	8.773088535	40.81464	52.75131	52.02452
Mean test time CV(sec)	0.056613684	0.068188	0.06898	0.069628
Mean cross validation accuracy	100%	99.98%	70.83%	54.38%
Classification accuracy	100%	100%	69.46%	54.38%

7.2 Experimental Results with Datasets of Type 1 – SVM

Characteristics of the SVM:

- Parameter C: the parameter that penalized the control term in SVM model is chosen by grid search method between 1,10,100,1000.
- Kernels are chosen by grid search method between RBF kernel, sigmoid kernel and linear.

For the evaluation of the model's prediction performance, the same as in the case of ANN, we use cross validation accuracy and the prediction accuracy on test sets.

Table 2. Characteristics and results for best SVM models trained on datasets with different noise levels.

	SVM 20	SVM 40	SVM 60	SVM 80
Kernel function	RBF	RBF	RBF	RBF
	$\gamma = 0.001$	$\gamma = 0.0001$	$\gamma = 0.001$	$\gamma = 0.001$
C	1	10	10	10
Mean train time (sec)	87.86414	90.04825	412.6743	209.3323
Mean test time CV (sec)	28.01891	34.50823	108.3143	51.02146
Mean cross validation accuracy	100%	99.97%	72.13	%60.95%
Classification accuracy	100%	100%	73.06%	65.80%

The parameters of the best model after grid search and its performance's evaluation are given in Table 2. We denoted by SVM x the SVM model trained on a dataset with maximum noise level x%.

As in the case of ANN model, both cross-validation and classification accuracy decrease as maximum noise levels increase. The SVM model requires significantly more time for training and classification. Up to 40% noise, the accuracy for both ANN and SVM models is similar. When noise levels exceed 40%, the SVM model performs better both in training and classification.

8   Practical Evaluation

For practical evaluation of our module and models, we generate a dataset of training examples using an improved version of the watermarking system presented in [16], train our models on these datasets and compare the results with those reported in Sect. 7.

The total number of examples in the training set is $9 \times 93 \times 26 = 21762$. In the rest of the article we will refer to this dataset as WA (Watermarking Attacks). The dataset was divided into 80% for training and 20% for testing.

We used the same ANN model characteristics and grid search parameters as in Sect. 7.1. We denote by ANN WA the best ANN model trained on WA dataset and obtained after grid search. The ANN WA characteristics are presented in Table 3.

Table 3. Characteristics and results for ANN WA model.

Characteristic	Value
Activation function	Sigmoid
Learning rate	0.001
Mean train time (sec)	63.87561
Mean test time CV (sec)	0.126163
Mean cross validation accuracy	99%
Classification accuracy	99.12%

We observe that the classification accuracy is between the accuracy obtained by the ANN trained on the datasets with 40% maximum noise level (ANN 40) and 60% maximum noise level (ANN 60), respectively. In order to formulate a conclusion, we compute the classification accuracy of all the ANN x models from Sect. 7.1 on the WA dataset. The results are reported in Table 4.

Table 4. Classification accuracy of ANN models from Sect. 7.1 on the WA dataset.

Models	ANN 20	ANN 40	ANN 60	ANN 80
Accuracy on WA dataset	51.52%	94.14%	89.33%	61.60%

We use the same SVM model characteristics and grid search parameters as in Sect. 7.2. We denote by SVM WA the best SVM model trained on WA dataset and obtained after grid search. The SVM WA characteristics are presented in Table 5.

Table 5. Characteristics and results for SVM WA model.

Characteristic	Value
Kernel function	Linear
C	1
Mean train time (sec)	1000.294362
Mean test time CV (sec)	365.7832131
Mean cross validation accuracy	96.53%
Classification accuracy	97.31%

On the WA dataset, the SVM model resulted after grid search performed slightly worse both in training and classification than the ANN model. In order to compare the ANN x and SVM x models, we compute the classification accuracy

Table 6. Classification accuracy of SVM models from Sect. 7.2 on the WA dataset.

Models	SVM 20	SVM 40	SVM 60	SVM 80
Accuracy on WA dataset	58.48%	57.59%	55.84%	52.93%

of SVM x models from Sect. 7 on the WA dataset. The results are reported in Table 6.

We observe that the generalization performance of SVM x models on the WA dataset decreases as the value of x increases.

In order to make a comparison of our ANN WA and SVM WA models we computed the precision and recall for both models and all watermarks. The results are presented in Figures 4 and 5. On the X axis we considered the watermarks ID.

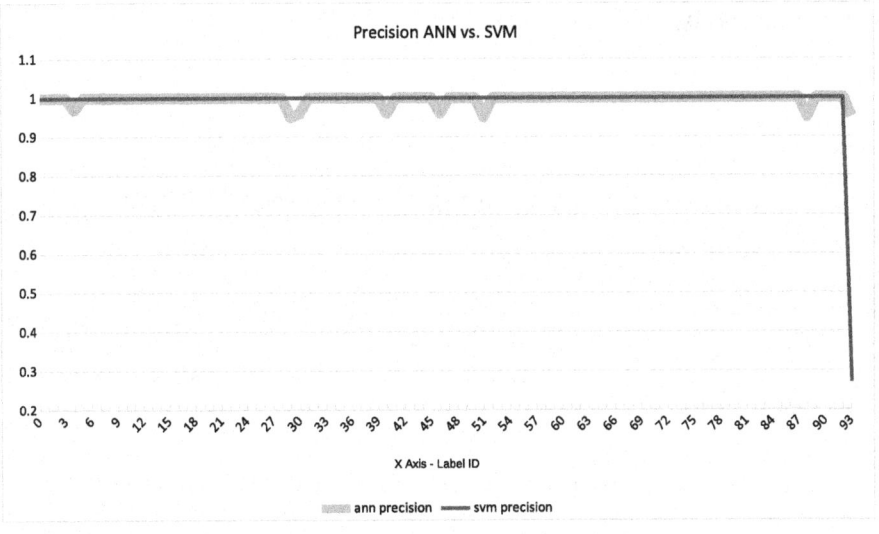

Fig. 4. Precision of ANN WA and SVM WA models

Excepting the last watermark image, the SVM WA model precision equals 1, but in terms of recall the SVM model behaves worst than the ANN one. The ANN WA model makes a complete recognition of all except two watermarks images and there are 7 watermarks images for which the identification is less relevant. Anyway, the precision for the ANN WA model is up than 0.95 for all watermarks.

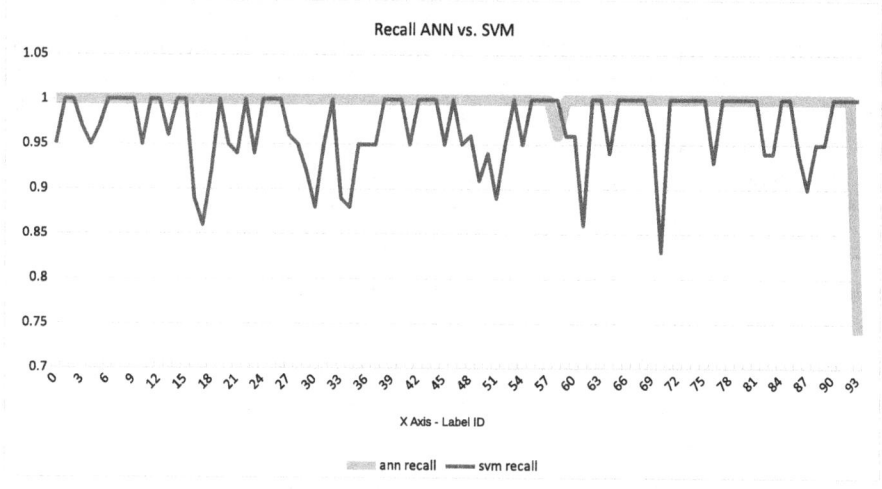

Fig. 5. Recall of ANN WA and SVM WA models

9   Conclusions and Discussions

The analysis of results from Sect. 8 lead us to the following conclusions.

The final experiment consists of 21762 data points (cases) and of 64×64 pixel images resulting in 4096 input feature dimensions. Given the high input dimension this is a relatively sparse problem.

The mean time for training/testing is significantly higher for SVMs than for ANNs. For a low level of noise x, up to 20%, both ANN x and SVM x models defined in Sect. 7 classify correctly all the instances from training and testing sets with artificially inserted noise.

The ANN WA model outperforms the SVM WA model in terms of recall. It has a maximum recall for all except two watermarks. The SVM WA model fails in complete identification of 44 watermarks with up to 17% (the worst recall is 0.83, for the watermark with ID 70).

The surprising result is that SVM is much worse than ANN on the WA dataset but not so on the preliminary exercise of Sect. 7 with 9300 examples and 93 outputs (classes). Clearly the 9 benchmark images used in the WA dataset lead to another type of (effectively even more sparse) problem, which destroys the SVMs ability to produce better separating hyperplanes. This seems to be confirmed by the fact that the winning SVMs remain on linear kernels. The ANNs being more monolithic in training seem to be also more robust against such structural training set variations as introduced into the WA dataset.

Both mean cross-validation accuracy and classification accuracy for ANN WA from Table 3 are situated between the accuracy of ANN 40 and ANN 60 models from Table 1. These results could suggest that the mean level of noise introduced by host image transformations in the WA dataset is between 40% and 60% (considered in a uniform distribution).

Analyzing the classification accuracy of the ANN x models computed on the WA dataset and reported in Table 1, we see that there is an ascending trend of accuracy between ANN 20 and ANN 40 and a descending trend between ANN 60 and ANN 80. This suggests that a maximum accuracy on WA dataset could be found for an ANN x model with x between 40% and 60%.

This could mean that an ANN model trained on a dataset with maximum level of uniform noise between 40% and 60% would give better accuracy on the WA dataset. The results obtained give support for the following conjecture.

Conjecture 1. Accuracy of ANN x models computed on a dataset of type WA has a maximum value in the neighborhood of the mean level of noise introduced by WA host image transformations.

Two important consequences of Conjecture 1 can be formulated:

Remark 1. If the WA dataset is obtained by applying a single type of transformation, and the maximum classification accuracy of an ANN x model on the WA dataset is obtained for x = x0, then the transformation applied to the images added a mean level of noise of x0%.

These results could be used for evaluating the robustness of a watermarking system against different transformations.

Remark 2. It is not possible to design a system based on ANNs completely independent of watermarking techniques for automatic identification of watermarks extracted from altered host images.

Analyzing the accuracy of ANN x models, we can observe that watermark identification accuracy is strongly dependent on the level of noise introduced by different types of transformations.

Conjecture 1 implies that by building a sequence of ANN x_i models with sufficiently low increments between x_i values, and testing these models on a WA dataset obtained using only one transformation applied to the host image, we can compute the robustness of the watermarking technique against this transformation. As future directions of study, we want to conduct more experiments to test Conjecture 1 and the proposed method for watermark robustness evaluation against existing techniques based on different metrics. We also want to use more benchmark host images and different existing watermarking systems.

Up to a certain level of noise introduced by host image transformations, ANN models trained on WA datasets generated using our proposed module from Sect. 3 can provide automatic watermark identification.

By studying the SVM models, we observe that their performances on the WA dataset are lower than the performances of ANN models, and they cannot be used to create a watermark robustness evaluator. Another drawback of SVM models is the long time required for training and classification, relative to ANN models. Therefore, in our next study we will focus on ANN models.

To the best of our knowledge, there are not reported similar studies on watermark robustness evaluation.

The article paves the way for further research in the direction of verifying and validating the proposed framework and the conjecture and conclusions formulated using others watermarking systems, host images and watermarks.

Acknowledgement. The first two authors, Dana Simian and Ralf D. Fabian were supported from the project financed from Lucian Blaga University of Sibiu & Hasso Plattner Foundation research action LBUS-RRC-2020-01.

References

1. Adam, S.P., Alexandropoulos, S.-A.N., Pardalos, P.M., Vrahatis, M.N.: No free lunch theorem: a review. In: Demetriou, I.C., Pardalos, P.M. (eds.) Approximation and Optimization. SOIA, vol. 145, pp. 57–82. Springer, Cham (2019). https://doi.org/10.1007/978-3-030-12767-1_5

2. Chang, C.Y., Wang, H.J., Su, S.J.: Copyright authentication for images with a full counter-propagation neural network. Expert Syst. Appl. **37**(12), 7639–7647 (2010)

3. Cox, I., Miller, M., Bloom, J., Fridrich, J., Kalker, T.: Digital watermarking and steganography morgan kaufmann publishers. Amsterdam/Boston (2008)

4. Deeba, F., Kun, S., Dharejo, F.A., Langah, H., Memon, H.: Digital watermarking using deep neural network. Int. J. Mach. Learn. Comput. **10**(2), (2020)

5. Fan, L., Ng, K.W., Chan, C.S.: Rethinking deep neural network ownership verification: embedding passports to defeat ambiguity attacks. In: Advances in Neural Information Processing Systems, pp. 4714–4723 (2019)

6. Fındık, O., Babaoğlu, İ., Ülker, E.: A color image watermarking scheme based on hybrid classification method: particle swarm optimization and k-nearest neighbor algorithm. Optics Commun. **283**(24), 4916–4922 (2010)

7. Jagadeesh, B., Kumar, P.R., Reddy, P.C.: Robust digital image watermarking scheme in discrete wavelet transform domain using support vector machines. Int. J. Comput. Appl. **73**(14), 1–7 (2013)

8. Khan, A., Siddiqa, A., Munib, S., Malik, S.A.: A recent survey of reversible watermarking techniques. Inf. Sci. **279**, 251–272 (2014)

9. Kingma, D., Ba, L.: Adam: a method for stochastic optimizations. In: 3rd International Conference on Learning Representations, ICLR 2015, San Diego, CA, USA, 7–9 May, 2015, Conference Track Proceedings (2015)

10. Knerr, S., Personnaz, L., Dreyfus, G.: Single-layer learning revisited: a stepwise procedure for building and training a neural network. Neurocomputing **68**, 41–50 (1990)

11. Lusson, F., Bailey, K., Leeney, M., Curran, K.: A novel approach to digital watermarking, exploiting colour spaces. Sig. Process. **93**, 1268–1294 (2013)

12. Peng, H., Wang, J., Wang, W.: Image watermarking method in multiwavelet domain based on support vector machines. J. Syst. Softw. **83**(8), 1470–1477 (2010)

13. Ramamurthy, N., Varadarajan, D.S.: Robust digital image watermarking scheme with neural network and fuzzy logic approach. Int. J. Emerg. Technol. Adv. Eng. **2**(9), 555–562 (2012)

14. Sharma, P., Kaur, M.: Classification in pattern recognition: a review. Int. J. Adv. Res. Comput. Sci. Softw. Eng. **3**, 555–562 (2013)

15. Shih, F.Y.: Image processing and pattern recognition: fundamentals and techniques. John Wiley & Sons (2010)

16. Simian, D., Fabian, R.: Ownership tracking with dynamic identification of water-mark patternss. In: Modeling and Development of Intelligent Systems. Proceeding of the International Conference MDIS 2015, pp. 113–124. Lucian Blaga University Press, Sibiu (2016)

17. Simian, D., Stoica, F.: A General Frame for Building Optimal Multiple SVM Kernels. In: Lirkov, I., Margenov, S., Waśniewski, J. (eds.) LSSC 2011. LNCS, vol. 7116, pp. 256–263. Springer, Heidelberg (2012). https://doi.org/10.1007/978-3-642-29843-1_29

18. Szyller, S., Atli, B.G., Marchal, S., Asokan, N.: Dawn: Dynamic adversarial watermarking of neural networks. arXiv preprint arXiv:1906.00830 (2019)

19. Tsai, H.H., Lai, Y.S., Lo, S.C.: A zero-watermark scheme with geometrical invariants using SVM and PSO against geometrical attacks for image protection. J. Syst. Softw. **86**(2), 335–348 (2013)

20. Tsai, H.H., Liu, C.C.: Wavelet-based image watermarking with visibility range estimation based on HVS and neural networks. Pattern Recogn. **44**(4), 751–763 (2011)

21. Vapnik, V.N.: The Nature of Statistical Learning Theory. Springer, New York (1995)

22. Wang, X.y., Wang, C.p., Yang, H.y., Niu, P.p.: A robust blind color image watermarking in quaternion fourier transform domain. J. Syst. Softw. **86**(2), 255–277 (2013)

23. Yahya, A.N., Jalab, H.A., Wahid, A., Noor, R.M.: Robust watermarking algorithm for digital images using discrete wavelet and probabilistic neural network. J. King saud Univ. Comput. Inf. Sci. **27**(4), 393–401 (2015)

24. Zhao, J., Xu, W., Zhang, S., Fan, S., Zhang, W.: A strong robust zero-watermarking scheme based on shearlets' high ability for capturing directional features. Math. Problems Eng. **2016**, (2016)

Mathematical Models for Development of Intelligent Systems

Intelligent System to Support Decision Making Using Optimization Business Models for Wind Farm Design

Daniela Borissova[1,2]($\boxtimes$) (ID), Zornitsa Dimitrova[1] (ID),
and Vasil Dimitrov[1] (ID)

[1] Institute of Information and Communication Technologies at the Bulgarian
Academy of Sciences, 1113, Sofia, Bulgaria
dborissova@iit.bas.bg, zrn.dimitrova@gmail.com,
vasil.zdimitrov@gmail.com
[2] University of Library Studies and Information Technologies,
1784, Sofia, Bulgaria

Abstract. In the digital age, all successful business processes depend on well-motivated and effective decision-making. To be more precise, these effective decisions are to be based on properly formulated mathematical models. To achieve such decisions a framework of an intelligent system to support decision-making is proposed. The proposed framework relies on the effective integration of multi-attribute group decision making models (MAGDM) and mathematical optimization models (single or multi-objective). While the MAGDM models contribute to the determination of the most preferred alternative by aggregation different points of view of the group experts, the goal of using the optimization models aims to determine the effectiveness of the selected alternative. The described framework is applied for the selection of wind turbine types for designing a wind farm. The efficiency of the designed wind farm is evaluated by using an optimization model. It is shown that in some cases the selected preferred turbine type leads to less wind farm performance taking into account the particular farm area and wind conditions.

Keywords: Decision-making · MAGDM · Business optimization models · Renewable energy · Wind turbine selection

1 Introduction

Properly processed input information is the prerequisite to gain the maximum advantage in effective business decision-making [1]. Such kind of tools that benefit business decisions is the so-called decision support systems that utilize data and models to solve different problems. In the era of digital transformation, there is no area that is not influenced by management based on Information Technologies (IT) [2]. Therefore, different approaches are proposed to design custom software to cope with specific needs [3–6]. When modelling complex processes, the advantages of multi-criteria decision analysis (MCDA) and the advantages of multi-objective decision making (MODM) could be integrated. The methods of MODM consider decision variables as

© Springer Nature Switzerland AG 2021
D. Simian and L. F. Stoica (Eds.): MDIS 2020, CCIS 1341, pp. 287–301, 2021.
https://doi.org/10.1007/978-3-030-68527-0_18

determined in a continuous or integer domain with an infinite or large number of alternatives, while the multi-attribute decision making (MADM) methods consider discrete decision space where a set of decision alternatives are predetermined [7, 8].

Today, there is no area in which different decision-making models are not used to achieve better results. This is also true for the renewable area where huge amounts are invested. The success of using wind energy depends on careful selection of the appropriate wind turbine type while considering multiple parameters as annual output energy, capacity factor, rotor diameter, nominal wind speed, cut-out wind speed, cut-in-wind speed, hub height, etc. A close to reality approach is using group decision making for selection model is based on collecting and aggregating estimates from different decision-makers (DMs).

Different models involving MADM and MODM techniques have been proposed for selection of wind turbines types. MADM models based on analytic hierarchy process (AHP) are used to select the best turbine type among many alternatives. The wind turbines are evaluated through their effective properties considering different wind turbine brands under same rated output [9]. The group of wind energy experts expresses an opinion in determining parameter significance toward wind turbine assessment and the rated parameters for the efficiency of the turbines. The most important factors that are identified include: the capacity factor, wind power curves, tower height and the higher values of these factors increase the attractiveness of the turbine. MODM technique is applied in a multi-objective evolutionary algorithm [10]. The authors show the possibility of multi-objective evolutionary algorithms usage in context of wind energy production, by selecting a combination of two different wind turbine types along with given wind speeds distributed over different time spans of the day. Determining suitable region for wind farm building can be done by using optimal score-radar map of wind speed probability distribution [11]. With the constant development of wind energy technologies, non-expert stakeholders are faced with the challenge of choosing among the different wind turbines types. A method based on the statistical analysis of nominal specifications for estimating the cost of energy is proposed for the selection of wind turbines [12]. An important consideration is the cost of energy used as a comprehensive indicator for the estimation of wind farm profitability. The classical MADM methods like simple additive weighting (SAW), weighted product model (WPM), simple multi-attribute rating techniques (SMART), AHP, the technique of order preference similarity to the ideal solution (TOPSIS), preference ranking organization method for enrichment evaluation (PROMETHEE), etc., are focused on determining attribute weights and consider all DMs' opinions with equal importance [13, 14]. Some MADM methods, especially in the renewable energy sector, rely on experts from different areas. In practice, not all of the DMs have an equal level of knowledge and experience and respectively equal relations to the problems. This means that their opinions must be assessed differently. These different assessments of DMs' opinions can be aggregated by using different models. For example, a fuzzy multi-attribute group decision-making model allows evaluating every group member, including himself/herself by assigning a score according to the importance of the DMs' expertise [15]. Another approach is realized as a two-stage algorithm with multi-objective optimization and group decision making based on MADM techniques. On the first stage, a set of alternatives to wind farm layouts is determined by using multi-objective methods. The second stage uses group decision-making for evaluation and selection of the most suitable alternative layout design [16].

The article describes an intelligent system to support decision making via a two-stage process involving different optimization business models. In the first stage, modelling business processes aims to make a selection by group decision-making and includes experts with different domain expertise and background. The second stage of the proposed intelligent system is intended to evaluate the determined alternative as a result of group decision at the first stage with respect to its effectiveness.

The rest of the article is organized as follow: Sect. 2 introduces the proposed framework of an intelligent system to support decision-making by involving multi-attribute group decision making (MAGDM) and mathematical modelling of the effectiveness of the selected alternative; Sect. 3 demonstrates the numerical application in the selection of wind turbine type and determination of the effectiveness of the designed wind farm using this alternative. In Sect. 4 an analysis and discussion of obtained results are given, while the conclusions are drawn in Sect. 5.

2   Framework of Intelligent System to Support Decision Making

Different techniques and methods could contribute to the effective designing of decision making systems. The proposed framework relies on the effective integration of MADM models and mathematical modelling based on single or multi-objective optimization.

2.1   Architecture of an Intelligent Framework System to Support Decision Making by Integration of MADM and MODM

The business intelligence system for decision-making could contain different modules and sub-modules depending on the particular subject and processing data [17]. The proposed decision support system with basic modules that contribute to the effective integration of MADM/MAGDM and MODM is illustrated in Fig. 1.

The user interface (UI) should be designed in such a way to provide easy access to all resources required the active participation of DMs. It should provide the required information for entering the needed data about the parameters relevant to the considered domain area. All of the entered data are to be stored in a predefined database. The UI along with the created database with the input information about the particular problem are two essential parts of the preparation stage of the proposed intelligent system.

The next essential stage is the selection stage that is composed of module #3 and module #4. Module #3 is intended to keep different databases, related to the MADM models and MAGDM models. These MADM/MAGDM models could be used to provide the ranked lists of alternatives considering single or group decision-making. There are different and well-known models for MADM/MAGDM that could be utilized to rank a predefined set of alternatives in accordance with the DM preferences' or group of experts.

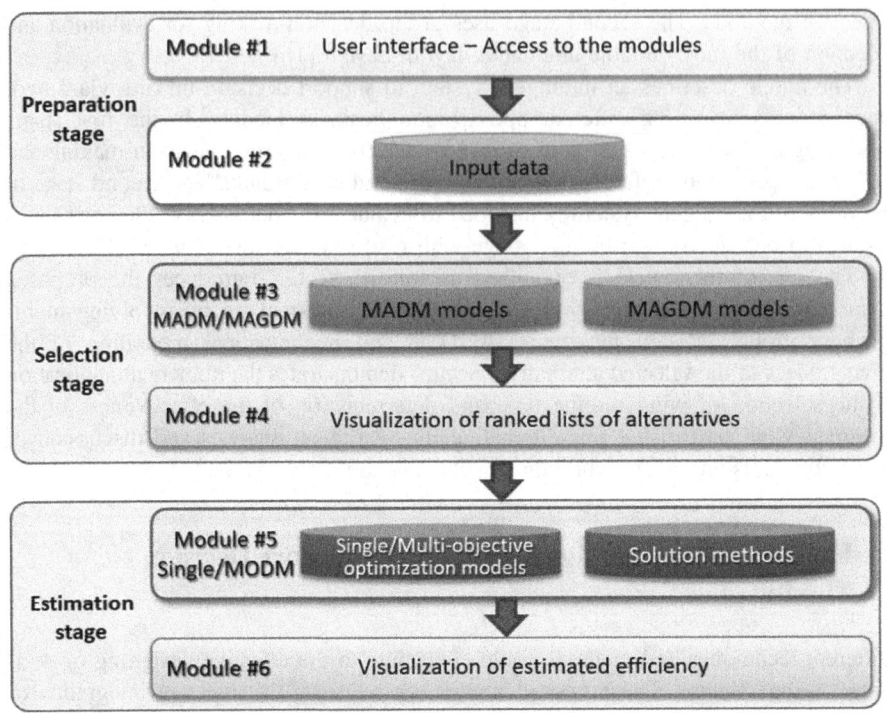

Fig. 1. A framework of an intelligent system to support decision making by effective integration of MADM/MAGDM and MODM

The usage of module #4 will provide a proper interface to visualize the ranked list of alternatives. Once the ranked list is known, the best alternative in the ranked list should be estimate toward determined criteria for effectiveness.

The third stage of the proposed intelligent system to support decision making is the estimation stage and involves two additional modules (module #5 and module #6). Module #5 is composed of additional two databases. The first database contains single and multi-objective optimization models, while the second is composed of different suitable methods for their solving. The database with single and multi-objective optimization models relies on the formulation of different optimization criteria expressed as objective function/s under given restrictions. The second database with different solution methods is intended to maintain a variety of exact or heuristic approaches to solve the transformed multi-objective problem into a single objective one. Using one of these solution methods will provide a Pareto-optimal solution of the investigated problem. The advantage of this approach is the fact that the obtained solution is surely optimal or Pareto-optimal for the formulated objective function and used restrictions.

In such a way, it is possible to estimate the ranked alternative from the selection stage using module #3 with respect to its effectiveness expressed via formulated single or multi-objective problems via module #5. Module #6 is intended to provide a proper interface to visualize the effectiveness of the alternative.

2.2   MADM Models for Selection via Group Decision Making

One of the essential stages of the proposed framework in Fig. 1 is the formulation of a mathematical model for the selection of turbine type via group decision making considering differences in DMs opinions importance. For the goal, the classical SAW model is modified by introducing of binary integer variables corresponding to the number of given alternative and using additional weighted coefficients to reflect the background experience and knowledge of the experts within-group (DMs). The resulting formulation leads to the determination of aggregated group decision by modified simple additive weighting model (MSAW) as follows:

$$max \sum_{i=1}^{M=1} \sum_{k=1}^{K} x_i \lambda^k A_i^k \tag{1}$$

subject to

$$\forall i = 1, 2, \ldots, M \; : \; \left(\forall k = 1, 2, \ldots, K : A_i^k = \sum_{j=1}^{N} w_j^k e_{i,j}^k \right) \tag{2}$$

$$\sum_{j=1}^{N} w_j^k = 1, \forall k = 1, 2, \ldots, K \tag{3}$$

$$\sum_{i=1}^{M} x_i = 1, x_i \in \{0, 1\} \tag{4}$$

$$\sum_{K=1}^{K} \lambda^k = 1, \lambda^k \in (0, 1) \tag{5}$$

The aggregate assessment of the i-th alternative against evaluation criteria, according to the k-th expert (DM) in the group is expressed by A_i^k. The weighted coefficient w_j^k represents the relative importance among criteria from DMs point of view. Each DM determines corresponding evaluation scores $e_{i,j}^k$ to express his opinion to the performance of i-th alternative toward j-th criterion. The binary integer variables x_i is used to perform a selection of the single alternative (turbines' type). To distinguish the importance of alternatives assessment between DMs, another type of coefficients λ^k is involved to indicate the differences in the expertise of each group's members. The values of these coefficients make it possible to differentiate the point of view of the experts within the group. The sum of corresponding coefficients λ^k should be equal to 1. It should be noted that acceptable values of coefficients λ^k are from the open interval (0,1). This restriction guarantees that all DMs opinions are taken into account. The relation (1) expresses the assessment of each alternative as a linear combination of the relative weight of criteria importance multiplied by corresponding evaluation scores for each criterion.

The assessment of alternatives via classical WPM uses multiplication of evaluation scores for each criterion on the degree of the relative weight of criteria importance. The proposed modified model for the selection of turbines type via group decision making based on WPM uses binary integer variables assigned to each alternative and a weighted coefficient for background experience and knowledge of each DM. The resulting formulation leads to the determination of aggregated group decision by modified weighted product model (MWPM):

$$max \sum_{i=1}^{M} \sum_{k=1}^{K} x_i \lambda^k R(A_i)^k \qquad (6)$$

subject to (3)–(5) and additional restriction

$$\forall i = 1, 2, \ldots, M : \left(\forall k = 1, 2, \ldots, K : R(A_i)^k = \prod_{j=1}^{N} \left(e_{ij}^k \right)^{w_j^k} \right) \qquad (7)$$

The overall assessment of the i-th alternative against a set of the evaluation criteria, according to the k-th DM is denoted by $R(A_i)^k$. The rest of the notations are the same as described in (1)–(5) restrictions of GDM-SAW model.

2.3   Optimization Model for Determining the Efficiency of the Selected Wind Turbine Type for Building Wind Farm

An essential stage of the described framework is the definition of some metric to determine the efficiency of the designed wind farm. For this goal, a model that represents the relation of parameters of chosen turbine type to parameters of wind farm site is proposed. One of the most used formulations for wind farm efficiency is the ratio of costs of wind farm building toward annual energy production. The determination of costs is based on the function of the total number of turbines that defines costs per year as a non-dimension number [18]. Using of rated turbine annual output is insufficiently justified for the determination of wind farm annual energy production (AEP). The determination of this parameter is complex and depends on both turbines parameters and wind conditions. The AEP could be estimated by considering the number of turbines and their rated power and corresponding utilization coefficient η that expresses the utilization of wind power over the year [19].

When building some wind farm is important to take into consideration the wake effect. This effect is manifested in the creation wake behind the turbine, i.e. a long trail of wind which is quite turbulent and slowed down. The result is decreasing of the effectiveness of the following turbine. A new concept for layout optimization through selective deactivation of wind turbines for fixed foundation wind farms' layout through purposefully deactivating selected wind turbines to improve wind farms' total power output is proposed [20]. To cope with the negative wake effect when building a wind farm, the turbines are to be located at a certain minimum distance from each other. In reality, it is impossible to determine exactly the needed distances between turbines to overcome the wake effect due to the stochastic nature of wind parameters. The values of separation distance SD between turbines (typically as values expressing the number of rotor diameters) can be taken from some practically determined interval [21–23]. Using these separation distances it is possible to determine the number of turbines for particular wind site dimensions and wind conditions. Taking into account all of these considerations, the efficiency of a particular turbine for the building of wind farm in accordance with the last stage of the proposed methodology is determined by the following single objective optimization model:

$$min\left(\frac{Costs}{Energy}\right) = \frac{N\left(\frac{2}{3} + \frac{1}{3}e^{-0.00174N^2}\right)}{NP_{wt}h_y\eta} \tag{8}$$

subject to

$$SD_x = k_xD_{wt} \tag{9}$$

$$SD_y = k_yD_{wt} \tag{10}$$

$$N = N_xN_y \tag{11}$$

$$N_x = \frac{L_x}{SD_x} + 1 \tag{12}$$

$$N_y = \frac{L_y}{SD_y} + 1 \tag{13}$$

$$k_x^{min} \le k_x \le k_x^{max} \tag{14}$$

$$k_y^{min} \le k_y \le k_y^{max} \tag{15}$$

The objective function (8) expresses dimensionless costs per unit of energy over a year as a function of integer variable N expresses the number of all wind turbines to be installed in the wind farm. The integer variables N_x and N_y are the number of turbines in columns and rows within the farm area. The rated power P_{wt} and rotor diameter D_{wt} of the chosen turbine type are considered as constant. The dimensions of the wind farm area are denoted by L_x and L_y, and SD_x, SD_y are separation distances between turbines to overcome the wake effect. The separation distances are calculated by involving separation coefficients k_x and k_y expressing the number of rotor diameters on y and x axes for the rectangular shape of the wind site. Because of the stochastic nature of wind parameters these coefficients could be taken within some boundaries for which practice shows that the wake effects can be negligible [21–23]. Furthermore, the investigations show that differences in turbine separation distances (e.g. 10.5 or 7 rotor diameters) do not affect significantly wake-related power losses [24]. The coefficient η expresses the nominal wind power utilization and is statistically defined [19]. The overall hours over the year are denoted by h_y and are equal to 8760 h.

3 Numerical Application

The numerical testing of the described intelligent system to support decision making using different optimization business models is applied for a case of the selection and evaluation of wind turbines for building a wind farm.

3.1   Input Data

For numerical calculations, six types of turbines with rated power equal to 2 MW are used and their corresponding parameters are shown in Table 1 [9].

Table 1. Wind turbines' parameters with rated power of 2.0 MW

Turbines	Annual output (8 m/s) in MWh	Capacity factor	Rotor diameter (m)	Hub height (m)	Cut-out wind speed (m/s)	Nominal wind speed (m/s)
T-1	9269	32.4	114	80/93/125/	25	12.0
T-2	9327	32.9	116	80/94	25	10.0
T-3	8498	30.2	103	80/90	22	10.0
T-4	9074	32.0	110	80/95	20	11.5–12
T-5	8825	31.1	98.0	60/80/98	25	15.0
T-6	8676	30.7	92.5	64/80/100	24	12.5

The turbines parameters can be found in the datasheet of any particular wind turbine type specifications. The most important parameter is annual output energy and it is a function of many parameters including air density, wind velocity, rotor swept area, etc. [25]. Usually, the turbines AEP is calculated based on the steady-state turbine output power and Weibull distribution of wind speed [25,26].

The capacity factor of a particular turbine represents the actual annual energy output divided by the theoretical maximum output if the turbine is running at its maximum during certain hours of the year. If this relation is bigger, more energy output will be produced. The parameter of hub height is related to the temperature, air density, and air pressure. The wind speed is increasing with heights above ground level. This is due to the decrease of the temperature and the pressure of the air when the altitude is increased. The usage of turbines with higher hub increases the extracted wind power and respectively turbines energy output. Therefore, this is better if its value is greater. Cut-out wind speed is the highest wind speed value at which wind turbines should stop in order to prevent overloading and stresses of the blades [27] to avoid damaging the turbine. If turbines are able to work at a bigger range of wind velocities, more energy output will be produced. Output energy depends on the rotor swept area i.e. on rotor diameter. The bigger rotor diameter will generate more energy. The nominal wind speed represents the lowest speed at which the nominal (rated) power output of the turbine is reached.

Following all of the above considerations, the DMs from the group should evaluate turbines types by their parameters and should determine corresponding weights for each of these parameters. According to the proposed methodology, the most preferred wind turbine type will be determined as a result of the aggregating of all DMs' opinions by using the proposed GDM-SAW and GDM-WPM models.

3.2 Selection of the Best Wind Turbines Type via Proposed MAGMD Models

The group of experts is formed by four DMs with different expertise and background who should make a decision for selection of the best from 6 wind turbines types (Table 1). Each DM evaluates a given set of turbines toward their parameters (performance criteria). The alternatives of turbines types are evaluated toward these criteria by direct scoring using values in the range between 0 and 10. Value 0 means that the criterion in question (alternative) does not entail any benefit; while the value of 10 means that the alternative best meets the efficiency requirements for a wind farm. The sum of the weights of all turbines parameters should be equal to 1. The summarized results of DMs evaluation scores of different turbines type alternatives and weights of the criteria of each expert from the group are shown in Table 2.

Table 2. The summarized evaluations from each DM

#	Evaluation parameters (criteria)						DMs
	C_1	C_2	C_3	C_4	C_5	C_6	
	$w_1^1 = 0.1$	$w_2^1 = 0.21$	$w_3^1 = 0.25$	$w_4^1 = 0.12$	$w_5^1 = 0.22$	$w_6^1 = 0.10$	DM-1
T-1	9	9.8	6	10	10	8.4	
T-2	9.7	9.8	5	7	9.9	9.6	
T-3	4.5	5.8	8.3	5.3	8	10	
T-4	7.8	8.5	8.9	8	8.2	8.6	
T-5	6.6	7.3	9	7.3	10	6	
T-6	5.9	6	10	9	9.4	7.3	
	$w_1^2 = 0.16$	$w_2^2 = 0.18$	$w_3^2 = 0.22$	$w_4^2 = 0.12$	$w_5^2 = 0.1$	$w_6^2 = 0.22$	DM-2
T-1	8.4	7	8.4	9	6.3	8.1	
T-2	9.5	7.8	9.7	9.8	7.4	9	
T-3	5.8	5.3	9.6	5.6	8	8.2	
T-4	8.4	8.3	9.2	8.4	7.3	7.5	
T-5	8.2	7.8	6.5	7	9	9.1	
T-6	6.3	9.1	8.4	7.6	10	8.9	
	$w_1^3 = 0.25$	$w_2^3 = 0.1$	$w_3^3 = 0.1$	$w_4^3 = 0.1$	$w_5^3 = 0.23$	$w_6^3 = 0.22$	DM-3
T-1	9.3	9.9	8	6.4	9	9.3	
T-2	9.4	10	9.5	5	10	9.8	
T-3	5.4	8	9.8	8.3	5	5.8	
T-4	7.7	8.2	9.1	7.4	8.7	8	
T-5	8.8	10	7.4	9.1	8.8	7	
T-6	7.1	9	9.3	9.8	7.9	6.7	
	$w_1^4 = 0.15$	$w_2^4 = 0.20$	$w_3^4 = 0.20$	$w_4^4 = 0.18$	$w_5^4 = 0.15$	$w_6^4 = 0.12$	DM-4
T-1	8.8	8.9	9.5	6.7	9.2	9.5	
T-2	9.7	9.1	7.9	5	9.3	9.2	
T-3	8.9	8.1	5.4	8.2	5.8	6.2	
T-4	9.2	9.7	8.9	7.9	8.4	8.6	
T-5	7.4	9.2	9.7	9	8.6	7	
T-6	7.2	9	10	9.8	6.7	7.9	

Using the results from Table 2 and MSAW model (1)–(5), and MWPM with objective function (6) subject to (3)–(5) and (7), the best wind turbines types considering 3 different sets of weighting coefficients reflecting the members' expertise, are obtained and shown in Table 3.

Table 3. Selected turbines types by using of MSAW and MWPM models

Cases	Weighted coefficients for DMs expertise				Optimal selection of turbine	
	DM-1	DM-2	DM-3	DM-4	MSAW	MWPM
Case-1	0.25	0.25	0.25	0.25	T-2	T-2
Case-2	0.25	0.10	0.35	0.30	T-1	T-1
Case-3	0.20	0.30	0.15	0.35	T-2	T-4

The results of using both modified MSAW and MWPM models show that the T-1 and T-2 turbines are good enough alternatives for the building of wind farms. For example, in Case-1 where experts are considered to have equal background and knowledge (i.e. equal weighted coefficients) the objective function values for both models (MSAW and MWPM) determines T-2 as the most preferred turbine type. In Case-2, where more importance of evaluation scores for turbines performance is given on DM-3, followed by DM-4, and DM-1, and less importance of DM-2 evaluation scores, the preferable turbine type is T-1 also for both models (MSAW and MWPM). Case-3 expresses the importance of evaluation scores of the experts in the following order DM-4, DM-2, DM-1, and DM-3. For this case, modified MSAW determines turbine T-2 as most preferred while modified GDM-WPM defines turbine T-4 as the best one.

3.3 Determination of the Effectiveness of the Selected Turbines for Building of Wind Farm via the Proposed Optimization Model

The ultimate goal of using the optimal choices of turbines types is to build an efficient wind farm. To investigate the wind farm efficiency, a wind farm with rectangular shape and dimensions 4 km × 2 km is considered. The efficiency of the wind farm is defined by objective function (8) that expresses the costs per energy ratio over a year for 3 possible wind directions: 1) predominant wind direction perpendicular to the long side of the wind farm (predominant-1); 2) wind direction predominant perpendicularly to the short side of the wind farm (predominant-2) and 3) uniform wind directions. The value for the wind power utilization coefficient η is considered to be equal to 0.32. The separation coefficients k_x and k_y (used for determination of turbines separations) are chosen as number of rotor diameters limited as: 1) $1.5 \leq k_x \leq 3$ and $8 \leq k_y \leq 11$ for predominant-1 wind direction; 2) $8 \leq k_x \leq 11$ and $1.5 \leq k_y \leq 3$ for predominant-2 wind direction; and 3) $4.5 \leq k_x = k_y \leq 5.8$ for uniform wind directions. The wind farm efficiency is estimated by using the optimization model (8)–(15) for the determined wind turbines types in accordance with Table 3 and Table 4. The obtained results for the efficiency of the wind farm and for the number of turbines for its building under different wind directions are shown in Table 4.

Table 4. Wind farm efficiency for different wind directions

Turbines type	Predominant-1 wind direction		Predominant-2 wind direction		Uniform wind directions	
	Wind farm efficiency	Turbines number	Wind farm efficiency	Turbines number	Wind farm efficiency	Turbines number
T-1	0.1206654E-03	45	0.1192838E-03	54	0.1341084E-03	28
T-2	0.1206654E-03	45	0.1192838E-03	54	0.1341084E-03	28
T-4	0.1196791E-03	50	0.1191202E-03	57	0.1206654E-03	45

The optimization model (8)–(15) along with estimation for the efficiency of wind farm determines also the number of turbines and their separation distances needed to define the position of each turbine within the designed wind farm. Using of turbine T-1 or T-2 determines the overall turbines number as 45 for the predominant wind direction perpendicular to the longest farm site, 54 turbines for opposite wind direction, and 28 in case of uniform wind directions. The number of turbines for wind farm building using turbine T-4 for predominant-1 wind direction is 50 turbines, for predominant-2 wind direction the farm could contain 57 turbines, and for uniform wind directions, the number of turbines is equal to 45.

4 Results Analysis and Discussion

The determined best turbine types by using of MSAW are illustrated in Fig. 2a and corresponding selections when using of MWPM are shown in Fig. 2b.

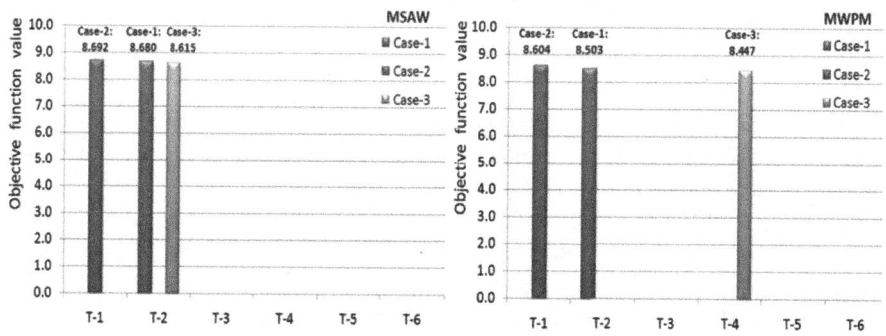

Fig. 2. a) The best turbines selection under different DMs weights and MSAW; b) The best turbines selection under different DMs weights and MWPM.

In Case-1 where all DMs have equal weighted coefficients show that turbine T-2 as the best selection for using both models MSAW and MWPM (Fig. 2) Based on the same input data from Table 1, other authors show that the optimal technical parameters are observed at wind turbine T-1 using of AHP techniques [9]. In contrast to the AHP, where the problem is decomposed hierarchically of more comprehended sub-problems which are analysed independently, the proposed approach integrates all evaluations, coefficients, and DMs competency in a single utility function. The most preferred alternative is obtained by a single run of optimizations task, based on using binary integer variables for selection.

In Case-2 both models determine as the best selection turbine T-1, while in Case-3 MSAW model determine T-2 as the most preferable one unlike to MWPM model where T-4 is determined as the most preferable turbine. This difference is due to the used different utility function and aggregation of evaluations weighs for criteria and weighed coefficients for DMs. Regardless of the model used to determine the best turbine, the final stage of the proposed framework of an intelligent system to support decision making requires evaluating the effectiveness of a selected turbine for wind farm building is to be done. The results for determining wind farm efficiency based on the optimization model (8)–(15) are illustrated in Fig. 3.

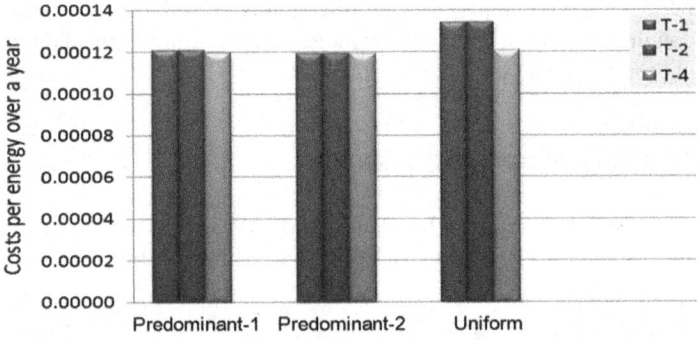

Fig. 3. Wind farm efficiency as function of selected turbine type

The selection of the most appropriate wind turbines is an important step in the building of wind farms, but the evaluation of the effectiveness of the designed wind farm should not be underestimated. Using the parameters of turbines T-1 and T-2 in the proposed model (12)–(21), obtained effectiveness for wind farm building are less than effective when parameters of turbine T-4 are used (Fig. 3). The number of turbines for the building of wind farms depends on available resources like wind speeds, wind directions, and determined site dimensions for building the farm. In this regard, the contemporary trend is to make turbines with a higher power while preserving the rotor diameter or make it smaller. For example, the 2 MW platform of Vestas Company offers wind turbines with rotor diameters of 90 m, 100 m, and 110 m. Using smaller rotor diameter allows building more compact wind farms and respectively more efficient.

As it is shown in Fig. 3, not always the best choice of the turbine (alternative) determines the best efficiency of the project for a wind farm building. Therefore, it is important to estimate somehow the expected efficiency before starting the wind farm building.

5 Conclusion

The proposed intelligent framework integrates techniques of MADM/MAGDM and MODM to evaluate the selected alternative with respect to its efficiency. Two new utility functions suitable for group decision making are proposed based on classical SAW and WPM models. The essence of the proposed MSAW and MWPM models is the fact that they are capable to aggregate not only evaluations and weights for evaluation criteria given by each DM but also incorporate the importance of the opinion of each group member into a final group decision. It is shown that using different weighted coefficients for DMs opinions importance influence on a group decision. Therefore, it is important carefully to identify the group members and properly determine their coefficients' expressing the importance of a formed group decision. The described approach could integrate other models based on the formulation of different utility functions. The second part of the described intelligent framework emphasizes the formulation of single- or multi-objective optimization models. The idea of this part is to estimate the effectiveness of the obtained selected alternative from the previous stage. These models should correspond to the goal of the company.

The main contribution of the proposed intelligent framework is the integration of proposed models (MSAW and MWPM) and verification by the use of an optimization model to evaluate the alternative' effectiveness. Future developments are related to the development of mathematical reasoned models for the determination of DMs' weights more precisely instead of using predetermined ones.

References

1. Borissova, D., Korsemov, D., Mustakerov, I.: Multi-criteria decision making problem for doing business: comparison between approaches of individual and group decision making. In: Saeed, K., Chaki, R., Janev, V. (eds.) CISIM 2019. LNCS, vol. 11703, pp. 385–396. Springer, Cham (2019). https://doi.org/10.1007/978-3-030-28957-7_32
2. Shalamanov, V.: Institution building for IT governance and management. Inf. Secur. Int. J. **38**, 13–34 (2017)
3. Mustakerov, I., Borissova, D.: Data structures and algorithms of intelligent web-based system for modular design. Int. J. Comput. Sci. Eng. **7**(7), 87–92 (2013)
4. Balabanov, T.: Distributed evolutional model for music composition by human-computer interaction. In: Proceedings of International Scientific Conference UniTech 2015, pp. 389–392, Gabrovo, Bulgaria (2015)
5. Stoyanova, K., Guliashki, V.: MOEAs for Portfolio Optimization Applications. LAP Lambert Academic Publishing, Riga (2018)

6. Borissova, D., Mustakerov, I.: A concept of intelligent e-maintenance decision making system. In: Innovations in Intelligent Systems and Applications (INISTA), Albena, Bulgaria, pp. 1–6. IEEE (2013). https://doi.org/10.1109/INISTA.2013.6577668 .

7. Climaco, J.: Multicriteria Analysis. Springer, New York (1997).https://doi.org/10.1007/978-3-642-60667-0

8. Kumar, A., Sah, B., Singh, A.R., Deng, Y., He, X., Kumar, P., Bansal, R.C.: A review of multi criteria decision making (MCDM) towards sustainable renewable energy development. Renew. Sustain. Energy Rev. **69**, 596–609 (2017)

9. Sagbansua, L., Balo, F.: Decision making model development in increasing wind farm energy efficiency. Renew. Energy **109**, 354–362 (2017)

10. Montoya, F.G., Manzano-Agugliaro, F., Lopez-Marquez, S., Hernandez-Escobedo, Q., Gil, C.: Wind turbine selection for wind farm layout using multi-objective evolutionary algorithms. Expert Syst. Appl. **41**, 6585–6595 (2014)

11. Miao, S., Gu, Y., Li, D., Li, H.: Determining suitable region wind speed probability distribution using optimal score-radar map. Energy Convers. Manage. **183**, 590–603 (2019)

12. Arias-Rosales, A., Osorio-Gomez, G.: Wind turbine selection method based on the statistical analysis of nominal specifications for estimating the cost of energy. Appl. Energy **228**, 980–998 (2018)

13. San Cristobal, J.R.: Multi-Criteria Analysis in the Renewable Energy Industry, Springer-Verlag, London (2012). https://doi.org/10.1007/978-1-4471-2346-0

14. Abu Taha, R., Daim, T.: Multi-criteria applications in renewable energy analysis, a literature review. In: Daim, T., Oliver, T., Kim, J. (eds.) Research and Technology Management in the Electricity Industry. Green Energy and Technology, pp. 17–30. Springer, London (2013). https://doi.org/10.1007/978-1-4471-5097-8_2

15. Elzarka, H.M., Yan, H., Chakraborty, D.: A vague set fuzzy multi-attribute group decision-making model for selecting onsite renewable energy technologies for institutional owners of constructed facilities. Sustain. Cities Soc. **35**, 430–439 (2017)

16. Borissova, D., Mustakerov, I.: A two-stage placement algorithm with multi-objective optimization and group decision making. Cybern. Inf. Technol. **17**(1), 87–103 (2017)

17. Borissova, D., Mustakerov, I.: An integrated framework of designing a decision support system for engineering predictive maintenance. Int. J. Inf. Technol. Knowl. **6**(4), 366–376 (2012)

18. Grady, S.A., Hussaini, M.Y., Abdullah, M.M.: Placement of wind turbines using genetic algorithms. Renew. Energy **30**, 259–270 (2005)

19. Mustakerov, I., Borissova, D.: Wind turbines type and number choice using combinatorial optimization. Renew. Energy **35**, 1887–1894 (2010)

20. Haces-Fernandez, F., Li, H., Ramirez, D.: Improving wind farm power output through deactivating selected wind turbines. Energy Convers. Manage. **187**, 404–422 (2019)

21. Pena, A., Rethore, P-E., Paul van der Laan, M.: On the application of the Jensen wake model using a turbulence-dependent wake decay coefficient: the Sexbierum case. Wind Energy **19**, 763–776 (2016)

22. Smith, G., Schlez, W., Liddell, A., Neubert, A., Pena, A.: Advanced wake model for very closely spaced turbines. In: Proceedings of EWEC 2006, Athens (2006)

23. Donovan, S.: Wind farm optimization. In: Proceedings of Annual Conference of the Operations Research Society, Wellington, New Zealand (2005)

24. Barthelmie, R.J., Pryor, S.C., Frandsen, S.T., Hansen, K.S., Schepers, J.G., Rados, K., Schlez, W., Neubert, A., Jensen, L.E., Neckelmann, S.: Quantifying the impact of wind turbine wakes on power output at offshore wind farms. J. Atmos. Oceanic Technol. **24**, 1302–1317 (2010)

25. Akdag, S.A., Guler, O.: Alternative moment method for wind energy potential and turbine energy output estimation. Renew. Energy **120**, 69–77 (2018)
26. Chen, J., Wang, F., Stelson, K.A.: A mathematical approach to minimizing the cost of energy for large utility wind turbines. Appl. Energy **228**, 1413–1422 (2018)
27. Hansen, A.D., Iov, F., Blaabjerg, F., Hansen, L.H.: Review of contemporary wind turbine concepts and their market penetration. Wind Eng. **28**, 247–263 (2004)

A Mixed Integer Program for Optimizing the Expansion of Electrical Vehicle Charging Infrastructure

Paul Brown[1], Marcello Contestabile[2], and Raka Jovanovic[2(✉)]

[1] Imperial College, Exhibition Rd, South Kensington, London SW7 2BU, UK
paul.brown19@imperial.ac.uk
[2] Qatar Environment and Energy Research Institute, Hamad Bin Khalifa University,
PO Box 5825, Doha, Qatar
{mcontestabile,rjovanovic}@hbku.edu.qa

Abstract. In recent years there has been a growing interest in the use of electric vehicles. This has resulted in the need to develop the necessary charging infrastructure. In this paper, the issue of optimizing the locations and capacity of charging stations is analyzed through the evaluation of its expansion. This is achieved using a model based on a mixed integer program. Since this is a highly practical problem, one of the main focuses of this work is in using real world data like population density and the state of the electrical distribution system. An efficient approach is proposed for acquiring such data and its integration into the model. The developed model is used for evaluating the expansion of EV charging infrastructure for the cities of Doha and San Francisco.

Keywords: Mixed integer programming · Electrical vehicles · Charging infrastructure

1 Introduction

In the recent years there has been an extensive increase in the use of electrical vehicles (EV). EVs can play a major role in the progressive decarbonisation of road transport and in the improvement of air quality in urban areas. For this reason governments have been strongly supporting their adoption through a combination of purchase subsidies, non monetary incentives (free parking and use of car pooling lanes), the provision of public charging networks, etc. This has resulted in rapid adoption of EVs. Public chargers are a key enabler of EV adoption [18,28] and therefore for the market to continue to grow the existing networks need to continue to expand as well.

On the other hand the adoption of electric cars has had a high growth and is steadily increasing. In contrast to fossil fuel vehicles, where the only way of refueling is visiting a station, EV's pose several possibilities due to different charger speeds [9]. To be more precise, it is possible to charge a vehicle using slow Level 1 chargers at home or using faster ones for out-of-home charging. For example, Level 2 (medium speed) chargers can be used at office parking lots with many positive impacts [12,19,31]. This type of infrastructure has the additional benefit of being suitable for use in demand management

© Springer Nature Switzerland AG 2021
D. Simian and L. F. Stoica (Eds.): MDIS 2020, CCIS 1341, pp. 302–314, 2021.
https://doi.org/10.1007/978-3-030-68527-0_19

systems [15,21,26]. The last approach is the use of Level 3 (fast) chargers, which can fully charge a typical EV in around 30 min. Level 3 chargers come at a high cost and are frequently close to 10 times more expensive than Level 2 ones, but are a necessity.

As the EV industry expands, drivers are becoming increasingly concerned about the comfort of usage, which is highly dependent on the available charging infrastructure. [20]. Due to the high cost of stations with Level 3 chargers, it is necessary to optimize their network. A large amount of research on EV charging infrastructure is modeled as location-allocation problems [22]. The high computational complexity of these problems, generally NP-Hard, has resulted in a large number of metaheuristics being developed for solving them [7,10,27].

A major body of this research is dedicated to optimizing infrastructure for long distance and regional travel due to range anxiety. In such systems one of the main goals is to make it possible for a majority of EVs to be able to reach EV stations with the minimal need for additional travel [14,23]. An essential part of such models is the use of origin-destination pairs and attempting to position charging stations at locations that catch the majority of traffic [17]. Another part of this research is focused on optimal locations of charging infrastructure within a city [6,13]. This research sites a more dense network of charging stations, over a city. Consequently, the models used for finding optimal locations of charging stations and their capacity has a high similarity to classical covering problems [10,25,32]. This is due to the assumption that an additional drive to reach a station is expected to be relatively short. An extended commute is acceptable for drivers if the waiting time is lower, this type of behavior is commonly observed for petrol car users.

In this paper, the focus is on developing a tool that makes it possible to assist in the decision making for the expansion of the EV charging network. This is done using a mixed integer program (MIP) for modeling the potential charging network of a city. The primary objective is to find the optimal locations of charging stations and their capacity. The optimality is observed as the minimization of the number of new stations that are installed. The second objective is to minimize the travel distances needed by an EV user to reach the charging stations.

The proposed model is used to assess the expansion strategies of charging infrastructure through case studies for two cities having different levels of existing infrastructure. This is done through using real world data on population density and existing charging infrastructure. In order to achieve a more realistic model, data regarding the electric distribution system is also used to specify the potential locations for new charging stations. The model is used to evaluate the expansion of the charging infrastructure for an increasing numbers of chargers per 10,000 population.

The paper is organized as follows. The second section introduces the mixed integer program for the problem of interest. The next section provides information how real world data is used to generate the instances for the model. The fourth section is dedicated to the case studies of the selected cities. The paper concludes with a discussion of the main insights derived from the study performed.

2   Mathematical Model

In this section, we present the mixed integer linear program for modeling the expansion of existing EV charging infrastructure on a city level. This type of problem is closely related to the capacitated facility location problem (CFLP) [30]. In the CFLP, it is expected that most facilities can satisfy demand from all locations and the distance only influences the cost. When modeling locations and capacities of EV charging stations this is not the case due to the fact that EV drivers are not willing to commute very far to use them. On the other hand, it is assumed that drivers will wish to avoid stations with long queuing times and in such cases they will go to a charging station that is not necessarily the closest. This common behavior of drivers often manages to balance the demand for charging stations over the whole system.

When setting up such infrastructure, the main question is how to select the best locations for new charging stations and their capacity. This should be done based on the population density of a city and the potential locations of charging stations based on the existing electrical distribution system. In practice, a complete EV charging systems for a city is not setup at once but through the expansion of an existing one with the goal of lowering the number of EVs per charger. The objective of the proposed MIP formulation is to be able to find the best locations and capacities of stations in this type of system.

In the proposed mathematical model several assumptions and simplifications are made. Firstly, it is assumed that the modeling is done on city scale and that the city is divided into sections. Each section has a fixed population and is considered as a single location. Each location has a potential capacity of EV charging that it can provided. More precisely, at each location some additional maximum charging capacity can be installed. The demand for EV charging for a location is proportional to its population. It is assumed that all the demand for EV charging must be satisfied. A demand can only be satisfied from a charging location that an EV driver is willing to travel to.

In addition, it is assumed that there is already some infrastructure in place which can provide charging. This is an additional property for each location. A location can at most satisfy the amount of charging demand that is less or equal to its capacity. It is expected that the majority of the cost for setting up the charging system is related to setting up a new locations and less to the number of chargers included. Because of this the objective will be to minimize the number of new locations that will be selected. It is assumed that a new location is being set up even if there are some existing charging facilities at the same area.

In formulating this model is it important to define the used parameters:

- Let N be the number of locations in the city. In relation let us define the set $V = \{1..N\}$ as the set of all locations.
- It is assumed that the distance between any two locations i and j is known. Let us define a real parameters d_{ij} for each pair $i, j \in V$ equal to the distance between the locations i and j. Note that the distance between location i to itself will be equal to zero.
- Let us define a parameter c_v for each $v \in V$ corresponding to the potential additional charging capacity of location v

- Let us define the parameter e_v for each $v \in V$ corresponding to the existing capacity of location v
- Let us define the parameter p_v for each $v \in V$ corresponding to the population of location v.
- Let W be the maximal distance a driver is willing to compute to the EV charging stations.
- Let α be the number of chargers needed per a 10 000 population.

Next, let us define the set of decision variables needed to fully specify the model.

- Let x_i be defined as a binary decision variable for each location $i \in V$. It has a value of 1 if at least one new charging stations will be set at location i and 0 otherwise.
- For each two locations $i, j \in V$, a real variables r_{ij} is defined. r_{ij} holds the information on how much of the population of location j is using the chargers at location i.

It is assumed that if a new charging facility is installed at location i that its total additional capacity can be used by EV drivers. The objective is to minimize the number of used locations for installing the additional chargers as follows:

$$\text{minimize} \sum_{v \in V} x_v \tag{1}$$

Let us define the constraints that these variables need to satisfy in the following way.

$$\sum_{j \in V} r_{vj} \leq x_i c_v + e_v \qquad v \in V \tag{2}$$

$$\sum_{i,v \in V} r_{iv} = \alpha p_v \qquad v \in V \tag{3}$$

$$r_{ij} = 0 \qquad i, j \in V \wedge d_{ij} > W \tag{4}$$

$$0 \leq r_{ij} \leq c_i + e_i \qquad i, j \in V \tag{5}$$

$$x_i \in \{0, 1\} \qquad i, j \in V \tag{6}$$

$$r_{ij} \in \mathbb{R} \qquad i, j \in V \tag{7}$$

The constraint given in (2) states that the maximal charge that can be provided from location v is less or equal to the sum of existing capacity and additional capacity. The constraints given in (3) guaranties that all the charging demand of a location is covered. Note that all the charging demand of a location is satisfied if it has α chargers per 10,000 population. The constraints given in Eq. (4) state that a location j can use chargers from location i only if the distance is less than a specified maximal distance.

The MIP with the objective (1) using constraints (2)–(6) provides us with the minimal number of locations needed to satisfy a specific number of chargers per 10 000 population. The MIP indirectly provides the additional capacity of location $v \in V$ and is equal to $\sum_{j \in V} r_{vj}$. More precisely, the capacity needed at location v is equal to number of chargers used by other locations in the city. Note that this sum is a real value and can be rounded up to the nearest integer value.

As previously stated, the majority of costs is related to the setup of new charging locations and significantly less to the number of chargers used at such a location. The setup of the new charging infrastructure is in essence a dual objective problem. Whilst the first one is related to the minimizing the cost, the second one is related to satisfaction of EV drivers. In case of the later, a good measure is to minimize the average distance the EV drivers need to reach the station. In practice, our goal is to find the optimal capacities for each new charging location. This can be done in two steps. The first one is finding the minimal number L of new locations using the previously presented model. Secondly, finding the capacities of the L locations. This can be achieved by a new model having the following objective function

$$\text{minimize} \sum_{i,j \in V} r_{ij} d_{ij} \tag{8}$$

Equation (9) states that the objective is to minimize the sum of distances traveled by the EV drivers. The constraints of the new MIP use the Eq. (2)–(6) but with an additional constraint that fixes the number of new stations as follows.

$$\sum_{i \in V} x_i = L \tag{9}$$

3   Use of Real World Data

The main goal of this work is to evaluate the expansion of existing EV charging networks in metropolitan areas, because of this, special attention is given to generating instance based on real word data. In this section an overview of the used information and how it is converted to parameters of the model are given.

In generating instances for the model, the following real world data has been used.

- Population density is essential for specifying the population of each of the locations in the model.
- Latitude and longitude data on existing charging infrastructure is used so that the model can build upon this.
- Data on transformer substations provide the information on the potential new locations for EV charging stations. This property has been selected since it is possible to add large energy consumers, like an EV charging stations, close to them without a high increase in capital expenditure (CAPEX)

The first step in converting this type of data to an instance is defining what is considered a location within the model. In the conducted work, a grid approach is used in the following way. The area corresponding to the city of interest is divided into a grid (rectangular subsections of the city) and each location resembles to a cell in the grid. The next step is specifying the methods for calculating the parameters in the model. The distance between locations i and j is equal to the distance between cell centers calculated using their latitudes and longitudes based on the Haversine formula [29]. The used measure unit for distances is kilometer.

The value of the population parameter p_v, for location v, is calculated based on the population density. To be exact, the used population density data provides the density for the region of a specific cell (location) and based on its area the corresponding total population can be trivially calculated. The parameter for the existing capacity e_v, for a location v, is calculated based on the locations of existing charging stations. It is equal to the total number of charging stations with a location that is inside the region of the corresponding cell.

The parameter c_v corresponding to the potential additional charging capacity has been calculated using the locations of transformer substations. New charging stations can be added in areas that are less than a specific maximal distance to transformer substations. To be more precise, if the distance from a charging cell is less than a value l it can provide an additional capacity of 1 to that cell. In the proposed instance generation method the value c_v of location v is be equal to the sum of all the additional charging capacity of all substations based on distance of the substation to the corresponding cell and the l value.

4   Case Studies

In this section the proposed model and the method for integrating real world data are used to evaluate the growth of EV charging infrastructure. This has been done through two case studies for Doha, Qatar and San Francisco, USA. Firstly, the specifics for data collection for these two cities are presented. Next, an analysis of the results of the conducted computational results using this data and the proposed model are given.

4.1   Instance Generation

To represent the versatility of the model, it was important to select cities with a variety of economic, demographic, and geographical landscapes. The cities selected were Doha and San Francisco. These exemplified urban areas with ambitious EV targets but differing EV markets, demographic characteristics, and geographical landscapes. Doha is an example of a very small EV Market, it currently only hosts 9 charging stations but looks to expand its charging network to reach targets of 10% EV penetration by 2030 [24]. Comparatively, San Francisco has a comprehensive network of 155 charging stations, but also seeks an expansion of its charging facilities to reach 100% EVs by 2030 [11]. Moreover, whilst both cities are considered densely populated, much of Doha's population is concentrated in a relatively small area towards the east, whereas San Francisco's population is more equally dispersed throughout the city. Furthermore, whilst San Francisco is surrounded by sea and hosts several lakes and large parks, Doha has sea on one side and has relatively little water bodies or parks in the city but large areas of desert. Hence, by selecting urban areas with differing geographical landscapes, demographic characteristics and EV markets, the adaptability of the model can be represented.

The real-world data for both cities (population density, pre-existing charging facilities and transformer substation) used for generating instances, was gathered from freely accessible online resources. For San Francisco pre-existing charging data was accessed from 'Open Charge Map' [5], transformer substation data from 'The Homeland Infrastructure Foundation' [2], and population data from 'WorldMap' [1]. For Doha, pre-existing charging information was extracted from online news articles and from a case

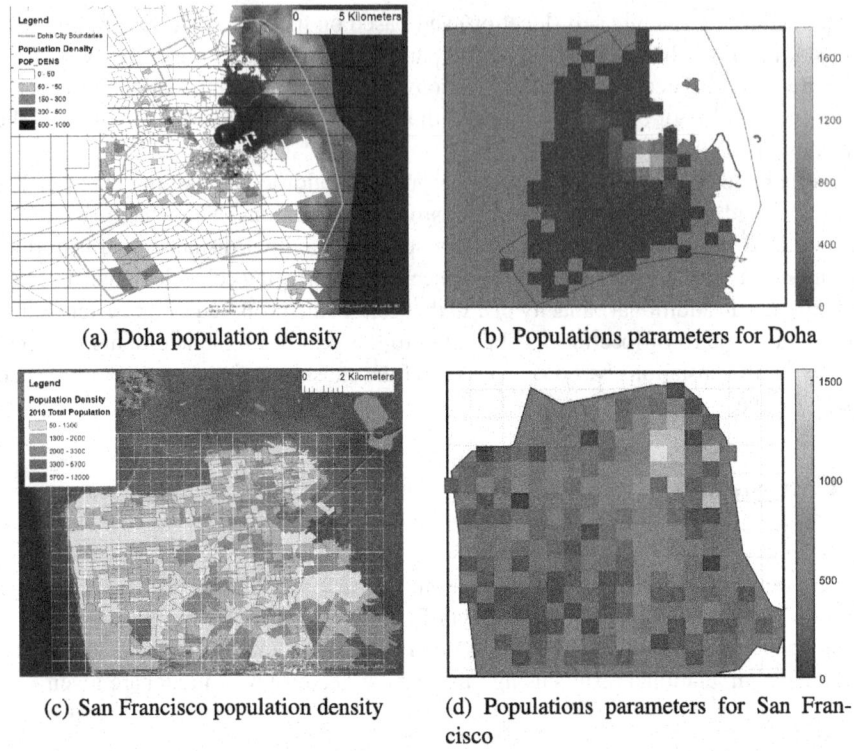

(a) Doha population density

(b) Populations parameters for Doha

(c) San Francisco population density

(d) Populations parameters for San Francisco

Fig. 1. Illustration of the conversion of population density to values of parameters within the model for Doha and San Francisco. The input data is on the left and the values of parameters corresponding to the population at a location (p_v) are on the right.

study regarding the hindrances of EV adoption in Qatar [16]. Moreover, transformer substation data was collected from 'Overpass Turbo' [4] and population data from 'WorldMap'. As sources such as 'Overpass Turbo' and 'Open Charge Map' used public contributions as part of their information, the reliability of their data was ensured through reviewing sites on Google Earth.

The data was extracted and reviewed, and adjustments were made to ensure it was suitable to be collected in the grid format and read into the optimization model. The accessible population data (density) for all the urban areas was only available online via census block data. Illustration of it's transformation to the model parameters can be seen in Fig. 1. The location of each transformer substation was assigned a 1-mile buffer zone (Fig. 2) to indicate the reach of its capacity, as recommended by [8]. Hence, any new charging station had to be located within a buffer zone of a Transformer Substation. The pre-existing charging data did not require any adjustments and so could be directly read into the grid data collection format (Fig. 3).

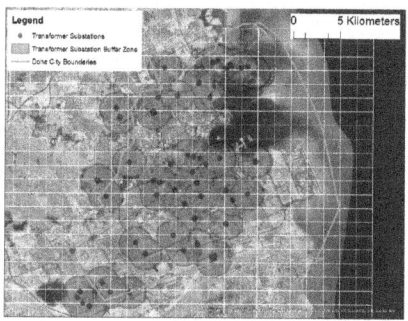

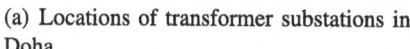

(a) Locations of transformer substations in Doha

(b) Additional capacity parameters for Doha

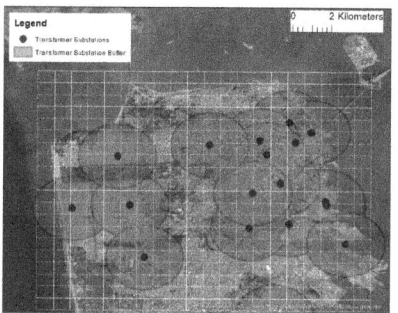

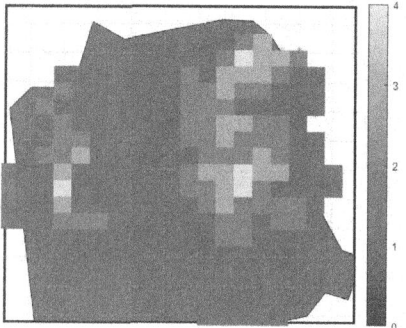

(c) Locations of transformer substations in San Francisco

(d) Additional capacity parameters for San Francisco

Fig. 2. Illustration of the conversion of transformer substation location data to values of parameters within the model for Doha and San Francisco. Red circles are used to indicate the 1-mile buffer zone for a substations. The input data is on the left and the values of parameters corresponding to the potential additional capacity that can be installed at a location (c_v) are on the right.

4.2 Computational Experiments

In this section the results of the conducted computational experiments are presented. The main objective was to analyze the expansion of the charging infrastructure based on the proposed model and real world data.

In the case studies for Doha and San Francisco, the selected maximal distance that a driver is willing to travel to a charging station (W) has 10 km. This value has been chosen to be an improvement to the current distance of EV drivers to the charging stations. For example is San Francisco, an EV drivers has a charging station with-in 16 km from his home [3].

In case of Doha and San Francisco there was a total of 127 and 213 potential locations (grid cells), respectively. The evaluation was done for several values of the number of chargers per 10 000 population (α). The selected lower bound was equal to the value of α_l for the current infrastructure, simply as the number of chargers divided by the total

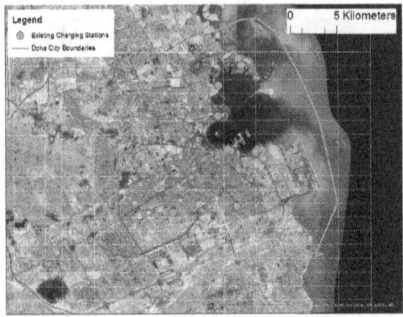
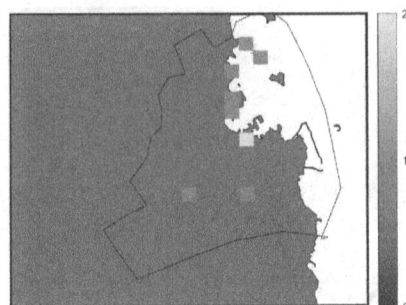

(a) Locations of existing charging stations in Doha

(b) Existing capacity parameters for Doha

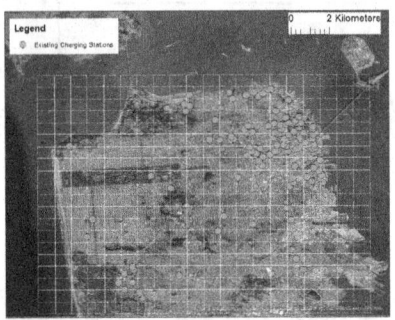
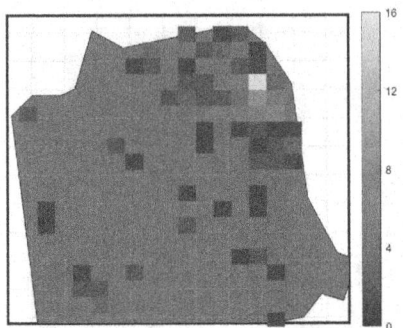

(c) Locations of existing charging stations in San Francisco

(d) Existing capacity parameters for San Francisco

Fig. 3. Illustration of the conversion of existing charging stations positions to values of parameters within the model for Doha and San Francisco. The green circles indicate the locations of the stations. The input data is on the left and the values of parameters corresponding to the existing capacity at a location (e_v) are on the right. (Color figure online)

population scaled by 10 000. The upper bound α_u was calculated in the same way using the sum of existing number of chargers and the maximal number of potential additional ones. It should be noted that due to the constraints related to parameter W, in practice these values could not be reached. The tests have been done for a 10 intermediate values α_i calculated as follows:

$$\alpha_i = \alpha_l + i\frac{\alpha_u - \alpha_l}{10} \tag{10}$$

The propose MILP has been implemented using OPL in IBM ILOG CPLEX Optimization Studio Version: 12.6.1.0, and executed using the default solver settings. A single problem instance could be solved to optimality within less than five seconds. The main results of the case studies for Doha and San Francisco are summarized in Figures 4 and 5, respectively. In addition in Table 1, the growth in the number of new charging locations that are needed to achieve different levels of charger availability, corresponding to values of α_i, are given.

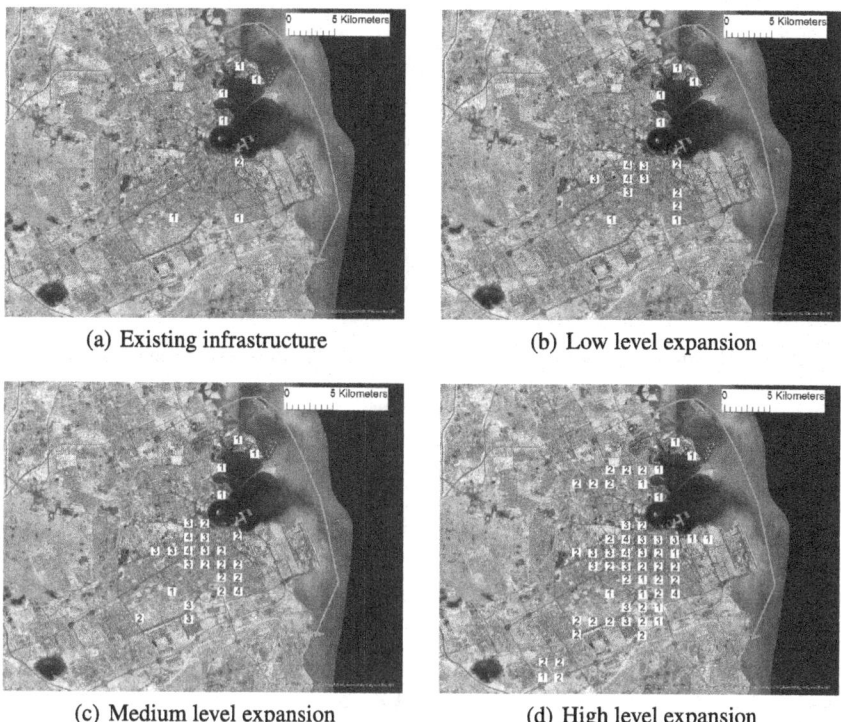

(a) Existing infrastructure

(b) Low level expansion

(c) Medium level expansion

(d) High level expansion

Fig. 4. Illustration of the expansion of charging infrastructure for Doha from current state to high levels of adoption. The numbers indicate the amount of charging capacity that is inside a location (cell in the grid) in the city.

In the case of Doha, where the EV charging infrastructure is at a very early stage, the initial stations have been deployed as test beds. Therefore, they are not positioned at the most densely populated areas of the city. The model indicates that the expansion of the infrastructure should be done initially at the most densely populated areas. In addition it can be observed that it is advantages to have the earlier stages of expansion focused in areas that can provide a high level of new charging capacity. It should be noted that for most areas the initial amount of charging provided would be equal to the maximal potential capacity.

This type of behavior is less noticeable in case of San Francisco, were an extensive EV charging network already exists. In this case, it is common that the charging capacity of an area would be expanded at later stages. The additional capacity is generally added mostly in densely populated areas. In case of both cities, in later stages of infrastructure development the network becomes more evenly distributed over the areas were installation is possible. It can also be observed that even at later stages of infrastructure expansion some areas do not have near by charging stations. One reason for this is that the objective in the model is minimal average distance from a charging station. Some of the disadvantages of this is that a small number of EV drivers will need to travel

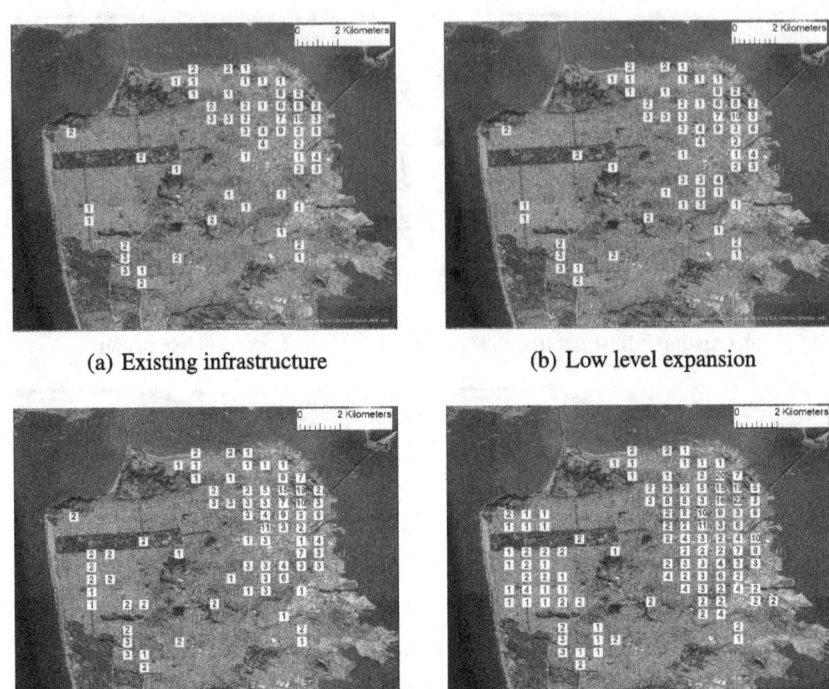

(a) Existing infrastructure   (b) Low level expansion

(c) Medium level expansion   (d) High level expansion

Fig. 5. Illustration of the expansion of charging infrastructure for San Francisco from current state to high levels of adoption. The numbers indicate the amount of charging capacity that is inside a location (cell in the grid) in the city.

Table 1. Number of new locations need to achieve different levels of chargers per 10 000 population for the cities of Doha and San Francisco.

Doha										
Chargers per 10 000 (α)	0.024	0.102	0.126	0.176	0.227	0.278	0.329	0.380	0.430	0.481	0.532
Num. new locations	0	4	9	14	21	27	34	41	52	65	78
San Francisco										
Chargers per 10 000 (α)	1.688	1.846	2.005	2.164	2.323	2.481	2.640	2.799	2.958	3.116	3.275
Num. new locations	0	7	15	25	37	49	61	81	104	128	151

further and there is a potential that the high density of charging stations in some areas will result in potential traffic congestion.

5   Conclusion

In this paper, a new approach has been presented for analyzing the expansion of EV charging infrastructure. The method uses information related to existing charging and electrical distribution infrastructure in combination with population density. The proposed method is based on a mixed integer linear program. One of the main objectives of the conducted research was to develop a method that can be used for real world application, consequently a large amount of effort has been dedicated to data collection. The analysis has been done through two case studies for the cities of Doha and San Francisco. It should be noted, that the data sources used for this have information for a wide range of other cities and the presented approach can easily be applied to them. The proposed MIP can be used to solve this problem for real world cities for a low computational cost. The two case studies indicate that by optimizing each stage of the infrastructure growth can results in a non-optimal final solution.

In the future we plan to extend this model to include additional parameters for the selection of charging stations like land area costs, points of interest, traffic and similar.

References

1. HarvardWorldMap. http://worldmap.harvard.edu. Accessed 10 Aug 2020
2. Homeland Infrastructure Foundation-Level Data (HIFLD). https://hifld-geoplatform.opendata.arcgis.com. Accessed 10 Aug 2020
3. Open Charge Map: The Global Public Registry Of Electric Vehicle Charging Locations. https://chargehub.com/en/countries/united-states/california/san-francisco.html. Accessed 10 Aug 2020
4. OverpassTurbo (2019). https://overpass-turbo.eu. Accessed 10 Aug 2020
5. ChargeHub: San Francisco, California EV Charging Stations Info. https://openchargemap.org. Accessed 10 Aug 2020
6. Choi, B.G., Oh, B.C., Choi, S., Kim, S.Y.: Selecting locations of electric vehicle charging stations based on the traffic load eliminating method. Energies 13(7), 1650 (2020)
7. Chraibi, K., Chraibi, A., Chaker, I., Zahi, A., Bekrar, A.: Hybrid metaheuristic to solve location problem for electric vehicles charging stations. In: 2018 IEEE International Conference on Technology Management, Operations and Decisions (ICTMOD), pp. 12–17. IEEE (2018)
8. Corey, J., Metcalf, M.: Direct Current (DC) Fast Charging Mapping. Technical report, Pacific Gas and Electric Company, Electric Program Investment Charge (EPIC) (2016)
9. Daina, N., Sivakumar, A., Polak, J.W.: Electric vehicle charging choices: Modelling and implications for smart charging services.Transp. Res. Part C: Emerg. Technol. 81, 36–56 (2017)
10. Efthymiou, D., Chrysostomou, K., Morfoulaki, M., Aifantopoulou, G.: Electric vehicles charging infrastructure location: a genetic algorithm approach. Eur. Transp. Res. Rev. 9(2), 27 (2017)
11. Environment, S.: San Francisco's Electric Vehicle Ready Community Blueprint. Tech. rep, The Department of the Environment City and County of San Francisco (2019)

12. Ferguson, B., Nagaraj, V., Kara, E.C., Alizadeh, M.: Optimal planning of workplace electric vehicle charging infrastructure with smart charging opportunities. In: 2018 21st International Conference on Intelligent Transportation Systems (ITSC). pp. 1149–1154. IEEE (2018)
13. Guo, C., Yang, J., Yang, L.: Planning of electric vehicle charging infrastructure for urban areas with tight land supply. Energies 11(9), 2314 (2018)
14. Jochem, P., Szimba, E., Reuter-Oppermann, M.: How many fast-charging stations do we need along European highways? Transp. Res. Part D Transp. Environ. 73, 120–129 (2019)
15. Jovanovic, R., Bayram, I.S.: Scheduling electric vehicle charging at park-and-ride facilities to flatten duck curves. In: 2019 IEEE Vehicle Power and Propulsion Conference (VPPC). pp. 1–5 (2019)
16. Khandakar, A., et al.: A case study to identify the hindrances to widespread adoption of electric vehicles in qatar. Energies 13(15), 3994 (2020)
17. Kong, C., Jovanovic, R., Bayram, I.S., Devetsikiotis, M.: A hierarchical optimization model for a network of electric vehicle charging stations. Energies 10(5), 675 (2017)
18. Langbroek, J.H., Franklin, J.P., Susilo, Y.O.: The effect of policy incentives on electric vehicle adoption. Energy Policy 94, 94–103 (2016)
19. Levinson, R.S., West, T.H.: Impact of convenient away-from-home charging infrastructure. Transp. Res. Part D Transp. Environ. 65, 288–299 (2018)
20. Mehrjerdi, H., Hemmati, R.: Electric vehicle charging station with multilevel charging infrastructure and hybrid solar-battery-diesel generation incorporating comfort of drivers. J. Energy Storage 26, 100924 (2019)
21. Meyer, D., Wang, J.: Integrating ultra-fast charging stations within the power grids of smart cities: a review. IET Smart Grid 1(1), 3–10 (2018)
22. Motoaki, Y.: Location-allocation of electric vehicle fast chargers–research and practice. World Electric Vehicle J. 10(1), 12 (2019)
23. Napoli, G., Polimeni, A., Micari, S., Andaloro, L., Antonucci, V.: Optimal allocation of electric vehicle charging stations in a highway network: Part 1. methodology and test application. J. Energy Storage 27, 101102 (2020)
24. Onat, N.C., Kucukvar, M., Aboushaqrah, N.N., Jabbar, R.: How sustainable is electric mobility? a comprehensive sustainability assessment approach for the case of qatar. Appl. Energy 250, 461–477 (2019)
25. Răboacă, M.S., Băncescu, I., Preda, V., Bizon, N.: An optimization model for the temporary locations of mobile charging stations. Mathematics 8(3), 453 (2020)
26. Rahman, I., Vasant, P.M., Singh, B.S.M., Abdullah-Al-Wadud, M., Adnan, N.: Review of recent trends in optimization techniques for plug-in hybrid, and electric vehicle charging infrastructures. Renew. Sustain. Energy Rev. 58, 1039–1047 (2016)
27. Ren, X., Zhang, H., Hu, R., Qiu, Y.: Location of electric vehicle charging stations: a perspective using the grey decision-making model. Energy 173, 548–553 (2019)
28. Sierzchula, W., Bakker, S., Maat, K., Van Wee, B.: The influence of financial incentives and other socio-economic factors on electric vehicle adoption. Energy Policy 68, 183–194 (2014)
29. Van Brummelen, G.: Heavenly mathematics: The forgotten art of spherical trigonometry. Princeton University Press (2012)
30. Verter, Vedat: Uncapacitated and capacitated facility location problems. In: Eiselt, H.A., Marianov, Vladimir (eds.) Foundations of Location Analysis. ISORMS, vol. 155, pp. 25–37. Springer, New York (2011). https://doi.org/10.1007/978-1-4419-7572-0_2
31. Wu, X.: Role of workplace charging opportunities on adoption of plug-in electric vehicles-analysis based on gps-based longitudinal travel data. Energy Policy 114, 367–379 (2018)
32. Zhang, Y., Qiu, Z., Gao, P., Jiang, S.: Location model of electric vehicle charging stations. J. Phys. Conf. Ser. 1053, 012058 (2018)

TabbyLD: A Tool for Semantic Interpretation of Spreadsheets Data

Nikita O. Dorodnykh[iD] and Aleksandr Yu. Yurin[✉][iD]

Matrosov Institute for System Dynamics and Control Theory, Siberian Branch of the
Russian Academy of Sciences, 134, Lermontov Street, Irkutsk 664033, Russia
`iskander@icc.ru`

Abstract. Spreadsheets are one of the most convenient ways to structure and represent statistical and other data. In this connection, automatic processing and semantic interpretation of spreadsheets data have
become an active area of scientific research, especially in the context of
integrating this data into the Semantic Web. In this paper, we propose a
TabbyLD tool for semantic interpretation of data extracted from spreadsheets. Main features of our software connected with: (1) using original
metrics for defining semantic similarity between cell values and entities of
a global knowledge graph: string similarity, NER label similarity, heading similarity, semantic similarity, context similarity; (2) using a unified
canonicalized form for representation of arbitrary spreadsheets; (3) integration TabbyLD with the TabbyDOC project's tools in the context of
the overall pipeline. TabbyLD architecture, main functions, a method for
annotating spreadsheets including original similarity metrics, the illustrative example, and preliminary experimental evaluation are presented.
In our evaluation, we used the T2Dv2 Gold Standard dataset. Experiments have shown the applicability of TabbyLD for semantic interpretation of spreadsheets data. We also identified some issues in this process.

Keywords: Semantic table interpretation · Annotation · Spreadsheet
data · Entity linking · Linked data · Knowledge graph · DBpedia

1 Introduction

Currently, a large volume of arbitrary tables presented in the spreadsheet-like
formats (for example, Excel, CSV, HTML) circulates in the world. The modern
estimations [1] show that a number of genuine tables in the Web reach hundreds
of millions. They can contain hundreds of billions of facts. A big variety and
heterogeneity of layouts, styles, and content, as well as a high rate of growth of
their volume, characterize the arbitrary tables. This information can be considered Big Data.

Spreadsheets are one of the most convenient ways to structure and represent
statistical and other data. Arbitrary spreadsheets are widely distributed, and
they can be used as a valuable data source in business intelligence and data-
driven research. However, difficulties that inevitably arise with the extraction

© Springer Nature Switzerland AG 2021
D. Simian and L. F. Stoica (Eds.): MDIS 2020, CCIS 1341, pp. 315–333, 2021.
https://doi.org/10.1007/978-3-030-68527-0_20

and integration of the tabular data often hinder the intensive use of them in the mentioned areas. Typically, spreadsheets are not accompanied by explicit semantics necessary for the machine interpretation of their content, as conceived by their author. Spreadsheets data is often unstructured and not standardized. Analysis of these data requires their preliminary extraction and transformation to a structured representation with a formal model. This research is aimed at solving this problem, which corresponds to the concept of five-star deployment of open data in the Web, proposed by Sir Tim Berners-Lee [2].

Semantic interpretation of spreadsheets data is a comparison of tabular data with a target knowledge graph (e.g., Wikidata, DBpedia, or YAGO) and selecting semantic tags for source spreadsheet elements. The complexity of this task results from incompleteness, ambiguity, or lack of metadata (e.g., spreadsheet heading names). A variety of approaches and software are proposed to solve this problem. However, most of them only target specific datasets or spreadsheet layouts (e.g., only statistical spreadsheets, etc.).

In this paper, we propose a web-based tool for semantic interpretation of spread-sheets data, namely, TabbyLD [3]. One of the features of TabbyLD is its versatility, achieved through the use of a canonical table form for a uniform representation of input data and preprocessing the input spreadsheets. Other features of our tool are original metrics for defining semantic similarity of cell values, and integration with the TabbyDOC project's software [4]. The TabbyLD architecture, main functions, method and metrics for annotating spreadsheets, and preliminary experimental evaluation are presented. In our evaluation, we used the T2D Gold Standard, version 2 [5].

The paper is organized as follows. Section 2 presents an analytical overview of related works. Section 3 describes the proposed software for semantic interpretation of spreadsheets data including a description of the method for annotating spreadsheets, the conceptual architecture, main functions, and key features. Section 4 presents the illustrative example for annotating a spreadsheet. Section 5 illustrates the preliminary experimental evaluation for our tool and discussion of obtained results, while Sect. 6 presents concluding remarks.

2   Related Works

Many approaches and tools aimed at the semantic interpretation of data extracted from web tables and spreadsheets have been proposed in recent years. In particular, the following works represented by research software or its prototypes can be distinguished in this area.

De Vos et al. [6] present several approaches for automatic annotation of natural science spreadsheets using a combination of structural properties of tables and external vocabularies (ontologies). Authors distinguish five structural properties of tables, and each one proposes its heuristic for semantic interpretation (annotation). The basic method for annotating natural science tables is a lexical comparison with concepts from the domain vocabulary. Annotation approaches for five properties are complementary to this basic method. Moreover, approaches are combined iteratively.

Zhang [7] proposes an approach and tool, namely, TableMiner+ for semantic interpretation of web tables. This method is capable of handling all tasks for semantic table interpretation. In particular, the method annotates individual values of table cells with instances from a knowledge graph; subject (named-entity) columns with the corresponding semantic type (class) or a property; relationships between subject columns and relationships between subject and literal columns with semantic properties from a target knowledge graph. The proposed method and tool improve the accuracy of annotations by using different contextual information both inside and outside a source table.

A new TAIPAN approach to restoring table semantics by extracting RDF triplets from web tables is proposed in [8]. The restoration of table semantics is divided into two tasks: identification of a subject column, which is a column containing resource labels that create an instance of the main object of a source table; comparison (mapping) of properties from an external knowledge base to a binary relationship between a subject column and another column. DBpedia is used as a target knowledge base, as well as Linked Open Vocabularies (LOV), which supports access to different ontologies based on rdfs:label and rdfs:comment.

Ritze et al. [9] present a T2K Match approach aimed at interpreting (annotating) web tables using general-purpose knowledge bases. The authors also describe the "Gold Standard T2D" dataset that was used to evaluate the performance of the proposed approach. This approach works only with entity-attribute tables, which describe a set of entities (rows in a table) that have a set of multivalued attributes (columns). A column heading name in a source table is considered by default to be some superficial form of a semantic attribute name. At the same time, different types of attributes (columns) are distinguished: strings and numeric values. The approach also requires tables to contain attributes only in natural language. The table normalization procedure (data cleaning, resolving abbreviations, etc.) is carried out before starting annotation. Data type in table columns is also determined using about 100 manually defined regular expressions.

A domain-independent framework, namely, "DIF-LDT" for extracting linked semantic data from tables is presented in [10]. The framework is based on graphical models and probabilistic reasoning for interpreting and inferring values linked with a table. The process of semantic table interpretation is divided into 4 main tasks: assigning to each column (or a row header) of a certain class from the corresponding knowledge graph (ontology); linking table cell values to corresponding instances from a knowledge graph; discovering relationships between table columns and linking them with properties from a knowledge graph; generating a linked data. The proposed approach and framework were tested on tables from the Internet (Google Squared and web tables from Wikipedia). DBpedia, YAGO, and Wikitology knowledge graphs were used.

Other successful examples for solving the problem of semantic table interpretation are presented in [11–19].

Analysis of existing approaches and tools showed:

1. DBpedia is the most popular open knowledge graph, that contains over 6 million concepts and over 1 billion RDF triplets [20].
2. As a rule, considered approaches are ad-hoc solutions, i.e. designed for processing relational tables with a certain structure. Such tables should describe data belonging to a single class (category), where each column is associated with a property from a knowledge graph, and all rows of a table are objects (instances) of this class.
3. Programming skills of end-users are required to configure tools for a new type of input tables.
4. The ubiquitous use of a table context is acceptable for web tables, but difficult for other table formats.
5. Mostly web tables (from wiki pages) are analyzed, CSV (Excel) tables processed less.

So, in this paper, we propose a prototype of a web-based tool called TabbyLD for semantic interpretation of spreadsheets data. TabbyLD implements a universal approach focused on processing spreadsheets with an arbitrary layout represented in the Excel format. TabbyLD versatility is achieved through using a unified canonicalized form for representation of a source arbitrary spreadsheet [21]. This form is one of the output formats for the TabbyXL tool. This form provides the integration of TabbyLD with the TabbyDOC project's software and builds a full chain of transformations from a source PDF-document to an annotated spreadsheet (linked to reference entities from a target knowledge graph). We also propose an aggregated method for defining similarity between candidate entities and cell value to disambiguation. This method consists of the sequential application of five metrics and combining ranks obtained by each metric.

3   TabbyLD

We created a prototype of a tool, namely TabbyLD to solve the task of semantic interpretation of spreadsheets data. Our software is implemented as a web application designed for non-programmers: domain experts, knowledge engineers, semantic technology specialists, system analysts, etc.

The main purpose of TabbyLD is annotating canonical spreadsheets represented in the Excel format (XLSX) by linking spreadsheet elements to entities of external knowledge graphs (global taxonomies) from the Linked Open Data (LOD) cloud, in particular, DBpedia [20]. Results of annotating are presented as marked up canonical XLSX-spreadsheets.

Basic principles and features of TabbyLD are as follows:

1. Using a unified canonicalized table representation form [21,22] for representing spreadsheets with arbitrary layouts, i.e. our tool can analyze spreadsheets with different layouts, header structures, and data areas.
2. Using an aggregated method for defining similarity between candidate entities and a cell value to disambiguation based on heuristic metrics.
3. Integrating TabbyLD into the software pipeline of the TabbyDOC project.

Next, let's consider each feature, and also architecture and main functions of TabbyLD in detail.

3.1 Canonicalized Form for Representing Arbitrary Spreadsheets

A table is a grid of cells arranged in rows and columns. Such tables are used as visual communication patterns, as well as data arrangement and organization tools. In this paper, we primarily focus on spreadsheets, i.e. tables embedded in various documents and reports. Such spreadsheets typically contain data from various dimensions or named entities for different domains and are presented in the MS Excel (XLSX) or CSV format. However, spreadsheets can have different variants of layouts, design styles, data representation logic, and others. Such spreadsheets are called arbitrary spreadsheets.

In this paper, we propose to use a rule-based approach and Cells Rule Language (CRL) [23] for the analysis of arbitrary spreadsheets and their transformation to a canonicalized (relational) form. A canonical spreadsheet is a formalized uniform representation of a subset of arbitrary spreadsheets [24]. Canonical spreadsheets contain high-quality relational data in the form of a set of entities, which could exist in rows (horizontal) or columns (vertical), the remainder of cells contains their descriptive attributes. Thus, using a unified canonicalized table representation form, TabbyLD can analyze spreadsheets with different layouts, header structures, and data areas.

Let's formally define the structure of the spreadsheet represented in a canonicalized form and processed by TabbyLD:

$$CS = \left\{ D, R^H, C^H \right\}, \tag{1}$$

where CS is a source spreadsheet in a canonicalized form; D is a data block that describes literal data values (entries) belonging to the same data type (e.g. numeric, date, time, or text); R^H is a set of labels of a row heading; C^H is a set of labels of a column heading. The values in cells for heading blocks can be separated by the "|" symbol to divide categories into subcategories. So, the canonical spreadsheet denotes hierarchical relationships between categories (headings). A heading label is a set of characters, for example, it can be one word, or a phrase, or a whole sentence.

3.2 Method for Annotating Spreadsheets

The main function of TabbyLD is semantic interpretation (annotation) of spreadsheets data using a target knowledge graph. This process is matching each cell values of a source canonical spreadsheet with entities from a target knowledge graph (in our case, DBpedia), and is implemented as a chain of the following stages:

1. Annotating canonical spreadsheet heading values (R^H and C^H).
2. Annotating cell values of a canonical spreadsheet in a data lock (D).

3. Generating annotated spreadsheets with links (tags) of concepts from a target knowledge graph (DBpedia).

Let's describe the annotation process for each stage in detail.

Stage 1: Annotating Canonical Spreadsheet Heading Values. The annotation strategy for heading values (R^H and C^H) is the same. Each spreadsheet header value is treated as one mention (concept). At the same time, this mention must already be converted to the required format (a preprocessing procedure has been performed). The following sequence of steps for each heading value must be performed to find reference entities from DBpedia for all spreadsheet heading values:

Step 1. Searching a class from an ontology segment of DBpedia for a heading value by making an exact match.

Step 2. Searching an instance from a resource segment of DBpedia for a heading value by making an exact match.

Step 3. Forming a set of candidate entities by compiling SPARQL (SPARQL Protocol and RDF Query Language) [25] queries against DBpedia for matching a string to a pattern. At the same time, we select only the first 100 candidates for performance reasons.

If no entity (classes or instances) from DBpedia has been linked to a heading value, then this heading value is marked as not annotated and the next steps for it are skipped.

Step 4. Next, a disambiguation procedure for all heading values is performed.

Since a heading value can refer to several entities from a set of candidates, it is rather difficult to choose a suitable reference entity from this set for linking. We propose an aggregated method for defining similarity between candidate entities and heading value to disambiguation. This method consists of the sequential application of two metrics and combining ranks obtained by each metric:

1) String similarity. This metric is used for selecting a reference entity from a set of candidates based on the maximum similarity of a sequence of characters in the accordance of the Levenshtein distance [26]:

$$f_1(e_i) = LevenshteinDistance\,(v_j, e_i)\,, v_j \in CS, e_i \in E, \tag{2}$$

where $f_1(e_i)$ is a function for calculating the Levenshtein distance; v_j is a string value of j-cell in a source canonical spreadsheet; e_i is a string name of i-entity from a set of candidates E.

2) NER label similarity. This metric is based on information about already recognized named entities and cell types form the Named Entity Recognition (NER) procedure. This procedure includes identifying named entities in cells of a canonical spreadsheet and assigning each spreadsheet cell of the corresponding type.

So, firstly, the procedure for extracting and recognizing named entities contained in canonical spreadsheet cells is carried out in this step. We use the library for natural language processing called Stanford CoreNLP [27] and, in particular, Java implementation of Stanford Named-Entity Recognizer [28]. Stanford NER marks words in the text that are object names such as peoples, companies, cities, or countries. Stanford NER defines many classes of named entities. In our work, we use 8 basic types (classes): *Location, Country, City, Person, Organization, Number, Percent, Date,* and *None* that is an undefined class. These classes are assigned to each cell in a source canonical spreadsheet, so one characterizes the data that it contains. We divide such specific typed cells into two groups: cells with named entities and cells with literal values.

Secondly, if the depth of classes nesting (a distance from this target class relative to a current candidate entity) is defined, then the rank is defined as follows:

$$f_2(e_i) = \frac{1}{distanceLevel(c_j)}, e_i \in E, c_j \in C, \tag{3}$$

where $f_2(e_i)$ is a function for calculating similarity to classes defined by NER labels for an entity e_i from a set of candidates E; c_j is a name of j-class from a set of classes C found by NER labels (see Table 1); $distanceLevel(c_j)$ is a class c_j distance (depth) in a transitive relationship with entity e_i. Moreover, if candidate entity e_i is an instance of the required class directly, then this distance is 1.

Step 5. Using (1) and (2), we calculate the aggregate rank for candidate entities linked to heading values from heading blocks (R^H and C^H):

$$f_{agg}^H = w_1 \times \left(1 - \frac{f_1(e_i)}{100}\right) + w_2 \times f_2(e_i), \tag{4}$$

where f_{agg}^H is an aggregate rating function for a candidate entity e_i; w_1 is a weighting factor that balances the importance of a rank obtained on the basis of the Levenstein distance calculation; w_2 is a weighting factor that balances the importance of rank obtained on the basis of the NER label similarity. By default, $w_1 = 1$ and $w_2 = 1$.

So, a reference entity e_i from a set of candidates E based on a combination of ranks f_{agg}^H is defined. Next, the interpretation (annotation) of cell values for a data block (D) is carried out.

Stage 2: Annotating Cell Values of a Canonical Spreadsheet in a Data Block. Each cell value for a data block is treated as one mention (concept). So, this mention must also already be converted to the required format (a prepro-cessing procedure has been performed). The following sequence of steps for each cell value for a data block (D) must be performed to find reference entities from DBpedia:

Step 1. Searching an instance from a resource segment of DBpedia for a cell value for a data block (D) by making an exact match.

Step 2. Forming a set of candidate entities by compiling SPARQL queries against DBpedia for matching a string to a pattern. We also select only the first 100 candidates for performance reasons. If no entity from DBpedia has been linked to a cell value for a data block (D), then this cell value is marked as not annotated and the next steps for it are skipped.

Step 3. The disambiguation procedure for all cell values from a data block (D) is also performed. We propose an aggregated method for defining similarity between candidate entities and cell value for a data block (D) to disambiguation. This method consists of the sequential application of five metrics and combining ranks obtained by each metric.

String similarity and *NER label similarity* from (1) and (2) are our basic metrics and they are applied to all cells of a source canonical spreadsheet. So, initially, these two metrics are calculated. The following three metrics are used only for cell values located in a data block (D):

1) Heading similarity. This metric is based on the hypothesis that heading values of a source canonical spreadsheet are some classes (types) which generalize entities in a data block (D). First, a set of classes C is formed for each entity e_i from a set of candidates E. Next, a set of all values L for heading blocks (R^H and C^H) is formed on the same row as v_j. The similarity between a heading value $l_k \in L$ and all classes from C is calculated. In this case, we also use the Levenshtein distance. So, an entity e_i is selected, where $c_j \in C$ for this entity will have maximum similarity for l_k:

$$f_3(e_i) = \arg\min HeadingSim\,(l_k, c_j),\qquad(5)$$

where $f_3(e_i)$ is a function for calculating header similarity; l_k is a string k-value of heading from a set of heading values L; c_j is a name of j-class from a set of classes C.

2) Semantic similarity. This metric is based on the hypothesis that data in a source spreadsheet column usually has the same type, i.e. entities usually belong to one class. This is typical for arbitrary spreadsheets with a common layout in the form of a set of columns with top headers. However, in this paper, we consider canonical spreadsheets in which all data from arbitrary spreadsheets is placed in a single column in the form of a block data (D). So, a candidate entity e_i for some cell value v_j from a data block (D) must be semantically similar to other entities selected for cell values in this block (column), and which are contained along with the same heading values for heading blocks (R^H and C^H).

Therefore, it is necessary to form a set of corresponding classes C for each entity e_i from a set of candidates E. Then, the task is to define the similarity of each class $c_k \in C_l, l = \overline{1,n}$ with classes from other sets of classes C^{all}. In this case, we also use the Levenshtein distance:

$$f_4(e_i) = SemSim\left(c_k, C^{all}\right),\qquad(6)$$

where $f_4(e_i)$ is a function for calculating semantic similarity; c_k is a name of k-class from a set of classes $C_l \in C^{all}$; C^{all} is a set of sets of classes for each entity candidate e_i, where $C^{all} = \{C_1, ..., C_n\}$.

3) Context similarity. This metric is based on the hypothesis that usually a cell value and a reference entity from a set of candidates have a shared context.

First, we select the neighboring cell values by a row and column for a certain cell to obtain its context. A context for a selected cell value will be a set of all collected cell values (mention context). Next, RDF triples are collected to define a context for each entity e_i from a set of candidates E. In this case, each subject (where a candidate entity e_i is an object) and object (where a candidate entity e_i is a subject,) from these RDF triples are selected. A context for a candidate entity e_i will be a set of all collected entities (entity context). Next, each item from a mention context is mapped to items from an entity context. We also use the Levenshtein distance for this mapping. If there is an exact match, then rank is increased by 1:

$$f_5(e_i) = \sum_{i=1}^{k} ContextSim\left(cn_i^v, cn_j^e\right), cn_i^v \in CN^v, cn_j^e \in CN^e, j = \overline{1, n}, \quad (7)$$

where $f_5(e_i)$ is a function for calculating context similarity; cn_i^v is a string value of i-mention from CN^v; cn_j^e is a name of j-entity from CN^e.

Our hypotheses for *heading similarity, semantic similarity* and *context similarity* metrics are based on the relational nature of source tables. Today, the relational form of table representation is the most common and widely used. A relational table is a set of data elements (values) that uses a model of vertical columns with a unique name and horizontal rows (tuples). Thus, we assume that such columns contain similar entities.

Step 4. Using all five similarity metrics (1)–(5) as a linear weighted convolution, we calculate the aggregate rank for candidate entities linked to cell values from a data block (D):

$$f_{agg}^D = w_1 \times \left(1 - \frac{f_1(e_i)}{100}\right) + w_2 \times f_2(e_i) \quad (8)$$
$$+ w_3 \times \left(1 - \frac{f_1(e_3)}{100}\right) + w_4 \times f_4(e_i) + w_5 \times f_5(e_i),$$

where f_{agg}^D is a function for calculating the aggregated rank for a candidate entity e_i; w_1, w_2, w_3, w_4, w_5 are weighting factors that balance the importance of ranks obtained using five similarity metrics for disambiguation. All weights are equal to one by default.

Step 5. Finding relationships (object properties in a property segment of DBpedia) between annotated non-literal cell values from a data block (D) and annotated heading values from heading blocks $(R^H$ and $C^H)$. This step is optional in the main algorithm for annotating spreadsheet cells.

Stage 3: Generating Annotated Spreadsheets. At this step, the procedure for generating an annotated spreadsheet with links (tags) to entities from the target DBpedia graph is carried out. An annotated spreadsheet is presented in the Excel format and is available for further processing.

3.3 Software Architecture

Our tool, namely TabbyLD [3], is used to annotate both a single spreadsheet and a set of spreadsheets presented in a canonicalized form. TabbyLD has a client-server architecture and is developed using the MVC (Model-View-Controller) pattern, PHP7, and Yii2 framework. The main components of TabbyLD are shown in Fig. 1:

1. *Canonical spreadsheets parser* – retrieving and analyzing source canonical spread-sheet structure in the Excel format (XLSX).
2. *Data cleaning module* – preparing (preprocess) data extracted from a source canonical spreadsheet for further qualitative entity recognition and linking, in particular, to generate correct SPARQL queries for a target knowledge graph (e.g., DBpedia). As a result, cell values are cleared from various "garbage" characters, except for letters and numbers. Multiple spaces are also removed, and single spaces are replaced by the underscore. All mentions in a cell are lowercase, with the first letter in the first word being capitalized.
3. *Canonical spreadsheets annotator* – the main software module that implements the proposed method of annotating canonical spreadsheets (see 3.2 section).
4. *LOD interaction module* – potentially different knowledge graphs can be used, we chose DBpedia.
5. *Generator of an annotated spreadsheet* in the Excel format (XLSX).
6. *Module for interacting with external tools via the REST API.*

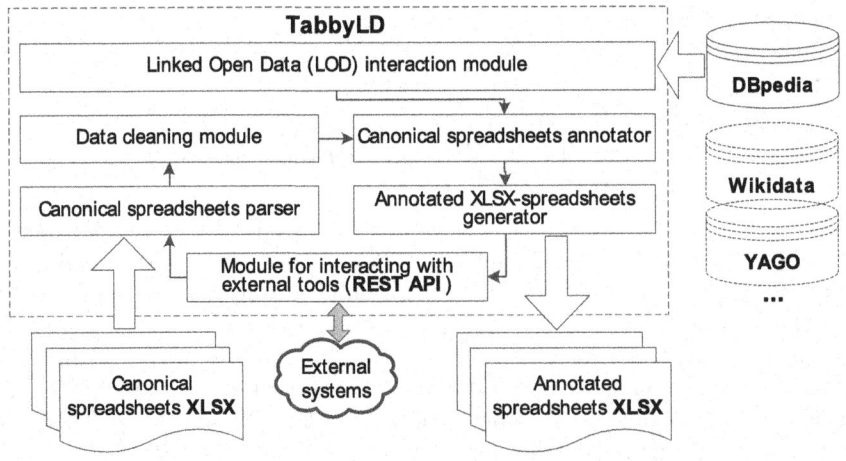

Fig. 1. Conceptual architecture of TabbyLD

3.4 Software Functions

The main functions of TabbyLD implemented by the considered architecture are the followings:

1. Importing (loading) and extracting (parsing) elements of a canonical spread-sheet (spreadsheets can be loaded in the form of an archive, or a single spread-sheet in the Excel format can be loaded).
2. Clearing extracted values and converting them to the required format for further annotation.
3. Annotating canonical spreadsheets using the DBpedia global ontology includ-ing:
   - defining the type (literal or subject, a.k.a. named-entity) for all cell values of a source canonical spreadsheet;
   - linking cell values from a data block (D) and heading blocks (R^H and C^H) to entities from a target knowledge graph (in our case, DBpedia);
   - searching for possible properties (relationships) between annotated cell values.
4. Generating the target annotated table in the Excel format (XLSX).

3.5   TabbyDOC Pipeline

One of the features of our tool is the integration into the overall pipeline of TabbyDOC project.

The TabbyDOC [4] project is aimed at creating a theoretical and instrumen-tal software platform for the accelerated development of systems for extracting data from various arbitrary spreadsheets. At the same time, this project is sup-posed to research and develop the following issues:

1. Algorithms for functional and structural analysis of spreadsheets (automatic restoration of their syntactic and semantic markup).
2. Algorithms for interpreting spreadsheets (cleaning and conceptualizing their natural language content).
3. An object model and a formal domain-specific language for transforming tab-ular data.
4. Tools for generating source code of tabular data transformation software.
5. Tools for generating relational and linked data from arbitrary spreadsheets.

Until today, the following tools are used as the main TabbyDOC project toolkit:

1. TabbyPDF [29] for extracting arbitrary spreadsheets from various PDF doc-uments.
2. TabbyXL [30] for transforming source arbitrary spreadsheets to a canonical-ized form includes stages: detection, recognition, role (functional) analysis, and structural analysis. TabbyXL is a Java console application that processes spreadsheet files in the Excel format (XLSX or CSV). Each Excel file can con-tain one or more spreadsheets with arbitrary layouts. TabbyXL uses rules to transform these spreadsheets data to the canonicalized form. Transformation rules can be presented in a domain-specific language called Cells Rule Lan-guage (CRL) and translated to the Java source code. Extracted data saved in separated files.

We suggest using TabbyLD in addition to these tools. So, we complete the full cycle of obtaining linked data based on arbitrary spreadsheets extracted from PDF documents (Fig. 2), which corresponds to the concept of five-star deployment of open data in the Web, proposed by Sir Tim Berners-Lee [2].

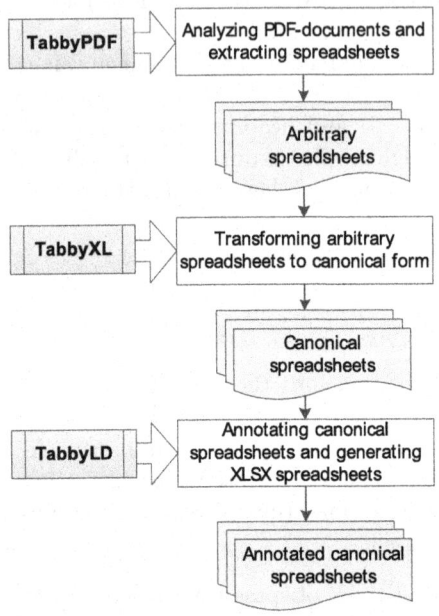

#	Media	MIX
1	Dainik Jagran	15.400
2	Dainik Bhaskar	14.000
3	CNN Editions (International)	14.000
4	CNN	12.000
5	NDTV	10.000
6	Times of India	4.800
7	Globo	4.500
8	Dailymail	4.500
9	Malayala Manorama	4.000
10	Dinamalar	3.500
11	WALL STREET JOURNAL USA	3.400
12	foxnews	3.300
13	New York Times	3.250
14	Gujarat Samachar	3.000
15	Telecinco	2.800
16	IBN live	2.800
17	USA Today	2.525
18	The Sun	2.500
19	Joong Ang Ilbo	2.250
20	AARP Bulletin	2.160

Fig. 2. TabbyLD in the TabbyDOC pipeline

Fig. 3. An example of a source arbitrary spreadsheet

4   Illustrative Example

Next, let's consider an example of forming an annotated spreadsheet in the Excel format using the proposed TabbyLD tool.

We used a spreadsheet with the media domain from the T2Dv2 Gold Standard set [5] as a source arbitrary spreadsheet (Fig. 3). This spreadsheet has only column headings (heading block C^H), and each row denotes some media objects. So, the "media" column is a subject column.

The following canonical spreadsheet is obtained (Fig. 4, a) with the use of TabbyXL by transforming a source arbitrary spreadsheet (Fig. 3). It should be noted that a resulting canonical spreadsheet does not contain row headings (heading block R^H), since it was not in a source arbitrary spreadsheet.

The NER procedure was also carried out. As a result, a canonical spreadsheet with NER labels was obtained (Fig. 4, b).

1	DATA	ColumnHeading
2	1	#
3	dainik jagran	media
4	15.400	mix
5	2	#
6	dainik bhaskar	media
7	14.000	mix
8	3	#
9	cnn editions (international)	media
10	14.000	mix
11	4	#
12	cnn	media
13	12.000	mix
14	5	#
15	ndtv	media
16	10.000	mix
17	6	#
18	times of india	media
19	4.800	mix
20	7	#
21	globo	media
22	4.500	mix
23	8	#
24	dailymail	media
25	4.500	mix
26	9	#
27	malayala manorama	media
28	4.000	mix
29	10	#
30	dinamalar	media
31	3.500	mix
32	11	#

a)

1	DATA	ColumnHeading
2	NUMBER	NONE
3	PERSON	NONE
4	NUMBER	NONE
5	NUMBER	NONE
6	PERSON	NONE
7	NUMBER	NONE
8	NUMBER	NONE
9	ORGANIZATION	NONE
10	NUMBER	NONE
11	NUMBER	NONE
12	ORGANIZATION	NONE
13	NUMBER	NONE
14	NUMBER	NONE
15	ORGANIZATION	NONE
16	NUMBER	NONE
17	NUMBER	NONE
18	NONE	NONE
19	NUMBER	NONE
20	NUMBER	NONE
21	PERSON	NONE
22	NUMBER	NONE
23	NUMBER	NONE
24	PERSON	NONE
25	NUMBER	NONE
26	NUMBER	NONE
27	PERSON	NONE
28	NUMBER	NONE
29	NUMBER	NONE
30	PERSON	NONE
31	NUMBER	NONE
32	NUMBER	NONE

b)

Fig. 4. Results of processing arbitrary spreadsheet on Fig. 3: a) a fragment of the resulted canonical spreadsheet; b) a fragment of the canonical spreadsheet with defined NER labels

Next, we used our tool to annotate a resulting canonical spreadsheet (Fig. 5).

Table 1 shows the result of semantic interpretation (annotation). This table contains extracted cell values and their corresponding referent entities from the DBpedia knowledge graph (from "http://dbpedia.org/resource/" segment). The result of each metric (LD is a string similarity from (1), NER is a NER label similarity from (2), HS is a heading similarity from (3), SS is a semantic similarity from (4), CS is a context similarity from (5)) of our aggregation method is also represented (Table 1). In this case, "+" is a marker indicating that a reference entity is defined correctly, "*" is a marker indicating that a reference entity is defined incorrectly, "−" is a marker indicating that a cell value has not been annotated.

1	DATA	ColumnHeading
2	http://dbpedia.org/resource/Number	#
3	http://dbpedia.org/resource/Dainik_Jagran	http://dbpedia.org/resource/Media
4	http://dbpedia.org/resource/Number	http://dbpedia.org/resource/Mixe
5	http://dbpedia.org/resource/Number	#
6	http://dbpedia.org/resource/Dainik_Bhaskar	http://dbpedia.org/resource/Media
7	http://dbpedia.org/resource/Number	http://dbpedia.org/resource/Mixe
8	http://dbpedia.org/resource/Number	#
9	cnn editions (international)	http://dbpedia.org/resource/Media
10	http://dbpedia.org/resource/Number	http://dbpedia.org/resource/Mixe
11	http://dbpedia.org/resource/Number	#
12	http://dbpedia.org/resource/CNN	http://dbpedia.org/resource/Media
13	http://dbpedia.org/resource/Number	http://dbpedia.org/resource/Mixe
14	http://dbpedia.org/resource/Number	#
15	http://dbpedia.org/resource/BroadbandTV_Corp	http://dbpedia.org/resource/Media
16	http://dbpedia.org/resource/Number	http://dbpedia.org/resource/Mixe
17	http://dbpedia.org/resource/Number	#
18	http://dbpedia.org/resource/The_Times_of_India	http://dbpedia.org/resource/Media
19	http://dbpedia.org/resource/Number	http://dbpedia.org/resource/Mixe
20	http://dbpedia.org/resource/Number	#
21	http://dbpedia.org/resource/Bojan_Globočnik	http://dbpedia.org/resource/Media
22	http://dbpedia.org/resource/Number	http://dbpedia.org/resource/Mixe
23	http://dbpedia.org/resource/Number	#
24	dailymail	http://dbpedia.org/resource/Media
25	http://dbpedia.org/resource/Number	http://dbpedia.org/resource/Mixe
26	http://dbpedia.org/resource/Number	#
27	http://dbpedia.org/resource/Malayala_Manorama	http://dbpedia.org/resource/Media
28	http://dbpedia.org/resource/Number	http://dbpedia.org/resource/Mixe
29	http://dbpedia.org/resource/Number	#
30	http://dbpedia.org/resource/Dinamalar	http://dbpedia.org/resource/Media
31	http://dbpedia.org/resource/Number	http://dbpedia.org/resource/Mixe
32	http://dbpedia.org/resource/Number	#

Fig. 5. A fragment of an annotated spreadsheet resulted from the canonical spreadsheet in Fig. 4

As a result, *string similarity (LD)*, *heading similarity (HS)*, and *semantic similarity (SS)* metrics performed well for this spreadsheet. However, *NER label similarity (NER)* and *context similarity (CS)* gave poor results. So, basically, $distanceLevel(c_j)$ was zero for *NER label similarity* metric. *Context similarity* metric, it was impossible to determine the context for table cell values based on the table's features (the table contains essentially only one column). The context for values of spreadsheet cells based on characteristics of this spreadsheet (spreadsheet contains only one named-entity column) could not be determined for context similarity metric.

Thus, the resulting annotated spreadsheet (Fig. 5), unlike the source spreadsheet (Fig. 3), contains explicit semantics in the form of URI tags from DBpedia, that can be used for the semantic search on tabular data and question-answering systems. In fact, such spreadsheets automate data mining and improve the general semantic interoperability of spreadsheets, for example, in the context of e-business. Generation of domain ontologies and linked data is also possible based on annotated spreadsheets.

Table 1. A result of semantic interpretation (annotation) of data extracted from a canonical spreadsheet (Fig. 4)

Cell value	DBpedia instance	LD	NER	HS	SS	CS
globo	Il_Globo	*	*	+	+	−
times of india	The_Times_of_India	+	−	+	+	−
foxnews	Fox_News_Channel	−	−	−	−	
dinamalar	Dinamalar	+	−	+	+	−
malayala manorama	Malayala_Manorama	+	−	+	+	−
dainik jagran	Dainik_Jagran	+	−	+	+	−
new york times	The_New_York_Times	+	−	+	+	−
wall street journal usa	The_Wall_Street_Journal	−	−	−	−	−
the sun	The_Sun_(United_Kingdom)	*	−	*	+	−
gujarat samachar	Gujarat_Samachar	+	−	+	+	−
dainik bhaskar	Dainik_Bhaskar	+	−	+	+	−
dailymail	Mail_Online	−	−	−	−	−
ibn live	CNN−IBN	−	−	−	−	−
joong ang ilbo	JoongAng_Ilbo	−	−	−	−	−
usa today	USA_Today	+	−	+	+	−

5   Preliminary Experimental Evaluation and Discussion

The T2Dv2 Gold Standard [5] is selected as the main dataset for experimental evaluation. The T2D Gold Standard provides a large set of human-generated correspondences between a public Web table corpus and the DBpedia knowledge graph. We selected all 287 spreadsheets that have correspondences with DBpedia instances, and also 150 random negative examples of non-relational spreadsheets that do not have annotations with DBpedia for our experiment.

We used standard measures of accuracy, precision, recall, and F1 score for preliminary experimental evaluation of semantic interpretation with TabbyLD:

$$Accuracy = \frac{|A|}{|S|}, Precision = \frac{|C \cap A|}{|A|}, Recall = \frac{|C \cap A|}{|S|}, \qquad (9)$$

$$F1 = \frac{2 \times Precision \times Recall}{Precision + Recall}, \qquad (10)$$

where C is a set of correctly annotated cell values; A is a set of annotated cell values (including both positive and negative annotation examples); S is a total number of cell values in a source canonical spreadsheet.

Following evaluations were obtained for annotating 237 positive examples of canonical spreadsheets from T2Dv2: *accuracy* = 0,74; *precision* = 0,73; *recall* = 0,5; *F1* = 0,58. The total runtime spent for semantic interpretation of spreadsheet data amounted to 70275,2 s (19,2 h).

The accuracy measure for 150 negative examples of non-relational spreadsheets turned out to be the same as for positive examples. The main purpose of the experiment with negative examples was to demonstrate that TabbyLD is suitable for annotating different spreadsheets including non-relational spreadsheets for which it is difficult to find annotations in a target knowledge graph.

Many studies have been conducted on the semantic table interpretation over the past decade [6–19]. Some of them provide an experimental evaluation with the T2Dv2 dataset. We selected several approaches with the highest results for comparison with our tool. Results of the quantitative comparison of accuracy, recall, and F1 values are presented in Table 2.

Table 2. A quantitative comparison of experimental evaluation for T2Dv2 dataset

Tool	Precision	Recall	F1
TACKO (2019) [11]	0,92	0,86	0,89
TableMiner+ (2017) [7]	0,96	0,68	0,80
T2KMatch (2015) [9]	0,94	0,73	0,82
TabbyLD	0,73	0,50	0,58

The result obtained turned out to be modest. This is mainly because we were not focusing on a specific table dataset, unlike other approaches. So, we have tried to take the first step towards a universal approach to the semantic interpretation of spreadsheet data. Existing approaches are also primarily aimed at the interpretation of tabular data extracted from web pages. Our tool aimed at the interpretation of spreadsheets data represented in the Excel format.

We defined the main issues that resulted to the current score:

1. Drawbacks of the algorithm for forming an initial set of candidate entities, which makes recall low. In the future, we plan to use a more flexible algorithm for generating candidate entities, for example, based on Elasticsearch or Solar. We also plan to optimize SPARQL queries against DBpedia for better matching a string to a pattern.
2. Drawbacks of the algorithm for annotating data cells with large text (cell values in data and heading blocks are interpreted as a single value). In the future, we plan to divide cell value into parts and annotate these parts separately.
3. Inaccurate calibration of metric ranks using weighting factors. In the future we plan to use special models, for example, graph models, Markov chains for this purpose.
4. Minor errors during the analysis and transformation of source arbitrary spreadsheets led to the absence of some values in cells for canonical spreadsheets. In our evaluation, we identified six such spreadsheets.

Note that our tool is aimed at processing canonical spreadsheets such as Excel. Such spreadsheets can be obtained from arbitrary spreadsheets with different formats and layouts. This opened up new vistas for processing different types of spreadsheets. However, our method for annotating spreadsheets is focused on processing canonical spreadsheets with headers (R^H and C^H) and do poorly with spreadsheets where there are no headers. TabbyLD will not work for such spreadsheets. In the future, we plan to improve our method and tool to determine the semantic of cell data without column or row headings or both.

6   Conclusions

The semantic interpretation of spreadsheets data stays an area of active scientific research. Existing methods, approaches, and tools in this area have limitations and drawbacks both for the considered spreadsheet layouts and for the domains covered. In this paper, we describe a tool, namely TabbyLD, for semantic interpretation of spreadsheets data presented in the Excel format.

The main features of our proposal are to use a canonicalized form for arbitrary spreadsheets representation and annotation and use a named entity recognition procedure for each cell of a transformed spreadsheet. Also, we use five heuristic-based similarity metrics based on the recognized types of named entities for disambiguating and identifying the most suitable (reference) DBpedia entities to linking them with cell values. TabbyLD integrates with the TabbyXL tool for getting canonical spreadsheets.

We used the T2Dv2 Gold Standard for preliminary experimental evaluation of our tool. Experiments have shown their applicability for semantic interpretation of spreadsheets data.

In the future, we plan to eliminate the identified shortcomings and make our tool publicly available in the form of SaaS to solve the problem of annotating various spreadsheets. We also plan to implement support for generating RDF triplets based on the transformation of annotated spreadsheet data.

Acknowledgement. This work was supported by the Russian Science Foundation, grant number 18-71-10001.

References

1. Lehmberg, O., Ritze, D., Meusel, R., Bizer, C.: A large public corpus of web tables containing time and context metadata. In: Proceedings of the 25th International Conference Companion on World Wide Web, pp. 75–76 (2016). https://doi.org/10.1145/2872518.2889386
2. Star Open Data. https://5stardata.info. Accessed 19 Oct 2020
3. TabbyLD. https://github.com/tabbydoc/tabbyld. Accessed 19 Oct 2020
4. Shigarov, A.O., et al.: Towards End-to-End Transformation of Arbitrary Tables from Untagged Portable Documents (PDF) to Linked Data. CEUR Workshop Proceedings for the 2nd Scientific-practical Workshop Information Technologies: Algorithms, Models, Systems, vol. 2463, pp. 1–12 (2019)
5. T2Dv2 Gold Standard for Matching Web Tables to DBpedia. http://webdatacommons.org/webtables/goldstandardV2.html. Accessed 19 Oct 2020

6. de Vos, M., Wielemaker, J., Rijgersberg, H., Schreiber, G., Wielinga, B., Top, J.: Combining information on structure and content to automatically annotate natural science spreadsheets. Int. J. Hum. Comput. Stud. **130**, 63–76 (2017). https://doi.org/10.1016/j.ijhcs.2017.02.006

7. Zhang, Z.: Effective and Efficient Semantic Table Interpretation using TableMiner+. Semantic Web **8**(6), 921–957 (2017). https://doi.org/10.3233/sw-160242

8. Ermilov, I. Ngomo, A.-C.N.: TAIPAN: automatic property mapping for tabular data. In: Proceedings of the 20th International Conference on European Knowledge Acquisition Workshop, EKAW, pp. 163–179 (2016). https://doi.org/10.1007/978-3-319-49004-5_11

9. Ritze, D., Lehmberg, O., Bizer, C.: Matching HTML tables to DBpedia. In: Proceedings of the 5th International Conference on Web Intelligence, Mining and Semantics (WIMS'15), pp. 1–6 (2015). https://doi.org/10.1145/2797115.2797118

10. Mulwad, V., Finin, T., Joshi, A.: A Domain Independent Framework for Extracting Linked Semantic Data from Tables. In: Ceri, S., Brambilla, M. (eds.) Search Computing. LNCS, vol. 7538, pp. 16–33. Springer, Heidelberg (2012). https://doi.org/10.1007/978-3-642-34213-4_2

11. Kruit, B., Boncz, P., Urbani, J.: Extracting novel facts from tables for knowledge graph completion. In: Proceedings of the 18th International Semantic Web Conference (ISWC 2019), pp. 364–381 (2019). https://doi.org/10.1007/978-3-030-30793-6_21

12. Efthymiou, V., Hassanzadeh, O., Rodriguez-Muro, M., Christophides, V.: Matching web tables with knowledge base entities: from entity lookups to entity embeddings. In: Proceedings of the 16th International Semantic Web Conference (ISWC 2017), pp. 260–277 (2017). https://doi.org/10.1007/978-3-319-68288-4_16

13. Ell, B., et al.: Towards l. In: Proceedings of the 5th International workshop on Linked Data for Information Extraction (LD4IE), pp. 1–12 (2017)

14. Wu, T., Yan, S., Piao, Z., Xu, L., Wang, R., Qi, G.: Entity linking in web tables with multiple linked knowledge bases. In: Proceedings of the 6th Joint International Semantic Technology Conference (JIST), pp. 239–253 (2016). https://doi.org/10.1007/978-3-319-50112-3_18

15. Venetis, P., et al.: Recovering semantics of tables on the web. Proc. VLDB Endowment, 528–538 (2011). https://doi.org/10.14778/2002938.2002939

16. Wang, J., Wang, H., Wang, Z., Zhu, K.Q.: Understanding tables on the web. In: Proceedings of the 31th International Conference on Conceptual Modeling (ER), pp. 141–155 (2012). https://doi.org/10.1007/978-3-642-34002-4_11

17. Shen, W., Wang, J., Luo, P., Wang, M.: LIEGE: link entities in web lists with knowledge base. In: Proceedings of the 18th ACM SIGKDD Conference on Knowledge Discovery and Data Mining, pp. 1424–1432 (2012). https://doi.org/10.1145/2339530.2339753

18. Muñoz, E., Hogan, A., Mileo, A.: Using linked data to mine RDF from wikipedia's tables. In: Proceedings of the 7th ACM International Conference on Web Search and Data Mining, pp. 533–542 (2014). https://doi.org/10.1145/2556195.2556266

19. Bhagavatula, C.S., Noraset, T., Downey, D.: TabEL: entity linking in web tables. In: Proceedings of the 14th International Semantic Web Conference (ISWC 2014), pp. 425–441 (2015). https://doi.org/10.1007/978-3-319-25007-6_25

20. Bizer, C., et al.: DBpedia - a crystallization point for the web of data. J. Web Semantics **7**(3), 154–165 (2009). https://doi.org/10.1016/j.websem.2009.07.002

21. Dorodnykh, N.O., Yurin, A.Y., Shigarov, A.O.: Conceptual model engineering for industrial safety inspection based on spreadsheet data analysis. In: Simian, D., Stoica, L.F. (eds.) MDIS 2019. CCIS, vol. 1126, pp. 51–65. Springer, Cham (2020). https://doi.org/10.1007/978-3-030-39237-6_4

22. Yurin, A.Yu., Dorodnykh, N.O.: A reverse engineering process for inferring conceptual models from canonicalized tables. In: Proceedings of the 2019 International Multi-Conference on Engineering, Computer and Information Sciences (SIBIRCON), pp. 485–490 (2020). https://doi.org/10.1109/SIBIRCON48586.2019.8958458

23. Shigarov, A.O., Mikhailov, A.A.: Rule-based spreadsheet data transformation from arbitrary to relational tables. Inf. Syst. **71**, 123–136 (2017). https://doi.org/10.1016/j.is.2017.08.004

24. Tijerino, Y.A., Embley, D.W., Lonsdale, D.W., Ding, Y., Nagy, G.: Towards ontology generation from tables. World Wide Web Internet Web Inf. Syst **8**(8), 261–285 (2005). https://doi.org/10.1007/s11280-005-0360-8

25. SPARQL 1.1 Query Language. https://www.w3.org/TR/sparql11-query/. Accessed 19 Oct 2020

26. Levenshtein, V.I.: Binary codes capable of correcting deletions, insertions, and reversals. Technical report 8, Soviet Physics Doklady (1966)

27. Stanford CoreNLP, https://stanfordnlp.github.io/CoreNLP/. Last accessed 19 Oct 2020

28. Stanford CoreNLP - Named Entity Recognition. https://stanfordnlp.github.io/CoreNLP/ner.html. Accessed 19 Oct 2020

29. TabbyPDF. PDF table extraction tool. http://cells.icc.ru/pdfte/. Accessed 19 Oct 2020

30. Shigarov, A., Khristyuk, V., Mikhailov, A.: STabbyXL: software platform for rule-based spreadsheet data extraction and transformation. SoftwareX **10**, 100270 (2019). https://doi.org/10.1016/j.softx.2019.100270

In-Depth Insights into Swarm Intelligence Algorithms Performance

Eva Tuba[1(✉)] , Peter Korošec[2] , and Tome Eftimov[2]

[1] Faculty of Informatics and Computing, Singidunum University, Belgrade, Serbia
etuba@ieee.org
[2] Computer Systems Department, Jožef Stefan Institute, Ljubljana, Slovenia
{peter.korosec,tome.eftimov}@ijs.si

Abstract. Solving hard optimization problems is one of the most important research topics due to the countless applications in different areas. Since solving such problems is of great importance, numerous metaheuristics were developed, many of which belong to the group of swarm intelligence optimization algorithms. In recent decades, there has been an explosion in the number of the proposed swarm intelligence algorithms most commonly compared to other metaheuristics using one statistic such as average or median which can lead to putting algorithms in different rankings even though there are only small differences between them. In order to provide more insights into swarm intelligence algorithms' performance, a deep statistical comparison is used. Five representative swarm intelligence optimization algorithms are ranked based on the obtained solutions values and their distribution in the search space while solving the CEC2013 benchmark functions. The used analysis differentiates algorithms that have statistically significant performance and measure the qualities of the exploration and the exploitation abilities of the tested algorithms.

Keywords: Swarm intelligence algorithms · Statistical analysis · Deep Statistical Comparison · Ranking optimization metaheuristics

1 Introduction

Optimization is one of the most widely used techniques, not only in science but also in everyday life in general since numerous problems can be defined as optimization problems. These problems can be simple for solving and some deterministic methods can be used but in reality problems are becoming more complicated so they cannot be solved by these methods. In order to solve such problems that are important for real-life applications, scientists turned to nature and tried to imitate the processes and phenomena. The search was guided by simulating processes from nature. It was a truly revolutionary idea that provided good solutions to the difficult optimization problems, and most importantly, these results were achieved in a very short time. Nature inspired algorithms are

© Springer Nature Switzerland AG 2021
D. Simian and L. F. Stoica (Eds.): MDIS 2020, CCIS 1341, pp. 334–346, 2021.
https://doi.org/10.1007/978-3-030-68527-0_21

a subset of Swarm Intelligence (SI) algorithms where a search is done by using a simple agent and their interaction.

Numerous swarm intelligence and nature inspired optimization algorithms were proposed in the last two decades. At some point in time, the search for new and original inspiration from nature and environments for the optimization algorithms was a hot research topic. This resulted in numerous algorithms that have similar or even the same mathematical background but rather different inspiration thus they were declared as the different algorithms. Nowadays, it is necessary to provide some kind of classification, comparison and analysis of these algorithms. The common practice was to compare new SI algorithms with the existing meta-heuristics by using one statistic such as average or median. A comparative analysis and ranking of five swarm intelligence algorithms based on the average was presented in [6]. The problem with such analysis is that algorithms can be ranked differently even though the differences are not significant. In order to provide a more meaningful analysis and comparison of SI algorithms, Deep Statistical Comparison (DSC) was used in the paper [3]. Rankings obtained by the DSC that use the distributions of the performance measure from multiple runs are more resistant to outliers and minor differences compared to the rankings obtained by using one statistic.

In this paper, five SI algorithms, Particle Swarm Optimization (PSO), Artificial Bee Colony (ABC), Firefly Algorithm (FA), Flower Pollination Algorithm (FPA), and Bare Bones Fireworks Algorithm (BBFWA), have been compared and ranked by the DSC. The selected algorithms are defined in the Sect. 2, while short description of the DSC is given in Sect. 3. Evaluation of the used SI algorithms is presented in Sect. 4. The paper is concluded in Sect. 5.

2 Swarm Intelligence Algorithms

In this paper, five swarm intelligence algorithms were tested and compared. These algorithms were selected based on the exploration and exploitation processes as well as on their quality reported in the literature. The first algorithm is the particle swarm optimization (PSO), which is the first proposed SI algorithm. The second one is the artificial bee colony (ABC), which have a rather different search process in comparison with the majority of the SI algorithms. The third one is the firefly algorithm (FA) that has been proved to be very efficient. The main drawback of the FA is computational time. The fourth algorithm is the flower pollination algorithm (FPA), a simple algorithm that has only one parameter that needs to be tuned. The last analyzed algorithm is the bare bones fireworks algorithm (BBFWA), one of the newest and the simplest SI algorithms.

2.1 Particle Swarm Optimization

Particle swarm optimization (PSO) is among the first swarm intelligence algorithms that were proposed. The PSO was introduced by Kennedy and Eberhart in 1995 [10]. It is well known and widely studied algorithm applied to numerous hard optimization problems. In the PSO algorithm, each of n solutions in the population is updated in each iteration based on the best known solution of the population and by its own best solution through iteration. New solutions are generated by using velocity vector defined as follows:

$$x_i^{t+1} = x_i^t + v_i^t, \tag{1}$$

$$v_i^{t+1} = v_i^t + c_1 r_1(x_i^{best} - x_i^t) + c_2 r_2(gBest - x_i^t) \tag{2}$$

where c_1 and c_2 are the algorithm's parameters, r_1 and r_2 are random numbers, x_i^{best} is the best solution of the solution i and $gBest$ is the best solution in the whole population.

2.2   Artificial Bee Colony

Artificial bee colony algorithm (ABC) was proposed in 2005 by Karaboga [7] and further improved by Karaboga and Bastuk [8,9]. ABC algorithm utilizes three classes of artificial bees: employed bees, onlookers and scouts. Exploration and exploitation processes were done by different three different solution replacement procedures: so-called onlooker and employed bees are used for implementing the exploitation and scouts perform the exploration.

In the ABC algorithm, solutions x_i $(i = 1, 2, ...SN)$ in D-dimensional space, where D represents problems dimension are initialized randomly in the search space. The employed bees are used for exploitation and a new solution is generated by combining two neighborhood solutions:

$$v_i = \begin{cases} x_i + \phi * (x_i - x_k), R < MR \\ x_i, otherwise \end{cases} \tag{3}$$

where x_i the old solution, x_k is a neighbor solution k, ϕ and R represent random numbers while MR denotes a modification rate which is an algorithm's control parameter. A newly generated solution is kept only if it is better than the old one. If the solution is not improved in a certain number of iterations, the solutions is abandoned and a new random solution is generated. Number of iterations is defined as an algorithm's parameter called *limit* or *abandonmentcriteria*.

Onlookers will additionally search the neighborhood of a solution x_i with a probability proportional to the solution's fitness:

$$p_i = \frac{fitness_i}{\sum_{i=1}^{SN} fitness_i}, \tag{4}$$

Scouts are used for the exploration process where in each iteration a certain number of random solutions are generated.

2.3   Firefly Algorithm

The firefly algorithm was proposed in 2009 by Yang [20] and it was applied to various optimization problems [13,18]. Compared to the PSO, the FA algorithm's displacement vector is based on all better solutions for each of n solutions in the population, it is not just based on the current global and local best solution. The influence of the better solution depends on the distance from the current solution. The displacement vector toward the better solution x_j^t for the solution x_i^t where $i = 1, 2, \ldots, n$ and t is the current iteration number is obtained by the following equation:

$$\Delta x_i = \beta e^{\gamma r_{ij}^2}(x_j^t - x_i^t) + \alpha_t \epsilon_i^t \tag{5}$$

where β represents the parameter that controls the attractiveness of the better solution, α denotes a randomization parameter, ϵ_i^t is a vector of dimensions D which is the problem dimension and it is generated from the Gaussian or uniform distribution, and r_{ij} represents the Cartesian distance between the solutions i and j. The parameter γ is used to control the convergence speed and its value is usually between 0.01 and 100. Influence of the better solution on the displacement vector is obtained by the following equation:

$$\beta(r) = \frac{\beta_0}{1 + \gamma r^2}, \tag{6}$$

where β_0 is the parameter of the FA algorithm that represents maximal attractiveness (when the distance $r = 0$).

2.4   Flower Pollination Algorithm

Flower pollination algorithm (FPA) was introduced by Yang in 2012 [21]. Global search is performed with the probability p while the exploitation occurs with the $1 - p$ probability. The movement towards the best solution in the population, x^t is determined by the following equations:

$$L(s) \approx \frac{\lambda \Gamma(\lambda) \sin(\pi \frac{\lambda}{2})}{\pi s^{1+\lambda}}, \quad (s \gg s_0 > 0)$$

$$s \leftarrow Levy(s_0, \gamma)$$

$$x = x_i^t + s(x^t * -x_i^t) \tag{7}$$

where Γ is a standard gamma function while λ and s_0 are algorithms parameters, s is N-dimensional step vector from a L'evy distribution, $i = 1, 2, \ldots, N$ is the solution's index and t is the current iteration number. Local search is defined by using two randomly selected solutions. It is performed via movement towards them with randomly selected step size ϵ [21]:

$$\epsilon = rand(0, 1) \tag{8}$$

$$r, q \leftarrow rand\_integer(1, N)$$

$$x_i^{t+1} = x_i^t + \epsilon(x_q^t - x_r^t) \tag{9}$$

where N is the population size.

2.5   Bare Bones Fireworks Algorithm

The bare bones fireworks algorithm (BBFWA) represents a recent swarm intelligent algorithms with computationally inexpensive exploration and exploitation operators. It was proposed by Li and Tan in 2018 [11] as simplified version of the fireworks algorithm (FWA) initially proposed Tan and Zhu in 2010 [15]. It has been widely used in the last decade [17,19]. The BBFWA generates n new solutions in each generation and keeps the best one as a referent point for the next one. Solutions are always generated randomly in the D-dimensional hypercube around the current best solution where D is the dimension of the problem while the size of the hypercube's side is regulated by two algorithm's parameters, $C_a > 1$ and $0 < C_r < 1$. The side of the hypercube will be increased by the factor C_a if the new best solution is found in the current generation and it will be decreased by factor C_r otherwise.

3   Deep Statistical Comparison

Deep Statistical Comparison (DSC) is a recently proposed approach for making a statistical comparison of meta-heuristic stochastic optimization algorithms using a set of benchmark problems [3]. The idea behind it, is that for each single benchmark problem, the distributions of the performance measure from multiple runs obtained for each algorithm are compared and the algorithms are ranked. By performing this, the DSC rankings obtained for each single benchmark problem have been already tested for a statistical significance. These rankings are more robust on outliers and small differences that exist between the data values, which can be a problem when using either mean or median of the performance measure from multiple runs for each algorithm.

In our study, we compare the SI algorithms twice, once with regard to the obtained solution value for which we used the basic DSC approach with the two-sample Anderson-Darling test, and once with regard to the distribution of the obtained solutions in the search space for which we used the extended DSC (eDSC) approach [2]. The experiments were performed using the DSCTool, which is a web-service-based e-Learning tool [4].

4   Evaluation of the Swarm Intelligence Algorithms

4.1   Experimental Setup

Five swarm intelligence algorithms have been tested on the CEC 2013 benchmark data set that contains 28 functions, 5 unimodal, 15 modal, and 8 composite functions [12]. We tested all algorithms for dimensions 10 and 30. The maximal number of the fitness function evaluation was set to $10.000 * D$, where D is the problem dimension and each test was run 30 times with different random seeds. Search range in all tests was $[-100, 100]$.

Parameters for the algorithms are summarized in the Table 1. In this paper, we used algorithms' parameters from the literature that were proved to be good for the considered benchmark dataset. The ABC algorithm's parameters were set according to the recommendations in [14]. Parameters for the BBFWA were set based on the results presented in [11] were the same benchmark data set was used. Based on the experiments conducted in [11], it was established that the best values for the BBFWA are $C_a = 1.2$ and $C_r = 0.9$ when CEC 2013 benchmark data set is used. Parameters for the FA were set based on the results presented in [16]. The probability parameter for the FPA was set based on the study presented in [1] where $p = 0.8$. In this paper, the PSO version presented in [22] was used. The standard PSO was adjusted and applied to CEC 2013 benchmark data set and the results were submitted for the competition section [22]. The same PSO parameters were used in this paper as the one proposed in [22].

Table 1. Parameters for the algorithms ABC, BBFWA, FA, FPA, and PSO

Algorithm	Parameters
ABC [14]	Population size = 100
	Modification rate = 1
	Limit = 250
BBFWA [11]	Population size = 100;
	$C_a = 1.2$; $C_r = 0.9$;
FA [16]	Population size = 25
	$\alpha = 0.5$; $\beta_0 = 0.2$; $\gamma = 1$;
FPA [1]	Population size = 25
	$p = 0.8$
PSO [22]	Population size = 50;
	$w = 0.72984$; $c_1 = 2.05 * w$; $c_2 = 2.05 * w$;

4.2   Results and Discussion

First, the statistical analysis was performed with regard to the obtained solutions values for which we used the original DSC approach. The DSC rankings for each algorithm on each problem in 10D and 30D are presented in Table 2(a) and 2(b), respectively. To make a general conclusion using all 28 benchmark problems, the obtained rankings were further analyzed by applying a relevant omnibus statistical test, which in both, 10D and 30D, is the non-parametric Friedman test.

In the case of 10D, the obtained p-value from the Friedman test is 0.00, which is less than the predefined statistical significance 0.05, so the null hypothesis is rejected and there is a statistical significance between the performance of the algorithms. However, to see where this significance comes from, we continued with a post-hoc test by applying multiple comparison with a control algorithm. In this paper, the algorithm with the lowest average ranking was selected as a control algorithm. The average DSC rankings through all benchmark problems

Table 2. DSC rankings for the algorithms ABC, BBFWA, FA, FPA, and PSO

	(a) 10D						(b) 30D				
	ABC	BBFWA	FA	FPA	PSO		ABC	BBFWA	FA	FPA	PSO
f1	5.0	3.0	4.0	1.5	1.5	f1	5.0	1.5	3.0	4.0	1.5
f2	5.0	2.5	2.5	1.0	4.0	f2	5.0	2.5	2.5	1.0	4.0
f3	5.0	3.0	1.0	2.0	4.0	f3	4.0	5.0	1.0	2.5	2.5
f4	5.0	1.0	3.0	2.0	4.0	f4	5.0	1.0	2.0	3.0	4.0
f5	5.0	3.0	4.0	1.5	1.5	f5	5.0	2.0	3.0	4.0	1.0
f6	4.0	2.5	5.0	1.0	2.5	f6	5.0	3.5	1.0	3.5	2.0
f7	5.0	2.5	1.0	4.0	2.5	f7	5.0	2.5	1.0	4.0	2.5
f8	3.0	2.0	5.0	4.0	1.0	f8	3.0	3.0	3.0	3.0	3.0
f9	5.0	2.0	1.0	4.0	3.0	f9	5.0	2.0	1.0	3.5	3.5
f10	5.0	3.5	2.0	1.0	3.5	f10	5.0	1.5	1.5	4.0	3.0
f11	4.5	4.5	1.5	3.0	1.5	f11	5.0	3.5	1.0	3.5	2.0
f12	5.0	3.0	1.0	3.0	3.0	f12	5.0	3.0	1.0	4.0	2.0
f13	4.5	4.5	1.0	2.5	2.5	f13	5.0	3.0	1.0	4.0	2.0
f14	5.0	4.0	2.0	3.0	1.0	f14	5.0	4.0	1.0	3.0	2.0
f15	5.0	3.0	1.0	4.0	2.0	f15	5.0	2.5	1.0	4.0	2.5
f16	4.5	2.0	1.0	4.5	3.0	f16	4.5	2.0	1.0	4.5	3.0
f17	5.0	3.5	1.5	3.5	1.5	f17	5.0	3.0	1.0	4.0	2.0
f18	4.5	4.5	1.0	2.5	2.5	f18	4.5	2.5	1.0	4.5	2.5
f19	5.0	4.0	2.0	3.0	1.0	f19	5.0	2.5	1.0	4.0	2.5
f20	5.0	2.0	4.0	3.0	1.0	f20	5.0	2.0	4.0	3.0	1.0
f21	5.0	4.0	3.0	1.5	1.5	f21	5.0	3.0	2.0	4.0	1.0
f22	5.0	4.0	1.5	3.0	1.5	f22	5.0	4.0	1.0	3.0	2.0
f23	5.0	3.0	1.0	4.0	2.0	f23	5.0	2.0	1.0	4.0	3.0
f24	5.0	3.0	4.0	1.0	2.0	f24	5.0	2.0	1.0	4.0	3.0
f25	5.0	1.0	2.0	4.0	3.0	f25	4.0	2.0	1.0	5.0	3.0
f26	5.0	4.0	3.0	1.5	1.5	f26	5.0	2.0	4.0	1.0	3.0
f27	5.0	1.5	1.5	4.0	3.0	f27	5.0	2.0	1.0	4.0	3.0
f28	5.0	1.0	4.0	2.0	3.0	f28	5.0	3.0	2.0	4.0	1.0

are: 4.821 (ABC), 2.910 (BBFWA), 2.303 (FA), 2.678 (FPA), and 2.285 (PSO), thus the PSO will be used as a control algorithm. The PSO was compared with all other algorithms using a post-hoc test appropriate for the Friedman test and the obtained p-values were corrected using the Holm correction. The obtained p-values are: 0.00 (PSO, ABC), 0.20 (PSO, BBFWA), 0.48 (PSO, FA), and 0.35 (PSO,FPA). These results indicate that there is a statistical significant difference between the PSO and ABC, but the performance of all other algorithms (even if all pairwise comparisons are done) are not statistically significant.

In the case of 30D, the Friedman's p-value is again smaller than 0.05, so there is a statistical significance between the performance of the five algorithms. Next,

Table 3. eDSC rankings for the algorithms ABC, BBFWA, FA, FPA, and PSO

(a) 10D

	ABC	BBFWA	FA	FPA	PSO
f1	4.5	3.0	4.5	1.5	1.5
f2	5.0	3.0	2.0	1.0	4.0
f3	5.0	4.0	1.0	2.0	3.0
f4	5.0	1.0	3.0	2.0	4.0
f5	5.0	3.5	3.5	2.0	1.0
f6	5.0	3.5	2.0	1.0	3.5
f7	5.0	3.0	1.0	4.0	2.0
f8	5.0	3.0	1.0	4.0	2.0
f9	5.0	3.0	1.0	4.0	2.0
f10	5.0	4.0	2.0	1.0	3.0
f11	3.0	5.0	2.0	4.0	1.0
f12	3.0	5.0	1.0	2.0	4.0
f13	3.0	3.0	3.0	3.0	3.0
f14	5.0	3.0	1.0	4.0	2.0
f15	5.0	4.0	1.0	3.0	2.0
f16	5.0	2.0	1.0	4.0	3.0
f17	1.0	5.0	2.0	4.0	3.0
f18	2.0	5.0	1.0	3.0	4.0
f19	2.0	5.0	1.0	4.0	3.0
f20	5.0	2.0	4.0	3.0	1.0
f21	2.0	2.0	2.0	4.5	4.5
f22	5.0	3.0	1.0	4.0	2.0
f23	5.0	3.0	1.0	4.0	2.0
f24	5.0	3.0	1.0	3.0	3.0
f25	5.0	3.0	1.0	4.0	2.0
f26	4.5	4.5	1.5	1.5	3.0
f27	5.0	3.0	1.0	4.0	2.0
f28	2.0	4.0	1.0	4.0	4.0

(b) 30D

	ABC	BBFWA	FA	FPA	PSO
f1	5.0	1.0	3.0	4.0	2.0
f2	5.0	3.0	1.0	2.0	4.0
f3	5.0	4.0	1.0	3.0	2.0
f4	5.0	1.0	2.0	3.5	3.5
f5	5.0	2.5	2.5	4.0	1.0
f6	5.0	3.5	1.0	3.5	2.0
f7	5.0	3.5	1.0	3.5	2.0
f8	5.0	3.0	1.0	4.0	2.0
f9	5.0	2.5	2.5	2.5	2.5
f10	5.0	1.5	1.5	3.5	3.5
f11	4.5	4.5	1.5	3.0	1.5
f12	5.0	2.5	1.0	4.0	2.5
f13	5.0	3.0	1.5	4.0	1.5
f14	5.0	3.0	1.0	4.0	2.0
f15	5.0	3.0	1.0	4.0	2.0
f16	5.0	3.0	1.0	4.0	2.0
f17	5.0	3.0	1.0	4.0	2.0
f18	5.0	3.0	1.0	4.0	2.0
f19	5.0	2.0	1.0	4.0	3.0
f20	5.0	2.0	4.0	3.0	1.0
f21	5.0	2.0	1.0	3.0	4.0
f22	5.0	2.0	1.0	4.0	3.0
f23	5.0	2.5	2.5	2.5	2.5
f24	5.0	2.0	1.0	4.0	3.0
f25	4.0	2.5	1.0	5.0	2.5
f26	5.0	2.0	3.0	1.0	4.0
f27	2.0	4.0	1.0	4.0	4.0
f28	5.0	3.0	2.0	4.0	1.0

a post-hoc analysis was done and the FA has been used as a control algorithm (it has the lowest average ranking). The average DSC rankings are: 2.589 (BBFWA), 4.821 (ABC), 1.607 (FA), 3.571 (FPA), and 2.410 (PSO). The Holm adjusted p-values are: 0.00 (FA, ABC), 0.02 (FA, BBFWA), 0.00 (FA, FPA), and 0.02 (FA, PSO). These results show that the FA is superior in 30D, since it is statistically significant to other four algorithms.

Table 3 presents the eDSC rankings when the algorithms are compared with regard to the distribution of the solutions in the search space, in 10D and 30D. We should mention here that the obtained results from the eDSC ranking scheme prefer clustered solutions in the search space. If we are interested in sparse solutions (e.g., depending from the application) the eDSC rankings will be vice-versa. For both dimensions, the rankings are further analyzed by the Friedman test. In the case of 10D, the p-value is 0.00, which means that there is a statistical significance between the compared algorithms with regard to the distributions of the solutions in the search space. The lowest ranked algorithm, which means the best in our case, or the algorithm that has the most clustered solutions, is the FA. Next, by performing a post-hoc test using the FA as a control algorithm, it follows that FA has statistically significant performance than all other algorithms. The same result is also obtained for 30D, so the FA algorithm finds more clustered solutions than all other algorithms.

If the DSC and eDSC rankings are combined, in addition to the information about the statistical significance, more insights into the performance of the algorithms could be obtained by applying performance2vec approach [5]. By using this approach, the obtained rankings for each algorithm are transformed in their vector representations (i.e. embedding). These vector representations are further analyzed by hierarchical clustering in order to find clusters of algorithms with similar performances and clusters of problems that are similarly solvable by the selected algorithm portfolio. In this paper, the portfolio consists of five SI algorithms.

The results from the hierarchical clustering in 10D are presented in Fig. 1. Please note that the algorithms' order in the heatmap is not the same. This holds true for the problems' order as well. From the presented figures, it can be seen that the ABC has the worst exploration powers in the majority of problems, since the solutions provided by the ABC are not clustered (red markings in Fig. 1b) and for the same problems the ABC rankings in Fig. 1a are also marked as red. Since exploitation could be hindered by the poor exploration, it cannot be said

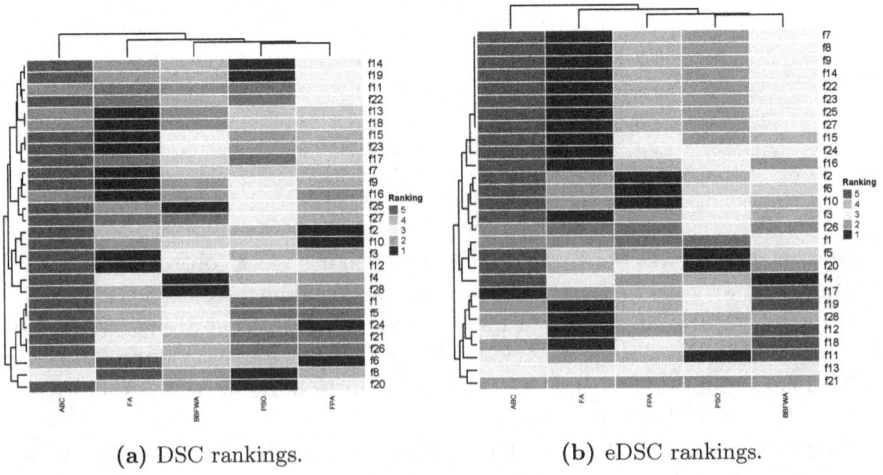

(a) DSC rankings.   (b) eDSC rankings.

Fig. 1. Heatmaps of the hierarchical clustering performed using the DSC and eDSC rankings in 10D. (Color figure online)

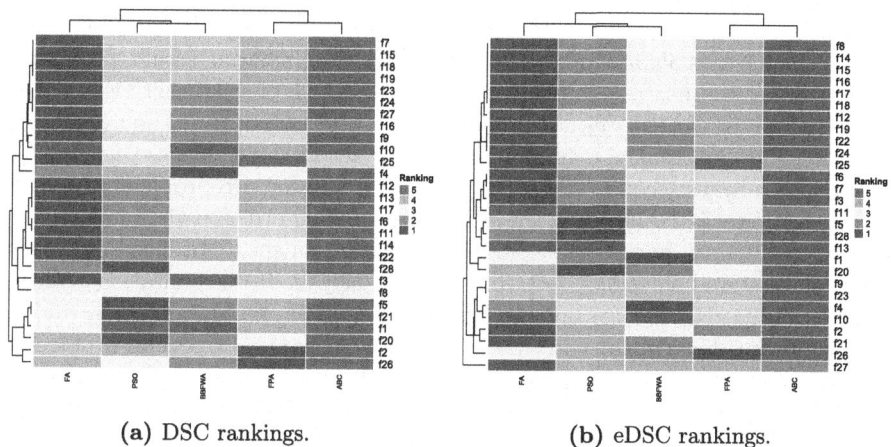

(a) DSC rankings.   **(b)** eDSC rankings.

Fig. 2. Heatmaps of the hierarchical clustering performed using the DSC and eDSC rankings in 30D.

much about the exploitation abilities. But if we check problems solved by the ABC that were marked with blue shades in Fig. 1b, we can see that they are also marked with red color in Fig. 1a, which indicates that ABC algorithm has poor exploitation power in comparison to other compared algorithms. On the other hand, the FA provides most clustered solutions in most problem cases that lead to good exploitation. This also indicates that the FA has good exploration abilities.

Figure 2 presents the results from the hierarchical clustering using the DSC and eDSC rankings in 30D. Using the results with regard to the obtained solutions values (Fig. 2a), two clusters of algorithms are obtained. The FA, PSO, and BBFWA, have more similar exploitation power. Within them, the PSO is more similar to the BBFWA, and then both of them are similar to the FA. The other cluster consists of the FPA and ABC. The same case is when looking at the results for the eDSC rankings (Fig. 2b). In general, clustered solutions are not an automatic guaranty of good exploration abilities. Let us look at the problems f_6 and f_8, where the FA achieved good clustering, but resulted in poor performance. From the results, one can conclude that the algorithm was stuck in the same local optima and the exploitation abilities could not be evaluated. The point is that having only information about clustered or sparse solutions is not enough to determine the quality of the exploration, but must always be considered together with the quality of values that were obtained. In multimodal problems with many similar solutions with regard to their quality, the sparse solutions are usually preferred, since this means that many, different solutions of similar quality were found. On the other hand, when dealing with problems with one obvious global optimum, the desire is to have clustered solutions where all of them are located in the basin of the global optimum. All mentioned scenarios can be seen in Figs. 1 and 2.

Additionally, we can also detect clusters of problems that are similarly solvable by the selected algorithm portfolio, by using the clustering results presented in the rows of the heatmaps. For example, in the case of 10D, there is a set of problems, f_7, f_8, f_9, f_{14}, f_{22}, f_{23}, and f_{25}, for which the exploration ability is the same. The first ranked algorithm is first for all problems, the second ranked algorithm is always second, and so on. However, only for two problems, f_{14} and f_{22}, the exploitation ability (i.e. the same DSC ranking order) is the same by the selected algorithms. For example, the FA algorithm on the f_8 problem obtained the most clustered solutions, but that did not help in the exploitation ability, where it is ranked as the last one (i.e. maybe it has poor exploration power and it is stuck in some local optima). In the case of 30D, there is also several problems' clusters that can be obtained with regard to the exploration and exploitation power. These results can further provide correlation of the algorithms behaviour.

5   Conclusion

To provide more insights into the swarm intelligence algorithms performance, we compared five algorithms using a recently proposed approach, known as Deep Statistical Comparison (DSC). The idea behind this comparison is that the algorithms are compared twice, once with regard to the obtained solutions values and once with regard to the distribution of the solutions in the search space. This information allows us not only to find groups of algorithms that have statistically significant performance, but also to find correlation between the exploration and the exploitation abilities of the compared algorithms. Based on the simulation results, it can be concluded that considering the obtained solutions values for 10D the statistically significant difference exists only between the PSO and ABC, while the FA algorithm has statistically significant performance when considering the distribution of the solutions. For 30D, the FA has been superior in both comparisons. This analysis clearly point to the drawbacks in the optimization algorithms, so the researchers can focus only on improving the exploration and exploitation operators and the balance between them. Future work will include tests on more benchmark and real-life problems by evolving larger algorithms portfolio.

Acknowledgement. The authors acknowledge the financial support from the Slovenian Research Agency (research core funding No. P2-0098, and project Z2-1867). We also acknowledge support by COST Action CA15140 "Improving Applicability of Nature-Inspired Optimisation by Joining Theory and Practice (ImAppNIO)".

References

1. Abdel-Basset, M., Shawky, L.A.: Flower pollination algorithm: a comprehensive review. Artif. Intell. Rev. **52**(4), 2533–2557 (2018). https://doi.org/10.1007/s10462-018-9624-4

2. Eftimov, T., Korošec, P.: A novel statistical approach for comparing meta-heuristic stochastic optimization algorithms according to the distribution of solutions in the search space. Inf. Sci. **489**, 255–273 (2019)

3. Eftimov, T., Korošec, P., Seljak, B.K.: A novel approach to statistical comparison of meta-heuristic stochastic optimization algorithms using deep statistics. Inf. Sci. **417**, 186–215 (2017)

4. Eftimov, T., Petelin, G., Korošec, P.: DSCTool: a web-service-based framework for statistical comparison of stochastic optimization algorithms. Appl. Soft Comput. **87**, 105977 (2020)

5. Eftimov, T., Popovski, G., Kocev, D., Korošec, P.: Performance2vec: a step further in explainable stochastic optimization algorithm performance. In: Proceedings of the Genetic and Evolutionary Computation Conference Companion. In Press (2020)

6. Hussain, K., Mohd Salleh, M.N., Cheng, S., Shi, Y.: Comparative analysis of swarm-based metaheuristic algorithms on benchmark functions. In: Tan, Y., Takagi, H., Shi, Y. (eds.) ICSI 2017, Part I. LNCS, vol. 10385, pp. 3–11. Springer, Cham (2017). https://doi.org/10.1007/978-3-319-61824-1_1

7. Karaboga, D.: An idea based on honey bee swarm for numerical optimization. Technical Report - TR06, pp. 1–10 (2005)

8. Karaboga, D., Basturk, B.: Artificial bee colony (ABC) optimization algorithm for solving constrained optimization problems. In: Melin, P., Castillo, O., Aguilar, L.T., Kacprzyk, J., Pedrycz, W. (eds.) IFSA 2007. LNCS (LNAI), vol. 4529, pp. 789–798. Springer, Heidelberg (2007). https://doi.org/10.1007/978-3-540-72950-1_77

9. Karaboga, D., Basturk, B.: A powerful and efficient algorithm for numerical function optimization: artificial bee colony (ABC) algorithm. J. Global Optim. **39**(3), 459–471 (2007)

10. Kennedy, J., Eberhart, R.: Particle swarm optimization. In: Proceedings of the IEEE International Conference on Neural Networks (ICNN '95), vol. 4, pp. 1942–1948 (1995)

11. Li, J., Tan, Y.: The bare bones fireworks algorithm: a minimalist global optimizer. Appl. Soft Comput. **62**, 454–462 (2018)

12. Liang, J., Qu, B., Suganthan, P., Hernández-Díaz, A.G.: Problem definitions and evaluation criteria for the CEC 2013 special session on real-parameter optimization. Computational Intelligence Laboratory, Zhengzhou University, Zhengzhou, China and Nanyang Technological University, Singapore, Technical Report 201212 (2013)

13. Senthilnath, J., Omkar, S., Mani, V.: Clustering using firefly algorithm: performance study. Swarm Evol. Comput. **1**(3), 164–171 (2011)

14. Shan, H., Yasuda, T., Ohkura, K.: A self adaptive hybrid enhanced artificial bee colony algorithm for continuous optimization problems. BioSystems **132**, 43–53 (2015)

15. Tan, Y., Zhu, Y.: Fireworks algorithm for optimization. In: Tan, Y., Shi, Y., Tan, K.C. (eds.) ICSI 2010, Part I. LNCS, vol. 6145, pp. 355–364. Springer, Heidelberg (2010). https://doi.org/10.1007/978-3-642-13495-1_44

16. Tuba, E., Mrkela, L., Tuba, M.: Support vector machine parameter tuning using firefly algorithm. In: 26th International Conference Radioelektronika, pp. 413–418. IEEE (2016)

17. Tuba, E., Strumberger, I., Bacanin, N., Jovanovic, R., Tuba, M.: Bare bones fireworks algorithm for feature selection and SVM optimization. In: 2019 IEEE Congress on Evolutionary Computation (CEC), pp. 2207–2214. IEEE (2019)

18. Tuba, E., Tuba, M., Beko, M.: Two stage wireless sensor node localization using firefly algorithm. In: Yang, X.-S., Nagar, A.K., Joshi, A. (eds.) Smart Trends in Systems, Security and Sustainability. LNNS, vol. 18, pp. 113–120. Springer, Singapore (2018). https://doi.org/10.1007/978-981-10-6916-1_10
19. Tuba, E., Tuba, M., Dolicanin, E.: Adjusted fireworks algorithm applied to retinal image registration. Stud. Inf. Control **26**(1), 33–42 (2017)
20. Yang, X.-S.: Firefly algorithms for multimodal optimization. In: Watanabe, O., Zeugmann, T. (eds.) SAGA 2009. LNCS, vol. 5792, pp. 169–178. Springer, Heidelberg (2009). https://doi.org/10.1007/978-3-642-04944-6_14
21. Yang, X.-S.: Flower pollination algorithm for global optimization. In: Durand-Lose, J., Jonoska, N. (eds.) UCNC 2012. LNCS, vol. 7445, pp. 240–249. Springer, Heidelberg (2012). https://doi.org/10.1007/978-3-642-32894-7_27
22. Zambrano-Bigiarini, M., Clerc, M., Rojas, R.: Standard particle swarm optimisation 2011 at CEC-2013: a baseline for future PSO improvements, pp. 2337–2344 (2013). https://doi.org/10.1109/CEC.2013.6557848

Modelling and Optimization of Dynamic Systems

A Numerical Parameter Estimation Approach of the Honeybee Population

Atanas Z. Atanasov[1] and Slavi G. Georgiev[2]([⊠])

[1] Department of Agricultural Machinery, AIF, University of Ruse, Ruse, Bulgaria
`aatanasov@uni-ruse.bg`
[2] Department of Applied Mathematics and Statistics, FNSE, University of Ruse, Ruse, Bulgaria
`sggeorgiev@uni-ruse.bg`

Abstract. In this study we aim to solve an inverse, or coefficient identification problem for honeybee population dynamics. The model we consider is a weakly coupled system of two nonlinear ordinary differential equations (ODEs) with dependent variables: the total number of bees and the number of bees that work outside the hive referred here as foragers. To recover the constant parameters, we minimize a quadratic discrete cost function, which expresses the difference between the computed and the measured numbers of bees. We present an efficient simple algorithm to solve the inverse problem using numerical optimization via the Trust-Region-Reflective method. To verify the convergence of the algorithm, we suggest particular statistical metrics to be analyzed. Ample computational results, demonstrating the capabilities of the approach, including the possibility to work with noisy measurements, are presented and discussed.

Keywords: Honeybee population dynamics · Parameter estimation · Cost function minimization

1 Introduction

The honeybee plays a vital role in the pollination of flowering plants, including crops. After all, one-third of the food we eat depends upon pollination, including almonds, apples, berries, cucumbers and melons.

While farmers are dealing with the challenge of providing more food for a growing population, the honeybee has been facing its own problems. The unexplained rise in honeybee deaths – mainly due to high winter losses in colonies in the EU, and 'Colony Collapse Disorder' (CCD) in the USA – has become an issue of concern, see e. g. [11], and for Canada [22].

But the greatest importance of honey bees to agriculture isn't a product of the hive at all. It's their work as crop pollinators. This agricultural benefit of honey bees is estimated to be between 10 and 20 times the total value of honey and beeswax. In fact, bee pollination accounts for about $ 15 billion in added crop value. Honey bees are like flying dollar bills buzzing over U.S. crops.

© Springer Nature Switzerland AG 2021
D. Simian and L. F. Stoica (Eds.): MDIS 2020, CCIS 1341, pp. 349–362, 2021.
https://doi.org/10.1007/978-3-030-68527-0_22

The life expectancy and mortality rate of the workers in a honeybee colony is influenced by dynamic interactions between them in the colony. The colony usually operates as a tidy, sanitary and well-supplied fortress with clearly distributed responsibilities, and thus the mortality rate of the bees working exclusively in the hive is very low, see e. g. [23]. On the contrary, foraging exposes bees to high level of metabolic stress and oxidative damage as well as a significant risk of predation, adverse weather conditions and getting lost. Forager mortality rates are very high: even in a well-organized and healthy colony forager mortality exceeds 15% per day, see e. g. [12,23]. The bee's total lifespan is, therefore, impacted by the age at which it becomes a forager, and it is determined by mechanisms of social feedback within the colony as pheromone-mediated systems of social inhibition.

The main benefit of the mathematical modeling is the ability of the models to describe real world phenomena in a quantitative manner and they enable us to derive conclusions about how the parameters influence the system. The dynamics of a honeybee colony population have been modeled under different assumptions and these models have been used to investigate the main causes of colony losses. In [18] a model has been constructed that links demographic dynamics of a colony to dynamics of infections and deceases within and outside the hive. A more general model has been set up in terms of uninfected hive/forage bees, infected hive bees, virusfree mites and virus-infected mites in [4,17]. In the studies [18,21], the factors influencing the colony failure have been explored and it has been concluded that the breakdown of food availability and the social inhibition are the main sources of colony collapse. A generalization of the model in [13], considering fractional-order derivatives, is developed in [24]. It preserves the memory property of the system. A modeling concerning the seasonal effects is presented in [20].

We follow the model proposed in [13,14], where the honey bee colony population dynamics is described with a weakly coupled system of ODEs. This model suggests the existence of a critical threshold of the forager death rate below which the colony maintains a stable population size. If the death rate appears to be beyond this threshold, it occurs a rapid population reduction and the colony failure is inevitable.

This paper also concerns the importance of honey bee population as pollinator of agriculture crops, fruit, vegetables, seeds etc. and losses in agriculture, honey bee and honey production due to use of chemical pesticides.

The next section introduces the model. The inverse problem is formulated in Sect. 3 and in Sects. 4 and 5, respectively, the numerical methods for the direct and inverse problems are developed, while in Sect. 6 numerical simulations with synthetic and real data are presented. Conclusions are drawn in the last section.

2 Model Description

Since 2006 in North America there has been observed a rapid decrease in the number of the honeybee colonies [12]. The syndrome, which is characterized by the disappearance of adult bees while the limited brood and honey bees work in the hive is known as Colony Collapse Disorder (CCD), which was first diagnosed in USA.

In [7] the authors presented a detailed review of mathematical models describing the bee population dynamics, where 31 models have been analyzed and classified as colony, varroa, and foraging models. In [8] a simple compartment model for the worker bee population of the hive was constructed. It only considers the population of the female workers since males (drones) do not contribute to the colony work. Let H be *the number of bees working in the hive* and F – *the number of bees who work outside the hive* hereafter referred to as *foragers*. We assume that all adult worker bees can be classified either as hive bees or as foragers, and that there is no overlap between these two behavioral classes [7,13]. Hence *the total number of adult worker bees in the colony* is $N = H + F$.

One limitation of the model in [13] is the fact that it does not consider the impact of brood diseases on colony failure. However, it is still useful since many cases of colony failure and the extreme case known as CCD are not caused by brood diseases [8,23], as we mentioned before. It is assumed that the death rate of the hive bees is negligible. Workers are recruited to the forager class from the hive bee class and die at rate m. Let t be the time, measured in days. Then, in [13] this process is described with the differential model that follows in terms of the *eclosion* function $\mathscr{E}(H,F)$ and the *recruitment* function $\mathscr{R}(H,F)$.

The rate of change in number of hive bee is represented by the equation

$$\frac{dH}{dt} = \mathscr{E}(H,F) - H\mathscr{R}(H,F), \tag{1}$$

where the change is explained with the eclosion, reduced with the proportional recruitment rate. The rate of change in the number of foragers is designed by

$$\frac{dF}{dt} = H\mathscr{R}(H,F) - mF, \tag{2}$$

and the change is constituted by the fresh recruits minus the deceased foragers.

The function $\mathscr{E}(H,F)$ describes *the way eclosion depends on the number of hive bees and foragers*. The recruitment rate function $\mathscr{R}(H,F)$ models *the effect of social inhibition on the recruitment rate*. The maximal rate of eclosion is assumed to be equivalent to *the queen's laying rate L* and that eclosion rate approaches its maximum as N (the total number of workers in the hive) increases. In the absence of other information, the simplest function that increases from 0 for no workers and tends to L as N becomes very large is

$$\mathscr{E}(H,F) = L\frac{H+F}{w+H+F} = L\frac{N}{w+N}. \tag{3}$$

Here w determines the rate of which $\mathscr{E}(H,F)$ approaches L as N gets large. On Fig. 2 in [13] $\mathscr{E}(H,F)$ as a function of N is plotted for a range of values of w.

The model suggests the following form for the recruitment function

$$\mathscr{R}(H,F) = \alpha - \sigma\frac{F}{H+F} = \alpha - \sigma\frac{F}{N}. \tag{4}$$

The first term α represents the maximal rate at which hive bees become foragers when there are no foragers in the colony. The second term $-\sigma\frac{F}{N}$ represents the *social inhibition* and, in particular, how the presence of foragers lowers the rate of recruitment of hive bees to foragers.

3 Inverse Problem

In the practice, the parameters m, α, σ and w are usually not known and they have to be recovered. After their «fair» values are found, the model could be used for reliable analysis.

We rewrite the problem (1)–(4) as follows:

$$\frac{dN}{dt} = L\frac{N}{w+N} - mF, \tag{5}$$

$$\frac{dF}{dt} = \alpha N - (\alpha + \sigma + m)F + \sigma\frac{F^2}{N}, \tag{6}$$

$$N(t_0) = N^0, \quad F(t_0) = F^0, \tag{7}$$

where $p = (p^1, p^2, p^3, p^4)$, $p^1 := m$, $p^2 := \alpha$, $p^3 := \sigma$, $p^4 := w$ and

$$p \in \mathbb{S}_{adm} = \left\{ p \in \mathbb{R}^4 : 0 < p^i < P^i, \ i = 1,2,3,4 \right\}. \tag{8}$$

Hereinafter all solutions $\{N(t;p), F(t;p)\}$, $p \in \mathbb{S}_{adm}$ are defined on the interval $t_0 \leqslant t \leqslant T$. When the parameters m, α, σ and w are known, the problem (5)–(8) is well-posed and it is called a *direct problem*.

We study the *inverse problem* of reconstructing the parameter $p \in \mathbb{S}_{adm}$ by means of the observed behaviour

$$\left\{N(t^i), F(t^i)\right\}, \ i = 1,\ldots,I_{obs}; \ t_0 = t^1 < \ldots < t^{I_{obs}} = T \tag{9}$$

of the dynamical system (5)–(8). Then, the inverse problem of the parameter reconstruction can be formulated in a variational setting as follows:

$$\min_{p \in \mathbb{S}_{adm}} \Phi(p), \qquad p = (p^1, p^2, p^3, p^4),$$

subjected to the solution of (5)–(8).

We seek the point $p = (p^1, p^2, p^3, p^4)$ of the local minimum of the functional $\Phi(p)$. The functional $\Phi(p)$ can be written as

$$\Phi(p) = \frac{1}{2}\sum_{i=1}^{I_{obs}} \left[\left(N(t^i;p) - N_{obs}(t^i)\right)^2 + \left(F(t^i;p) - F_{obs}(t^i)\right)^2 \right]. \tag{10}$$

Here, $\{N_{obs}(t^i), F_{obs}(t^i)\}$ are experimental data (9), and $\{N(t^i;p), F(t^i;p)\}$ is the solution to the problem (5)–(8).

The parameter admissible set $\mathbb{S}_{adm}$ follows the biology of the honey bee [23] as well as the conception of the concrete model [13] and experimental data.

4 Solution to the Direct Problem

In this section, we will briefly mention how to solve the direct problem (5)–(8). The solution to the direct problem is required to solve the inverse problem.

We introduce the piecewise-uniform mesh

$$\overline{\omega}_\tau = \left\{ t^1 = t_0,\ t^i = t^{i-1} + \tau_i J_i,\ t^{I_{obs}+1} = T \right\} \text{ for } i = 2, \dots, I_{obs}, \tag{11}$$

and the subinterval division follows

$$t_j^i = t^{i-1} + j\tau_i,\ j = 1, \dots, J_i,$$

where $\forall i = 2, \dots, I_{obs}$ t^i are the time instances at which observations are taken, t_j^i, $j = 1, \dots, J_i$ and τ_i are the time points and the time step corresponding to $(t^{i-1}, t^i]$ (see Fig. 1 for an example).

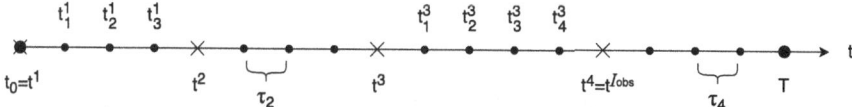

Fig. 1. Mesh ω_τ (11)

There are a number of numerical methods to solve the initial problem (5)–(7), see e. g. [1]. Here we solve the direct problem using the MATLAB ode45 subroutine [16], passing the mesh (11). The solver is based on an explicit Runge-Kutta-Dormand-Prince formula [10, 19], which achieves fourth- and fifth-order accurate solutions.

5 Solution to the Inverse Problem

The objective function defined in (10) corresponds to the sum of the squares of the residuals $N(t^i; \boldsymbol{p}) - N_{obs}(t^i)$ and $F(t^i; \boldsymbol{p}) - F_{obs}(t^i)$. These are essentially the differences between the theoretical quantities $N(t^i; \boldsymbol{p})$, $F(t^i; \boldsymbol{p})$ and the observed ones in practice, $N_{obs}(t^i)$, $F_{obs}(t^i)$ for all observation times t^i, $i = 1, \dots, I_{obs}$.

Let us denote the *argmin* of the objective function $\Phi(\boldsymbol{p})$ with $\check{\boldsymbol{p}}$. The vector $\check{\boldsymbol{p}}$ is called *nonlinear least squares estimator (LSE)*. There are a number of approaches how to find it, i. e. how to minimize the cost function $\Phi(\boldsymbol{p})$. In this paper, we will apply the *Trust Region Reflexive* algorithm [9], using the MATLAB subroutine lsqnonlin [15].

To explore the properties of the minimizer $\check{\boldsymbol{p}}$, we also use some metrics that allow us to ensure convergence. The first one is the *norm of the step* $\delta \boldsymbol{p}_k$, which measures the difference between two successive iterations of the minimizer $\boldsymbol{p}_k$ and $\boldsymbol{p}_{k+1}$ and is defined as

$$\delta \boldsymbol{p}_k := \|\triangle \boldsymbol{p}_k\| = \|\boldsymbol{p}_{k+1} - \boldsymbol{p}_k\|. \tag{12}$$

The next quantity is *the relative change in the cost function*, which is defined as

$$\delta \Phi_k := \frac{|\triangle \Phi_k|}{1 + |\Phi(\boldsymbol{p}_k)|}, \tag{13}$$

where $\triangle \Phi_k := \Phi(\boldsymbol{p}_{k+1}) - \Phi(\boldsymbol{p}_k)$.

The stopping criterion for the minimization algorithm is

$$\min \{\delta \boldsymbol{p}_k, \delta \Phi_k\} < \varepsilon_\delta, \tag{14}$$

where ε_δ is a user-prescribed tolerance. When this condition is met, the algorithm stops at iteration $k+1$ and returns an approximation of the nonlinear LSE $\boldsymbol{\check{p}} := \boldsymbol{p}_{k+1}$.

The implementation also provides the *first-order optimality measure*, which shows how close the approximation is to the actual minimum of (10). It is defined as the infinity norm of the gradient of the objective function, evaluated at the LSE, which in turn is given by the maximal absolute value of the partial derivatives of the objective function w. r. t. $\boldsymbol{p}$:

$$\|\nabla \Phi(\boldsymbol{\check{p}})\|_\infty = \max_{r=1,\dots,4} \left| \frac{\partial \Phi}{\partial p^r}(\boldsymbol{\check{p}}) \right|. \tag{15}$$

Based on the aforementioned quantities, which are provided at the end of the algorithm execution, we investigate the following statistical measures to quantify the *goodness-of-fit* performance. After we arrive at the nonlinear LSE, we compute the metrics as follows: *the variance of the residuals* $\tilde{\sigma}^2$, *the root mean squared error (RMSE)* $\hat{\sigma}$, and *the coefficient of determination* R^2 [5]:

$$\tilde{\sigma}^2 = \frac{\Phi(\boldsymbol{\check{p}})}{I_{\text{obs}}}, \tag{16}$$

$$\hat{\sigma} = \sqrt{\frac{1}{I_{\text{obs}} - 3} \Phi(\boldsymbol{\check{p}})}, \tag{17}$$

$$R^2 = 1 - \frac{\Phi(\boldsymbol{\check{p}})}{\sum_{i=1}^{I_{\text{obs}}} \left((N_{\text{obs}}(t^i) - \overline{N_{\text{obs}}})^2 + (F_{\text{obs}}(t^i) - \overline{F_{\text{obs}}})^2 \right)}, \tag{18}$$

where $\overline{B_{\text{obs}}} = \dfrac{\sum_{i=1}^{I_{\text{obs}}} B_{\text{obs}}(t^i)}{I_{\text{obs}}}$ denotes the mean value of the experimental data for $B \equiv N$ and $B \equiv F$, respectively. While $\tilde{\sigma}^2$ and $\hat{\sigma}$ are *the lower the better* metrics, R^2 is *the bigger the better*. Nevertheless, the coefficient of determination shows what part of the variability of the model is explained by the variability of the parameters $\boldsymbol{\check{p}}$, so it cannot be greater than 1.

If the residuals $(N(t^i; \boldsymbol{p}) - N_{\text{obs}}(t^i))$, $(F(t^i; \boldsymbol{p}) - F_{\text{obs}}(t^i))$ were normally distributed or if the number of the observations I_{obs} was sufficiently large, then the estimated *covariance matrix* of the nonlinear LSE $\boldsymbol{\check{p}}$ could be expressed as

$$\Sigma = \frac{\hat{\sigma}^2}{J^\top(\boldsymbol{\check{p}})J(\boldsymbol{\check{p}})}, \tag{19}$$

where $J(p)$ is the *sensitivity* or *Jacobian matrix* and it is defined as

$$J(p) = \begin{pmatrix} \frac{\partial N^\top(p)}{\partial p} \\ \frac{\partial F^\top(p)}{\partial p} \end{pmatrix} = \begin{pmatrix} J_N(p) \\ J_F(p) \end{pmatrix}. \tag{20}$$

In explicit form, the sensitivity matrix is written as

$$J_N(p) = \begin{pmatrix} \frac{\partial N(t^1)}{\partial p^1} & \frac{\partial N(t^1)}{\partial p^2} & \frac{\partial N(t^1)}{\partial p^3} & \frac{\partial N(t^1)}{\partial p^4} \\ \vdots & \vdots & \vdots & \vdots \\ \frac{\partial N(t^{I_{obs}})}{\partial p^1} & \frac{\partial N(t^{I_{obs}})}{\partial p^2} & \frac{\partial N(t^{I_{obs}})}{\partial p^3} & \frac{\partial N(t^{I_{obs}})}{\partial p^4} \end{pmatrix},$$

$$J_F(p) = \begin{pmatrix} \frac{\partial F(t^1)}{\partial p^1} & \frac{\partial F(t^1)}{\partial p^2} & \frac{\partial F(t^1)}{\partial p^3} & \frac{\partial F(t^1)}{\partial p^4} \\ \vdots & \vdots & \vdots & \vdots \\ \frac{\partial F(t^{I_{obs}})}{\partial p^1} & \frac{\partial F(t^{I_{obs}})}{\partial p^2} & \frac{\partial F(t^{I_{obs}})}{\partial p^3} & \frac{\partial F(t^{I_{obs}})}{\partial p^4} \end{pmatrix}.$$

The sensitivity matrix $J(p)$ measures the variation of the variables $N(t^i; p)$, $F(t^i; p)$ w. r. t. changes in p. Sensitivity analysis provide information about the relevance of the observations to the parameter identification, see e. g. [6]. Using the covariance matrix (19), we compute the *standard error (SE)* and the *normalized standard error (NSE)* associated to the nonlinear LSE $\check{p}$, which further help us to quantify the accuracy of the parameter estimate. They are defined by

$$\widehat{SE} = \sqrt{\mathrm{diag}(\Sigma)},$$

$$\widehat{NSE} = \frac{\widehat{SE}}{\check{p}} \times 100, \tag{21}$$

where the division should be interpreted elementwisely.

The quantities (19)–(21) are *asymptotic*, i. e. they are valid only if the number of observations I_{obs} is sufficiently large, because the LSEs are asymptotically normally distributed [5].

Now we are ready to present the algorithm for determining the parameters concerning the honeybee population dynamics:

Numerical Algorithm

1. Choose an initial approximation and denote it as p_0: $(k = 0)$.
2. Execute the `lsqnonlin` solver with starting point p_0. When the stopping criterion (14) is satisfied, the procedure returns the nonlinear LSE $\check{p}$.
3. Check the quality of the minimizer. The output of the subroutine consists also of the first-order optimality measure (15), the number of iterations $k + 1$, the norm of the step (12), the relative change in the cost function (13) and the numerical approximation of the Jacobian (20), evaluated at $\check{p}$.

  – if $\|\nabla\Phi(\breve{p})\|_\infty$, δp_k and $\delta\Phi_k$ are small and the number of iterations is not large, then the solver has converged to the optimal solution $\breve{p}$;
  – if not, then go to Step 1. and choose a different starting point.
4. Compute the goodness-of-fit metrics: the variance of the residuals $\tilde{\sigma}^2$ (16), the RMSE $\hat{\sigma}^2$ (17), the coefficient of determination R^2 (18) and the normalized standard errors $\widehat{NSE}$ (21).

6 Numerical Results

In this section we present computational results in order to verify the proposed algorithm. In our quasi-real test framework, we first solve the direct problem to explore the properties of the model and, more importantly, to use the results as measurements to solve the inverse problem.

First, we solve the direct problem (5)–(7) using *real world parameter values* from [13]. Let us set $L = 2000$, $\alpha = 0.25$, $\sigma = 0.75$, $w = 27000$. We assume two types of colonies: a small honey bee colony with $N^0 = 4500$ and an average honey bee colony with $N^0 = 9000$. In both cases at the initial time t_0 there are no foragers $F^0 = 0$. Firstly we conduct our simulations up to $t_1 = 1$ year in order to see whether, when and how the colony approaches an equilibrium state.

When we assume a moderate mortality rate $m = 0.24$, the bee colonies thrive, see Fig. 2.

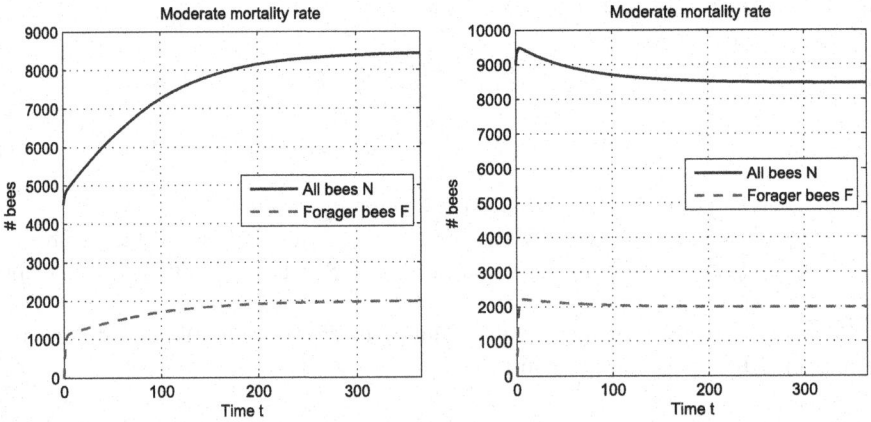

Fig. 2. The number of bees in the colony for $m = 0.24$: $N^0 = 4500$ (left) and $N^0 = 9000$ (right)

If the mortality rate is above a certain threshold, e. g. $m = 0.4$, then the colonies collapse (Fig. 3).

The last conclusions are confirmed by the phase plane diagrams (Fig. 4). The blue curves track particular solutions development, where the stars denote the starting points, and the circle – the end point. All drawn solutions consider the case with no foragers

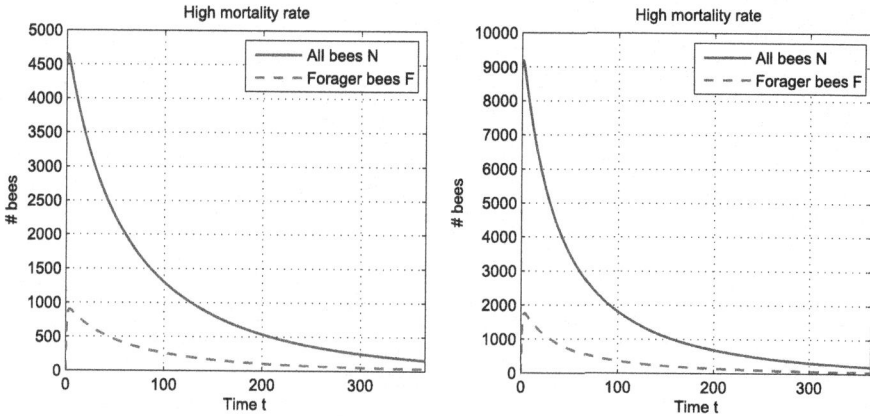

Fig. 3. The number of bees in the colony for $m = 0.40$: $N^0 = 4500$ (left) and $N^0 = 9000$ (right)

in the beginning. The populations in the moderate mortality case tend to a stable equilibrium state. When the mortality is high, the bee colonies collapse. We used the aforementioned set of parameters and the results well agree with those in [13].

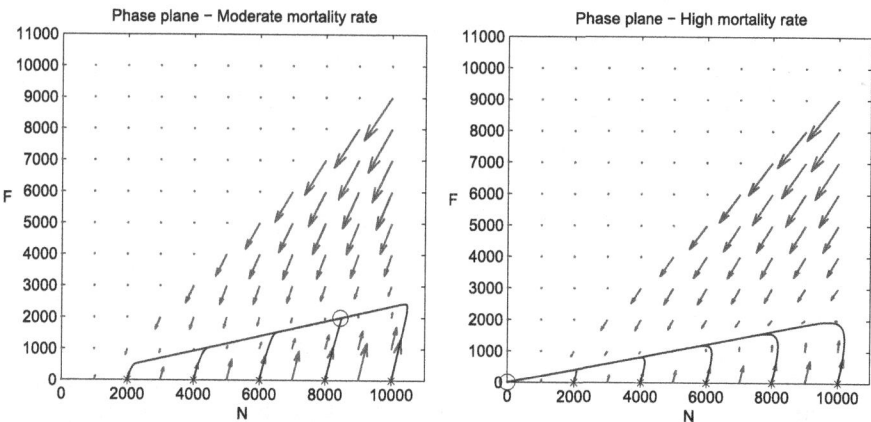

Fig. 4. Phase plane diagrams of the solution for $m = 0.24$ (left) and $m = 0.40$ (right) (Color figure online)

Now we proceed to the inverse problem. We again set $L = 2000$ and consider an average honeybee colony with $N^0 = 9000$ and $F^0 = 0$. We are going to identify the parameter values $\boldsymbol{p} = (m, \alpha, \sigma, w)^\top = (0.24, 0.25, 0.75, 27000)^\top$ and consider the problem (5)–(8) on the interval $t_0 = 0$, $t_1 = 100$ days. Furthermore, we set $\varepsilon_\delta = 1e - 16$ in (14) and the lower and upper bounds of the parameters are $\underline{\boldsymbol{p}} = (0, 0, 0, 0)^\top$ and $\overline{\boldsymbol{p}} = (1, 1, 1, 1e5)^\top$, respectively.

We assume $I_{obs} = 10$ measurements of type (9), distributed equidistantly from $t^2 = 5^{th}$ day to $t^{11} = 95^{th}$ day. Firstly, we test with the initial vector $p_0 = (0.2, 0.3, 0.8, 30000)^\top$, which is interpreted as close to the true values p. The results could be viewed in Table 1.

Table 1. Test with p_0 close to p

Parameter	p_0	p	$\check{p}$	$\|p - \check{p}\|$	$\dfrac{\|p - \check{p}\|}{p}$	$\widehat{NSE}$
m	0.3	0.24	0.24	8.5765e-15	3.5735e-14	2.3744e-14
α	0.3	0.25	0.25	1.1380e-14	4.5519e-14	1.9468e-14
σ	0.7	0.75	0.75	3.2196e-14	4.2929e-14	2.7920e-14
w	30000	27000	27000	1.4807e-09	5.4839e-14	2.3635e-14

The next test is performed with the initial values $p_0 = (0.5, 0.5, 0.5, 50000)^\top$, which are considered far from p. Since p^i for $i = 1, 2, 3$ are rates (between 0 and 1), these are the farthest possible values. The results follow in Table 2.

Table 2. Test with p_0 far from p

Parameter	p_0	p	$\check{p}$	$\|p - \check{p}\|$	$\dfrac{\|p - \check{p}\|}{p}$	$\widehat{NSE}$
m	0.5	0.24	0.24	5.0793e-15	2.1164e-14	2.7736e-14
α	0.5	0.25	0.25	7.1054e-15	2.8422e-14	2.2741e-14
σ	0.5	0.75	0.75	1.9096e-14	2.5461e-14	3.2615e-14
w	50000	27000	27000	9.3132e-10	3.4493e-14	2.7609e-14

It is obvious that the values of the implied parameters are extremely close to the real ones. The relative error for all the parameters is of order $1e - 14$. What is more, this is the case also with the normalized standard errors, i. e. their smallness implies the accuracy of the estimation. This is true for both cases, when the starting point p_0 is close to and far from the real values, and these results confirm the method is *stable*.

Table 3. Test with $I_{obs} = 10$ observations of type (9)

p_0 to p?	$\|\nabla\Phi(\check{p})\|_\infty$	δp_k	$\delta\Phi_k$	$k+1$	$\Phi(\check{p})$	$\tilde{\sigma}^2$	$\hat{\sigma}$	R^2
close	9.9428e-7	5.6963e-12	4.76e-21	22	7.9078e-22	7.9078e-23	1.0629e-11	$1-1.2968e-27$
far	1.8473e-6	1.2853e-12	1.17e-21	52	1.0791e-21	1.0791e-22	1.2416e-11	$1-1.7695e-27$

The performance of the algorithm is very successful (Table 3), which is reflected by the fact that the variance ($\tilde{\sigma}^2$) and the RMSE ($\hat{\sigma}$) are rather low, while the coefficient of determination R^2 is practically 1. The minimization subroutine has terminated its execution, because the relative change of the cost function $\delta\Phi_k$ has become smaller than the threshold ε_δ. The main difference in the results concerning the two initial guesses regards the number of iterations needed for convergence. The number in the «far» case is more than twice as the number in the «close» case, which is quite normal, since both numbers are not large.

To quantify the effect of measurement errors on the parameter estimate, we conduct a test with *perturbed* observations. We again consider the case with initial point far from the true values, and we add a Gaussian noise to the measurements $\{N_{\text{obs}}(t^i), F_{\text{obs}}(t^i)\}$, $i = 1, \ldots, I_{\text{obs}}$ (9). More precisely, we add white noise $\varepsilon_B \sim \mathcal{N}(0, \sigma_B^2)$ to all data points in B_{obs} for $B \equiv N$ and $B \equiv F$, respectively. The assumption follows the paradigm that with 95% confidence the bias in a particular measurement is not greater than 1% of the value itself. This means $\sigma_N = 46.0609$ and $\sigma_F = 10.8154$. The results are presented in Table 4.

Table 4. Test with p_0 far from p and perturbed data (9)

Parameter	p_0	p	$\check{p}$	$|p - \check{p}|$	$\dfrac{|p - \check{p}|}{p}$	$\widehat{NSE}$
m	0.5	0.24	0.2482	0.0082	0.0342	0.1020
α	0.5	0.25	0.2496	3.6628e-4	0.0015	0.0829
σ	0.5	0.75	0.7378	0.0122	0.0163	0.1207
w	50000	27000	25795	1204.6	0.0446	0.1015

We can conclude that the method approximately converges, i .e. the LSE $\check{p}$ is relatively close to the actual extremum. The absolute errors are small except that for w, which is simply because the order of w is significantly different from those of m, α and σ. The relative errors are less than 5% and the normalized standard errors are about 0.1%, which provide moderate accurate parameter estimation.

We present the relative errors (RE) and other aggregated metrics in Table 5. They are defined as follows:

$$RE_p := \frac{\|\check{p} - \check{p}^{\text{pert}}\|_\infty}{\|\check{p}\|_\infty},$$

$$RE_B := \frac{\|B(t; \check{p}) - B(t; \check{p}^{\text{pert}})\|_\infty}{\|B(t; \check{p})\|_\infty},$$

where $B(t; p)$ are the implied solutions for $B \equiv N$ and $B \equiv F$ using p, and $\check{p}^{\text{pert}}$ is the nonlinear LSE, derived with the perturbed data. The value of RE_p is around 4%, while RE_N and RE_F are far below 1%. This is why the implied solutions regarding the non-perturbed data $\{N(t, \check{p}), F(t, \check{p})\}$ are not plotted on Fig. 5 – because they practically coincide with the solutions regarding the perturbed data $\{N(t, \check{p}^{\text{pert}}), F(t, \check{p}^{\text{pert}})\}$.

This cannot be surprising because of the *ill-posed* nature of the inverse problem: the algorithm recovers the parameters imperfectly, but the implied solution is indistinguishable from the true one (or from the implied solution according to non-perturbed data).

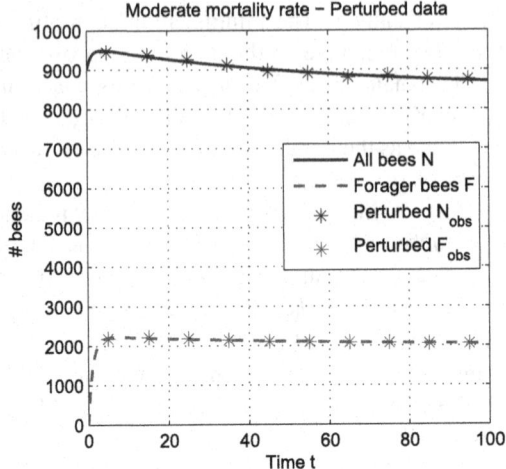

Fig. 5. The plotted solution $\{N(t, \check{p}^{\text{pert}}), F(t, \check{p}^{\text{pert}})\}$, constructed from the perturbed data (9). The perturbed measurements themselves are also shown.

Table 5. Test with $I_{\text{obs}} = 10$ observations of type (9) and perturbed data (9)

RE_p	RE_N	RE_F	$\|\nabla\Phi(\check{p})\|_\infty$	δp_k	$\Phi(\check{p})$	$\tilde{\sigma}^2$	$\hat{\sigma}$	R^2
0.0446	0.0018	0.0017	0.0545	4.2208e-16	14433.1	1443.31	45.4079	0.9772

The convergence metrics $\|\nabla\Phi(\check{p})\|_\infty$ and δp_k correspond to the first-order optimality measure (15) and the norm of the step (12), as before. Their small values confirm the convergence of the procedure. The exit condition is the reached maximum number of iterations, because the condition (14) is not met. Nevertheless, the optimal value is achieved on $k + 1 = 90$, which expectedly is higher that the «non-perturbed» case. The value of the cost function $\Phi(\check{p})$, the variance $(\tilde{\sigma}^2)$ and the RMSE $(\hat{\sigma})$ look large at first glance, but this is due to the magnitude of w, as we mentioned before. The value of the coefficient of determination R^2 is excellent and we could eventually approve the *robustness* of the suggested algorithm.

7   Conclusion and Future Work

In this paper we have proposed and described an efficient algorithm for identification of parameters in a honeybee population dynamics modeling. The model is introduced and reformulated to a system of two nonlinear ODEs, where the dependent variables consist of the number of foragers and the total number of bees. The problem of recovering the parameters that cannot usually be measured is formulated in a minimization setting, as the cost function is the sum of the squares of the difference between the theoretical and the experimental data. The solution to the direct problem is briefly commented, while the solution to the inverse problem is discussed in detail and a numerical algorithm is presented in a simple and easy to understand manner.

The computational results empirically demonstrate that the algorithm is stable and robust. The parameters are effectively reconstructed even if the initial guess is far from the real values. What is more, the recovery is satisfactory even if the observed data contain noise – to a some extend, of course. The statistical metrics, which are part of the proposed methodology, shed light on its performance in the particular environment, as the numerical examples convincingly show that the method is capable of identifying the unknown parameters in a straightforward way. The algorithm works well with parameters in the whole range of biologically reasonable values.

We hope that the results of this research give insight of the dynamics of the honey bee population and could help the beekeepers to manage the bee hive colony in a right way. For example, the professional beekeepers could investigate the implied parameters for their colonies and compare them with the benchmark thresholds. If it appears that an apiary is endangered, the beekeeper could take certain actions, e. g. to provide more food, to keep the colony warm in the winter or even to move the hive to another place. Furthermore, using the current and past observations, the beekeeper could simulate and forecast the development of the colony, which in turn would help them to take precautionary measures. The algorithm is easy to reproduce since the only prerequisite is an ordinary PC having an optimization software installed.

There are many ways in which this study could be further developed. One possible direction is to use more complex models (f. i. [2,3]), which, for example, may take into account some factors of the population dynamics, as food collection, infections, or others. The models, on the other hand, may involve time-dependent parameters, which are more difficult to recover. In the future we plan to consider such problems. What is more, a bifurcation analysis of the biologically relevant (healthy or disease) equilibria points seems interesting from both theoretical and practical points of view.

Acknowledgement. The authors would like to thank the reviewers for their constructive comments and suggestions, which significantly improved the quality of the paper.

The study of the first author is supported by contract 2020-FNI-01, funded by the Research Fund of the University of Ruse. The second author is supported by the Bulgarian National Science Fund under Young Scientists Project KP-06-M32/2–17.12.2019 "Advanced Stochastic and Deterministic Approaches for Large-Scale Problems of Computational Mathematics".

References

1. Atanasov, A.Z., Georgiev, S.G.: A numerical parameter reconstruction in a model of a honey bee population. In: AIP Conference Proceedings. AIP Publishing, Melville (2020, in press)
2. Atanasov, A.Z., Georgiev, S.G., Vulkov, L.G.: Parameter identification of colony collapse disorder in honeybees as a contagion. In: Simian, D., Stoica, L.F. (eds.) MDIS 2020. CCIS, vol. 1341, pp. 363–377. Springer, Cham (2021). https://doi.org/10.1007/978-3-030-68527-0_23
3. Bagheri, S., Mirzaie, M.: A mathematical model of honey bee colony dynamics to predict the effect of pollen on colony failure. PLoS ONE **14**(11), e0225632 (2019)
4. Bailey, L.: The Isle of Wight Disease. Central Association of Bee-Keepers, Poole (2002)
5. Banks, H.T., Davidian, M., Samuels, J.R.: An inverse problem statistical methodology summary. In: Chowell, G., Hyman, J.M., Bettencourt, L.M.A., et al. (eds.) Mathematical and Statistical Estimation Approaches in Epidemiology, pp. 249–302. Springer, Dordrecht (2009). https://doi.org/10.1007/978-90-481-2313-1_11

6. Banks, H., Dediu, S., Ernstberger, S.: Sensitivity functions and their uses in inverse problems. J. Inverse Ill-Posed Probl. **15**(7), 683–708 (2008)

7. Becher, M.A., Osborne, J.L., Thorbek, P., Kennedy, P.J., Grimm, V.: Review: towards a systems approach for understanding honeybee decline: a stocktaking and synthesis of existing models. J. Appl. Ecol. **50**, 868–880 (2013)

8. Booton, R.D., Iwasa, Y., Marshall, J.A.R., Childs, D.Z.: Stress-mediated Allee effects can cause the sudden collapse of honey bee colonies. J. Theoret. Biol. **420**, 213–219 (2017)

9. Coleman, T.F., Li, Y.: An interior, trust region approach for nonlinear minimization subject to bounds. SIAM J. Opt. **6**, 418–445 (1996)

10. Dormand, J.R., Prince, P.J.: A family of embedded Runge-Kutta formulae. J. Comp. Appl. Math. **6**, 19–26 (1980)

11. Dornberger, L., et al.: Death of the bees: a mathematical model of colony collapse disorder. Technical Report 2012–12, Mathematics Preprint Series, University of Texas at Arlington Mathematics Department (2012)

12. Fact Sheet: The economic challenge posed by declining pollinator populations. Office the Press Secretary, The White House (2015)

13. Khoury, D.S., Myerscough, M.R., Barron, A.B.: A quantitative model of honey bee colony population dynamics. PLoS ONE **6**(4), e18491 (2011)

14. Khoury, D.S., Barron, A.B., Meyerscough, M.R.: Modelling food and population dynamics honey bee colonies. PLoS ONE **8**(5), e0059084 (2013)

15. MathWorks: Solve nonlinear least squares (nonlinear data-fitting) problems. https://www.mathworks.com/help/optim/ug/lsqnonlin.html. Accessed 23 Aug 2020

16. MathWorks: Solve nonstiff differential equations - medium order method. https://www.mathworks.com/help/matlab/ref/ode45.html. Accessed 23 Aug 2020

17. Ratti, V., Kevan, P.G., Eberl, H.J.: A mathematical model of forager loss in honeybee colonies infested with Varroa destructor and the acute bee paralysis virus. Bull. Math. Biol. **79**(6), 1218–1253 (2017)

18. Russel, S., Barron, A.B., Harris, D.: Dynamics modelling of honeybee (Apis mellifera) colony growth and failure. Ecolog. Model. **265**, 138–169 (2013)

19. Shampine, L.F., Reichelt, M.W.: The MATLAB ODE suite. SIAM J. Sci. Comp. **18**, 1–22 (1997)

20. Switanek, M., Crailsheim, K., Truhetz, H., Brodschneider, R.: Modelling seasonal effects of temperature and precipitation on honey bee winter mortality in a temperate climate. Sci. Tot. Environm. **579**, 1581–1587 (2017)

21. Torres, D.J., Ricoy, V.M., Roybal, S.: Modelling honey bee populations. PLoS ONE **10**(7), e0130966 (2015)

22. Van der Zee, R., et al.: Managed honey bee colony losses in Canada, China, Europe, Israel and Turkey for the winters of 2008–2009 and 2009–2010. J. Appl. Res. **51**(1), 100–114 (2012)

23. Winston, W.L.: The Biology of the Honey Bee. Harvard University Press, Cambridge (1991)

24. Yıldız, T.A.: A fractional dynamical model for honeybee colony population. Int. J. Biomath. **11**(5), 1850063 (2018)

Parameter Identification of Colony Collapse Disorder in Honeybees as a Contagion

Atanas Z. Atanasov[1] , Slavi G. Georgiev[2(✉)] , and Lubin G. Vulkov[2]

[1] Department of Agricultural Machinery, AIF, University of Ruse, Ruse, Bulgaria
aatanasov@uni-ruse.bg
[2] Department of Applied Mathematics and Statistics, FNSE, University of Ruse, Ruse, Bulgaria
{sggeorgiev,lvalkov}@uni-ruse.bg

Abstract. The Colony Collapse Disorder (CCD) is a major problem of honeybee farms because of the massive decline in the colony numbers. The CCD model considered is described by a system of three ordinary differential equations that account for multiple hive population behaviour patterns including colony collapse, environmental issues and Allee effects. The population dynamics is studied by numerical algorithms to solve the parameter identification problem of the model. Computational experiments demonstrate the capabilities of the numerical approach. Moreover, the numerical analysis suggests the farmer about the role of accelerated forager recruitment in employing hives during a colony collapse.

Keywords: Honeybee population dynamics · Colony Collapse Disorder · Allee effect · Parameter identification · Cost function minimization

1 Introduction

The honeybee Apis mellifera is of great importance to the agriculture, therefore to the economics. The primary purpose of the bees is to pollinate plants. This agricultural benefit of the honey bees is estimated to be between 10 and 20 times the total value of honey and beeswax. In 2000, the bee pollination accounts for about $ 15 billion in added crop value [24]. Clearly, honeybees produce honey, beeswax, royal jelly and other products used frequently by the people. Much of the food production in the world is dependent on honeybees. With expanding nutrition industry, there is a further increase in the need for honeybees.

In a usual honeybee hive, the bees can be categorized in three classes. The queen lays eggs which constitute the brood. During their larval development, the maggots of the brood class are reared by the adult worker bees in a capped honeycomb compartment [10]. After maturing out of the larval stage, the bees join the hive worker class. They are responsible for taking care of the brood, build the combs of the hive, and clean, repair, ventilate and when needed – cool or heat the hive [34]. This includes maintaining the temperature of the brood chamber in the hive. If the brood is raised in abnormal temperature (the typical temperature of the brood nest is $34.5 \pm 0.5\,^\circ\mathrm{C}$), they become physically normal adult bees, but show deficiencies in learning and memory as well as they tend to get lost in the fields [24]. After between 7 and 21 days, the hive bees mature

© Springer Nature Switzerland AG 2021
D. Simian and L. F. Stoica (Eds.): MDIS 2020, CCIS 1341, pp. 363–377, 2021.
https://doi.org/10.1007/978-3-030-68527-0_23

and join the forager class [34]. The forage bees make between 5 and 20 flights per day, during which they collect nectar and pollen from the plants. The food supplies are stored by the hive bees. After foraging for between 14 and 21 days during the foraging season, or up to 4 months in the winter, a bee eventually dies [34].

The first common record of a ubiquitous massive loss of honeybee colones dates from the winter of 2006–2007. Some apiaries have lost about 90% of their colonies and this phenomenon is observed in USA, Europe and other destinations. There were another events concerning honeybee losses in the far and near history at different places [13,29], but in the early 2007, there were beekeepers who experienced 80–100% losses. Such an extraordinary event, besides its environmental and economic impact, attracts a lot of scientific interest and is a topic of a vast body of research.

There are conditions described by the absence of adult bees [19], but particular losses in the winters of 2007, 2008 and other recent years share similar symptoms and the causal syndrome is called 'Colony Collapse Disorder' (CCD). It is characterized by three distinguishable traits: 1) a rapid loss of adult bees in colonies while no dead bodies are found in and around the hive; 2) the presence of the queen and the capped brood; 3) the presence of a food stores which, as well as the hive, are not robbed by pests or scavengers for an extended period of time. Also, some authors [33] add up to the definition of CCD that 4) at the time of collapse, varroa mite and nosema populations are not at levels known to cause typical population decline. Another interesting sign is that the bee cluster is reluctant to consume food provided by the beekeeper [11].

A lot of research aimed to find the cause of the CCD, which is *still mostly unclear*. In [24] some hypotheses about the reasons for CCD are presented, amongst which are diseases and parasites, in- and out-hive chemicals, genetically modified crops, even narrow genetic base and cold brood. It is concluded that CCD has a multifactorial cause, and the bees are immunosuppressed by different factors. What is more, a short list of the primary hypotheses is given in [11], including chemical environmental pollution. The authors of [8] has used a metagenomic approach to study the microflora in CCD hives, and examined possible contribution of different pathogens in colonies collapse. They found that two *dicistroviruses*, Kashmir bee virus (KBV) and Israeli acute paralysis virus (IAPV) of bees present in almost all of the CCD-affected hives and in very few of the control hives. The latter virus is found to be strongly correlated with a CCD occurrence. Nevertheless, in none of the control hives a combination of pathogens is detected, in contrast to their presence in the majority of the affected hives. In [25], a special attention is paid to a particular parasite *Nosema ceranae*, which is believed to lead to colony collapse in Spain. Although it is dismissed as a sole factor for CCD, it is generally associated with heavily diseased colonies.

The common conclusion in the literature is that there is no single causative factor of CCD. Similarly, in [33] a descriptive epizootiological study is performed, where many quantitative variables, including bee body mass and protein analyses, morphometric measures, parasite and pathogen analyses as well as pesticide and genetic analyses are examined. One of the main findings is that, since affected colonies are more likely to neighbour such colonies and vice versa, it is plausible CCD to be either contagious condition or to result from exposure to a common risk factor. Interestingly, the bees from CCD colonies were more symmetrical from those from control colonies, due to the fact that only the fittest bees have remained in the hive. As mentioned before, the CCD colonies were co-infected with a greater number of pathogens than their control counterparts, suggesting either greater pathogen exposure or reduced defenses in CCD bees. Also, there are some stressor agents which quantity was higher in the control colonies, explained with a possible developed tolerance to these stressors, which might in turn protect the bees themselves.

In order to figure out the dynamics of a honeybee colony population, the mathematical models come to the rescue. They are able to delineate different processes and phenomena in a quantitative way and help drawing implications about the physical behaviour of the dynamic system and, in particular, the way a colony collapse progresses and the inevitable failure is reached. Such CCD model is introduced in the next section. In Sect. 3 the inverse problem is formulated and the numerical methods for the direct and inverse problems are developed in Sects. 4 and 5, respectively. Section 6 is devoted to computational experiments with realistic data. Discussion and summary are presented in the last section.

2 Mathematical Model

In this section we will describe a model of CCD in a honeybee colony. As known, it is a major problem which honeybee farms face due to the massive decline in the colony number, see e.g. [12,32].

In [5] different bee population models are considered and classified as colony, varroa, and foraging models. The authors of [27] created a model that finds a connection between the demographic dynamics of a colony and dynamics of honeybee infections and diseases. Models that considers uninfected hive/forage bees, infected hive bees, virusfree mites and virus-infected mites have been built in [2,21,26]. A bee as a projected simulation unit and the collective defence of honeybee colonies against predators is considered in [20]. Factors that influence the colony failure have been explored in [27,31] and it is found that the shortage of nutrition and the pheromone-driven social inhibition are the main causes of colony collapse. The seasonal effects are taken into account in a model constructed in [30].

A simple compartment model concerning the forage bee population was developed in [6]. The model proposed in [15,16] describes the honey bee colony population dynamics with two weakly coupled ODEs. It determines whether the colony would survive or die depending on whether the level of the forager death rate is below or above of a certain threshold, respectively. The latter model is extended in [35]. The common propery of these models is that they consider the population of the female bees because the males/drones do not contribute to the colony labour. So does the CCD model [10] we follow henceforward.

The model presented in this paper demonstrates the dynamics of a honeybee colony during an extended foraging season from early spring to late summer. The prominent feature of the CCD is the insufficient workforce to sustain the colony, thus urging the young hive bees to be recruited into the forager class. One possible way to explain the unusual absence of adult honey bees is that infected individuals emigrate from their hives in attempt to prevent the other bees from contamination [17]. We consider three separate classes of bees – hive, forager and infected forager class. The brood is not included in the model since they do not interfere in the infection of the hive [10]. Furthermore, the behavioral classes are distinct and do not overlap, since every bee has a clearly defined role in the hive [5, 15].

We denote *the number of bees which work inside the hive* with H, *the number of bees that work outside the hive* with F, which are called *foragers*, and *the number of infected foragers* with I. The early stage of the model contained a fourth variable, the *disease vector*. Under some biological assumptions, this extra dimension was incorporated in the model by the parameter k [10]. Then, the system of three ODEs follows:

The rate of change of the hive class size is modeled as

$$\frac{dH}{dt} = \varepsilon(H) - H\rho(F,I), \tag{1}$$

where the change consists of the eclosion, reduced with the proportional recruitment rate. The rate of change of the forager class size is designed by

$$\frac{dF}{dt} = H\rho(F,I) - \mu_1 F - F\delta(I), \tag{2}$$

where the change is constituted by the fresh recruits, reduced with the deceased foragers and the proportionally infected ones. Finally, the rate of change of the infected forager class size is given by

$$\frac{dI}{dt} = F\delta(I) - \mu_2 I, \tag{3}$$

where the change is explained by the *infection* portion of foragers minus the died infected bees.

The function $\varepsilon(H)$ shows how the eclosion rate depends on the number of hive bees. The maximal rate of eclosion is assumed to be equivalent to the queen's laying rate L and that eclosion rate approaches its maximum as H increases. The simplest function that increases from 0 for no hive workers and tends to L as H becomes very large is

$$\varepsilon(H) = L\frac{H}{H+\omega}. \tag{4}$$

Here ω defines the way how $\varepsilon(H)$ approaches L as H gets large.

The transition $\rho(F,I)$ from the hive to the forager class seems to be extremely affected by the disorder and its exploration is of great importance [10]. The modeled CCD appears to increase this rate, and that is why it is constituted by two rates – a *healthy* and an *unhealthy* rate. The unhealthy rate starts to influence the total rate when, for a reason, the recruitment from the hive to the forager class gets faster. In particular, the term $\alpha\frac{\varphi}{F+I+\varphi}$ works when the number of the healthy and the infected foragers approaches some low level. α is actually the maximal unhealthy recruitment rate. On the other hand, γ represents the healthy dynamics. It is the normal recruitment rate in terms of the average time a bee spends in the hive class. Then, the recruitment rate function $\rho(F,I)$ is

$$\rho(F,I) = \alpha\frac{\varphi}{F+I+\varphi} + \gamma. \tag{5}$$

φ is a constant like ω. Actually, ω represents the minimal number of bees needed to raise the brood for the hive to survive. φ is the threshold of the number of the total (healthy and infected) foragers below which a maximal recruitment rate would occur.

The disease rate $\delta(I)$ is composed by β, the infection rate of the susceptible foragers, and the rate of the transmission of the infection. The whole rate $F\delta(I)$ is represented as mass action and is the combined contact rate of the healthy and infected bees. k represents the infection of plants and has the form $k = \dfrac{pc}{b}$, where p is the total number of plants in the hive foraging region, c is the clearance rate of the infection from the plants, and b is the contamination rate of the bees from the plants [10]. The disease rate is as follows:

$$\delta(I) = \beta \frac{I}{I+k}. \tag{6}$$

Of course, μ_1 is the death rate of a healthy forager bee, representing the average life longevity of a forager. μ_2 is the average time that an infected bee remains a functional forager before leaving the hive or dying.

3   Inverse Problem Formulation

In this section we will define the inverse identification problem, or what and how we would be looking for. In the real world, the values of the parameters α, β, γ, μ_1 and μ_2 are typically unknown and their reconstruction plays a crucial role in the honeybee colony management.

The problem (1)–(6) is restated as follows:

$$\frac{dH}{dt} = L\frac{H}{H+\omega} - H\left(\alpha\frac{\varphi}{F+I+\varphi} + \gamma\right), \tag{7}$$

$$\frac{dF}{dt} = H\left(\alpha\frac{\varphi}{F+I+\varphi} + \gamma\right) - \mu_1 F - \beta\frac{FI}{I+k}, \tag{8}$$

$$\frac{dI}{dt} = \beta\frac{FI}{I+k} - \mu_2 I, \tag{9}$$

$$H(t_0) = H^0, \quad F(t_0) = F^0, \quad I(t_0) = I^0, \tag{10}$$

where $\boldsymbol{p} = (p^1, p^2, p^3, p^4, p^5)$, $p^1 := \alpha$, $p^2 := \beta$, $p^3 := \gamma$, $p^4 := \mu_1$, $p^5 := \mu_2$ and

$$\boldsymbol{p} \in \mathbb{S}_{\text{adm}} = \left\{\boldsymbol{p} \in \mathbb{R}^5 : 0 < p^i < P^i, \ i = 1, \dots, 5\right\}. \tag{11}$$

Henceforward all solutions $\{H(t;\boldsymbol{p}), F(t;\boldsymbol{p}), I(t;\boldsymbol{p})\}$, $\boldsymbol{p} \in \mathbb{S}_{\text{adm}}$ are defined on the interval $t_0 \leq t \leq T$. When the parameters α, β, γ, μ_1 and μ_2 are known, the problem (7)–(11) is well-posed and it is called a *direct problem*.

Following the definition of the functions $H(t)$, $F(t)$ and $I(t)$, the positivity of the solution to problem (7)–(10) is required. We show that the solutions to all subpopulations $\{H(t), F(t), I(t)\}$ in the system (7)–(9) are nonnegative for all $t \geq 0$ given nonnegative initial conditions (10). The result follows from the application of the Theorem 7.1 in [14]. Consider the IVP for the ODE system

$$\frac{du}{dt} = f(t, u(t)), \quad t \geq t_0, \ u(t_0) = u_0, \ t_0 \in \mathbb{R}, \ u_0 \in \mathbb{R}^m, \tag{12}$$

where $f : \mathbb{R} \times \mathbb{R}^m \to \mathbb{R}^m$ has the following property: the system (12) is positive iff

$$v_i = 0, \ v_j \geq 0 \ \forall j \neq i \implies f_i(t, v) \geq 0$$

holds for all t and any vector $v \in \mathbb{R}^m$ and all $i = 1, 2, \ldots, m$.

Let us assume now that the coefficients p^i, $i = 1, \ldots, 5$ are unknown. We explore the inverse problem of reconstructing the parameters $p \in \mathbb{S}_{adm}$ through the observed behaviour

$$U(t^i) := \{H(t^i), \ F(t^i) + I(t^i)\}, \ i = 1, \ldots, I_{obs}; \ t_0 = t^1 < \ldots < t^{I_{obs}} = T \qquad (13)$$

of the dynamic system (7)–(11).

Here, a natural question arises concerning the form of the observations. The primary causative agent of CCD remains unknown. On one hand, the illnesses that make the bees to look sick (like the paralysis viruses, which disable the bees from flying [24]) would eventually result in many carcases in and around the hive, which in turn would reject the CCD hypothesis. On the other hand, the CCD-symptomatic bees do not look ill in appearance, while they still serve the hive. To put it together, it is extremely hard, if possible at all, to distinguish a healthy forager from an infected, but still functional forager. This leads to the practical requirement all the foragers to be measured together.

The original inverse coefficient problem is replaced by a minimization problem and can be formulated in a variational setting as follows:

$$\min_{p \in \mathbb{S}_{adm}} \Phi(p), \qquad p = (p^1, p^2, p^3, p^4, p^5),$$

subjected to the solution of (7)–(11).

We look for the point $p = (p^1, p^2, p^3, p^4, p^5)$ of the local minimum of the functional $\Phi(p)$. The functional $\Phi(p)$ can be written as

$$\Phi(p) = \frac{1}{2} \left(U(t^i; p) - U_{obs}(t^i) \right)^\top \left(U(t^i; p) - U_{obs}(t^i) \right), \qquad (14)$$

where $U(t^i; p) = \{H(t^i; p), \ F(t^i; p) + I(t^i; p)\}$ is the solution to the problem (7)–(11), and $U_{obs}(t^i) = \{H_{obs}(t^i), \ F_{obs}(t^i) + I_{obs}(t^i)\}$ are the experimental data (13).

The admissible set $\mathbb{S}_{adm}$ follows both the specification of the particular model [10] and the biology of the honey bees [34].

4   Solution to the Direct Problem

This section briefly shows how to solve the direct problem (7)–(11). It is of a particular importance, because we need to solve the direct problem a number of times in order to solve the inverse problem.

Let us introduce the piecewise-uniform mesh

$$\overline{\omega}_\tau = \{ \text{for } i = 2, \ldots, I_{obs} \text{ we set } t_j^i = t^{i-1} + j\tau_i, \ j = 1, \ldots, J_i,$$
$$t^1 = t_0, \ t^i = t^{i-1} + \tau_i J_i, \ t^{I_{obs}+1} = T \}, \qquad (15)$$

where t^i, $\forall i = 2,\ldots,I_{\text{obs}}$ are the time instances which measurements are taken at. Moreover, t^i_j, $j = 1,\ldots,J_i$ are the time point and τ_i are the time steps corresponding to $(t^{i-1}, t^i]$, respectively.

There are many approaches to solve the initial problem (7)–(10), including numerical, Monte Carlo or machine learning methods. However, not only the accuracy is important. The designed methods must preserve positivity of the solutions and the local behaviour. The computational results show positivity of the numerical solutions, see Sect. 6. We solve the direct problem using the MATLAB ode45 subroutine [23] using the mesh (15). The solver is based on an explicit Runge-Kutta-Dormand-Prince formula [9,28], which achieves fourth- and fifth-order accurate solutions.

In [10,18], a thorough equilibria analysis is performed. There are six equilibria states of the system (7)–(9), five of which are biologically relevant. The first one is the trivial equilibrium $H = 0$, $F = 0$, $I = 0$. This *extinction* equilibrium is stable provided that $L < \omega(\alpha + \gamma)$.

There are two different *disease-free* equilibria, which require

$$\alpha\omega > \mu_1\varphi, \quad 2\sqrt{\alpha\mu_1\varphi\omega} - \mu_1\varphi \leq L - \gamma\omega.$$

These inequalities condition the larger disease-free equilibrium. The smaller one requires meeting them, together with $L < \omega(\alpha + \gamma)$. Simply, when the conditions for the smaller equilibrium are satisfied, the conditions for the larger one are also met. They can exist at the same time [10].

There are also two *endemic* equilibria. They may exist if the basic reproductive number $R_0 = \dfrac{\beta F}{\mu_2 k} > 1$. In this case, the disease sticks on the hive. If they do not exist, the hive goes extinct due to the CCD. When $R_0 < 1$, if both disease-free equilibria exist, the larger one is stable and the smaller is unstable. On the contrary, if only one disease-free equilibria exists, it is always stable and the extinction equilibrium loses its stability [10]. For more details please refer to Table 1 in [18].

We will present some particular (and important) cases which will later guide us in performing the simulations. Henceforward, we assume that $L < \omega(\alpha + \gamma)$ is satisfied. First we take into consideration the disease-free case, i. e. when $I^0 = 0$. Its phase plane diagram is presented on Fig. 1 (left), compare with Fig. 3.

The blue curves follow concrete solutions development and the stars denote the starting points. All the solutions (except in the extinction case) tend to the disease-free equilibrium, which is normal for a healthy colony.

The more interesting case is the honeybee population dynamics in presence of infected foragers. The circles on Fig. 1 (right) designate the end points after one foraging season. All the solutions tend to an endemic equilibrium, which is characterized by low number of foragers and similar quantity of infected foragers. Such a pair of solutions are plotted on Fig. 4.

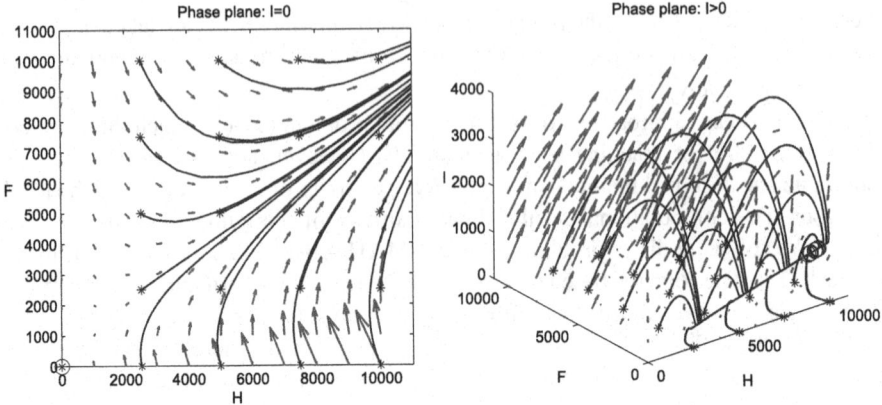

Fig. 1. Phase plane diagram of the solution in the case $I = 0$ (left) and $I > 0$ (right)

5 Solution to the Inverse Problem

In this section we describe the methodology to solve the parameter identification inverse problem. It results in finding those values of the parameters $\boldsymbol{p}$, which reproduce the system dynamics that is as closest as possible to the one observed in practice.

The cost functional defined in (14) represents the sum of the squares of the residuals $U(t^i; \boldsymbol{p}) - U_{\text{obs}}(t^i)$. They are formed by the differences between the theoretical quantities $U(t^i; \boldsymbol{p})$ and the measurements $U_{\text{obs}}(t^i)$ for all time instances t^i, $i = 1, \ldots, I_{\text{obs}}$.

The point of minimum of the objective function $\Phi(\boldsymbol{p})$ is designated with $\check{\boldsymbol{p}}$. It is called *nonlinear least squares estimator (LSE)* and it will play a central role in our analysis. From the many ways to minimize $\Phi(\boldsymbol{p})$, we chose to apply the *Trust Region Reflexive* method [7], employing the MATLAB subroutine lsqnonlin [22].

To further assess the quality of optimal solution $\check{\boldsymbol{p}}$, we also compute several statistical quantities to evaluate the convergence. Let us begin with the *norm of the step* $\delta \boldsymbol{p}_k$, which measures the change between two successive iterates of the minimizer $\boldsymbol{p}_k$ and $\boldsymbol{p}_{k+1}$ and is defined as

$$\delta \boldsymbol{p}_k := \|\triangle \boldsymbol{p}_k\| = \|\boldsymbol{p}_{k+1} - \boldsymbol{p}_k\|. \tag{16}$$

Moreover, *the relative change in the cost function* is defined as

$$\delta \Phi_k := \frac{|\triangle \Phi_k|}{1 + |\Phi(\boldsymbol{p}_k)|}, \tag{17}$$

where $\triangle \Phi_k := \Phi(\boldsymbol{p}_{k+1}) - \Phi(\boldsymbol{p}_k)$.

The minimization algorithm stops when the following condition is met:

$$\min \{\delta \boldsymbol{p}_k, \delta \Phi_k\} < \varepsilon_\delta, \tag{18}$$

where ε_δ is a user-defined tolerance. The implementation stops the algorithm at iteration $k+1$ and returns an approximation of the nonlinear LSE $\check{\boldsymbol{p}} := \boldsymbol{p}_{k+1}$.

To check the convergence of the algorithm, we provide the *the first-order optimality measure*, which describes how close the approximation is to the real minimum of (14). It is defined as the infinity norm of the gradient of the objective function, evaluated at the nonlinear LSE $\check{p}$, which in turn is given by the maximal absolute value of the partial derivatives of the objective function w. r. t. p:

$$\|\nabla\Phi(\check{p})\|_\infty = \max_{r=1,\dots,5}\left|\frac{\partial\Phi}{\partial p^r}(\check{p})\right|. \tag{19}$$

What is more, to quantify the *goodness-of-fit* performance, after deriving the non-linear LSE, we study the following statistics: *the variance of the residuals $\tilde{\sigma}^2$, the root mean squared error (RMSE) $\hat{\sigma}$, and the coefficient of determination R^2* [3]:

$$\tilde{\sigma}^2 = \frac{\Phi(\check{p})}{I_{obs}}, \quad \hat{\sigma} = \sqrt{\frac{1}{I_{obs}-3}\Phi(\check{p})}, \quad R^2 = 1 - \frac{\Phi(\check{p})}{\displaystyle\sum_{i=1}^{I_{obs}}\left((U_{obs}(t^i) - \overline{U_{obs}})^2\right)},$$

where $\overline{U_{obs}} = \left\{\dfrac{\displaystyle\sum_{i=1}^{I_{obs}} H_{obs}(t^i)}{I_{obs}}, \dfrac{\displaystyle\sum_{i=1}^{I_{obs}}\left(F_{obs}(t^i)+I_{obs}(t^i)\right)}{I_{obs}}\right\}$ designates the mean value of the

experimental data. The variance of the residuals $\tilde{\sigma}^2$ and the RMSE $\hat{\sigma}$ measure the model fit w. r. t. the data available, while the coefficient of determination R^2 shows the model fit to the data mean. A good fit would be indicated by small $\tilde{\sigma}^2$ and $\hat{\sigma}$ and a value of R^2 close to 1.

In case the residuals $\left(U(t^i;p) - U_{obs}(t^i)\right)$ are normally distributed or the measurements set size I_{obs} is sufficiently large, then the estimated *covariance matrix* of the nonlinear LSE $\check{p}$ is expressed as

$$\Sigma = \frac{\hat{\sigma}^2}{J^\top(\check{p})J(\check{p})}, \tag{20}$$

where $J(\check{p})$ is the *sensitivity* or *Jacobian matrix*, evaluated at $\check{p}$ and it is defined as

$$J(\check{p}) = \left(\frac{\partial H(\check{p})}{\partial p}, \frac{\partial F(\check{p})}{\partial p}, \frac{\partial I(\check{p})}{\partial p}\right)^\top. \tag{21}$$

The sensitivity matrix $J(p)$ quantifies the variation of the variables $U(t^i;p)$ w. r. t. changes in p. It favors gathering information about the appropriateness of the data measurements to the parameter recovery and is of a particular importance in the inverse problem studies [4]. The covariance matrix (20) helps further to compute the *standard error (SE)* and the *normalized standard error (NSE)*. They are used to quantify the accuracy of the parameter estimate and their definitions follow:

$$\widehat{SE} = \sqrt{\text{diag}(\Sigma)}, \quad \widehat{NSE} = \frac{\widehat{SE}}{\check{p}} \times 100, \tag{22}$$

where the division should be interpreted elementwisely.

The relevance of the metrics (20)–(22) could be validated only if the number of observations I_{obs} is sufficiently large since the nonlinear LSEs are asymptotically normally distributed [3].

The summarized methodology for the honeybee colony dynamics parameter reconstruction is schematically presented of Fig. 2.

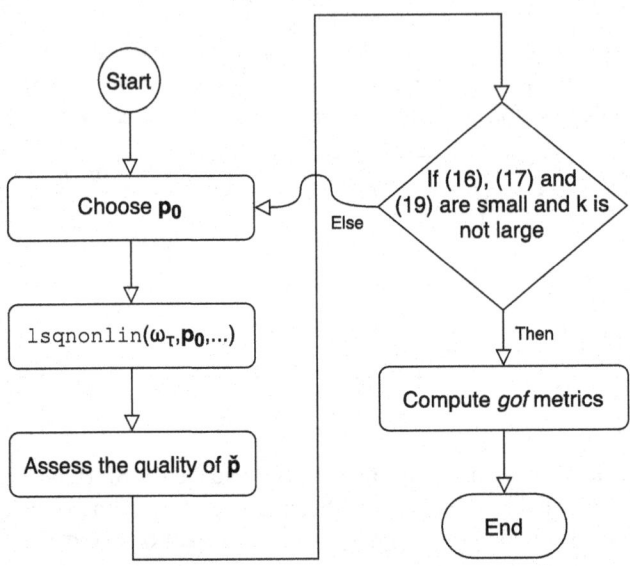

Fig. 2. Flowchart of the computational algorithm

First of all, we choose a suitable initial approximation, denoted by p_0. Then, we execute the lsqnonlin subroutine with the aforementioned initial point, the temporal mesh ω_τ and appropriate values for the remaining parameters. When the condition (18) is met, the nonlinear LSE $\check{p}$ is found. Then we read the quality of the current minimizer via the following check – if $\|\nabla\Phi(\check{p})\|_\infty$, δp_k and $\delta\Phi_k$ are small and the number of iterations is not large, then the solver has arrived at the optimal solution $\check{p}$. Otherwise, we have to choose another starting point and repeat the steps. When $\check{p}$ is accepted, then, in order to further evaluate it, we compute the considered goodness-of-fit metrics $\breve{\sigma}^2$, $\hat{\sigma}^2$, R^2 and $\widehat{NSE}$.

6   Computational Experiments

We devote this section to the numerical results, which confirm the robustness and efficiency of the proposed algorithm. Here we present quasi-real data test, using the information provided in [10].

6.1   Direct Problem

First, we solve the direct problem stated as (7)–(10). The number of the queen's maximum daily eggs laid is set to $L = 2000$. The minimum number of adult bees needed to rear the brood is $\omega = 15000$, while the threshold level of total foragers is $\varphi = 1000$. In normal circumstances, the bees work 21 days in the hive before becoming foragers [1], so $\gamma = 1/21$ d^{-1}. On the other hand, it takes at least 7 days for a bee to mature enough to forage [1]. The recruitment rate function ρ must take this into account, and finally $\alpha = 1/7 - 1/21$. The infection rate is assumed to be $\beta = 0.8$, together with $k = 100$. The healthy death normally occurs on the 21th day, thus $\mu_1 = 1/21$. The average time an infected forager remains functional is estimated to be 1.25 days, so $\mu_2 = 1/1.25$ d^{-1}.

To demonstrate the solution to the direct problem, we assume a colony with $H^0 = 15000$ hive bees at $t_0 = 0$. We plot two cases – with ($F^0 = 15000$) and without ($F^0 = 0$) foragers at the beginning at the foraging season. It's typical length is $T = 250$ days.

Firstly, we consider a «healthy» scenario, i. e. without infected bees. We can confidently conclude that the colony achieves a healthy equilibrium state with normal total number of bees, see Fig. 3.

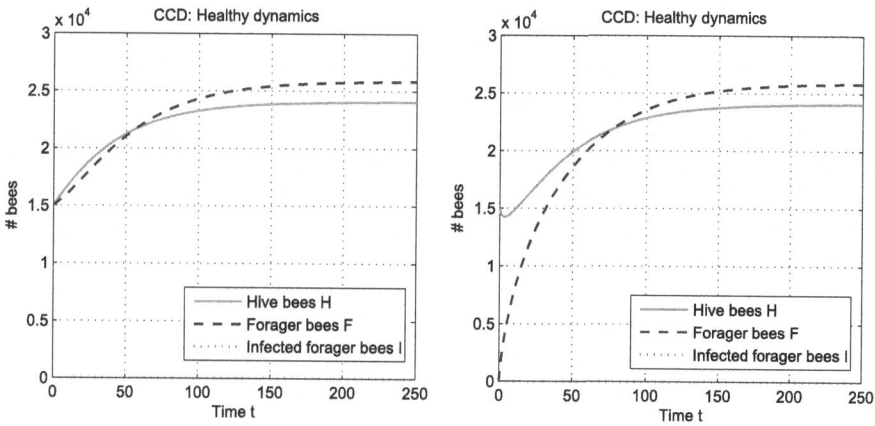

Fig. 3. Healthy dynamics: $F^0 = 15000$ (left) and $F^0 = 0$ (right)

Next, we simulate an «unhealthy» scenario, this time with $I^0 = 10$ infected bees at t_0. Note that we switch to a logarithmic abscissa, since the infection spreads extremely fast in the beginning. Some simulations reveal that the number of the initially infected bees is not crucial for the dynamics (Fig. 4). Also, it shows that the healthy equilibrium state is unstable.

6.2   Inverse Problem

Here we will present results concerning the inverse problem. Let us again set $L = 2000$, $\omega = 15000$, $\varphi = 1000$, $k = 100$ and $t_0 = 0$, $T = 250$. We aim to recover the parameter

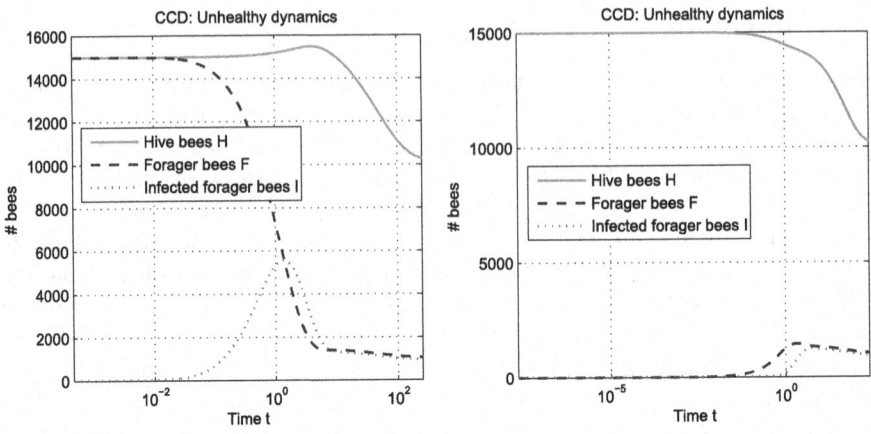

Fig. 4. Unhealthy dynamics: $F^0 = 15000$ (left) and $F^0 = 0$ (right)

set $p = (\alpha, \beta, \gamma, \mu_1, \mu_2)^\top = (2/21, 0.8, 1/21, 1/21, 0.8)^\top$ solving (7)–(11). We look for the parameters values in the unit hypercube $p \in \mathbb{S}_{\mathrm{adm}} \equiv (0, 1)^5$ with tolerance $\varepsilon_\delta = 1e - 16$ (18).

In the absence of any other information, we start the iterations with the initial vector $p_0 = (0.5, 0.5, 0.5, 0.5, 0.5)^\top$, which values are as neutral as possible. Surprisingly at first glance, we will not demonstrate parameter reconstruction of healthy dynamics. If one have a second look on the model (7)–(9), they might see that the coefficients β and μ_2 are impossible to recover if infection dynamics is not involved. Hence we will consider only the «unhealthy» case $H^0 = 15000$, $F^0 = 15000$, $I^0 = 10$.

The other crucial moment is the distribution of the observation time instances. As it can be seen on Fig. 4, we have switched to a logarithmic axis due to the vigorous development of the infection at the beginning of the foraging season. This is why we suggest observations to be taken in the first days and set $t^i = i^{\mathrm{th}}$ day, $i = 2, \ldots, I_{\mathrm{obs}}$. The results for $I_{\mathrm{obs}} = 10$ follow in Table 1.

Table 1. Unhealthy dynamics with $I_{\mathrm{obs}} = 10$ observations

Parameter	p_0	p	$\check{p}$	$|p - \check{p}|$	$\dfrac{|p - \check{p}|}{p}$	$\widehat{NSE}$
α	0.5	0.0952	0.0952	3.2008e-05	3.3608e-04	0.0188
β	0.5	0.8	0.7640	0.0360	0.0450	0.0027
γ	0.5	0.0476	0.0476	3.2252e-06	6.7728e-05	0.0040
μ_1	0.5	0.0476	0.0485	8.6462e-04	0.0182	0.0189
μ_2	0.5	0.8	0.8354	0.0354	0.0443	0.0027

The parameters are reconstructed relatively accurately. The 'farthest' guesses from their true values are of β and μ_2, since their relative errors are the highest. All the normalized standard errors are below 0.02%, which implies the efficiency of the recovery. Now we present the case of $I_{obs} = 15$ observations in Table 2.

Table 2. Unhealthy dynamics with $I_{obs} = 15$ observations

Parameter	p_0	p	$\check{p}$	$|p - \check{p}|$	$\dfrac{|p - \check{p}|}{p}$	$\widehat{NSE}$
α	0.5	0.0952	0.0952	2.9143e-16	3.0601e-15	1.0766e-13
β	0.5	0.8	0.8	6.6169e-14	8.2712e-14	3.1882e-14
γ	0.5	0.0476	0.0476	4.8572e-17	1.0200e-15	3.0501e-14
μ_1	0.5	0.0476	0.0476	3.3723e-15	7.0818e-14	2.3867e-13
μ_2	0.5	0.8	0.8	5.6510e-14	7.0638e-14	3.0533e-14

Here the unknown parameters are identified in an exact way. There are no practical differences between the implied and the real values, which is affirmed by the smallness of the error metrics.

Table 3. Error and goodness-of-fit indicators

I_{obs}	$\|\nabla\Phi(\check{p})\|_\infty$	δp_k	$\delta\Phi_k$	$k+1$	$\Phi(\check{p})$	$\tilde{\sigma}^2$	$\hat{\sigma}$	R^2
10	0.000153	5.580812e-17	–	50	0.471693	0.0472	0.2596	$1 - 3.0431e-09$
15	2.55e-08	1.77933e-10	2.37e-17	34	1.32762e-22	8.8508e-24	3.3262e-12	$1 - 5.0091e-31$

The proposed algorithm performed in an excellent manner, since the variance ($\tilde{\sigma}^2$) and the RMSE ($\hat{\sigma}$) are very small, while the coefficient of determination R^2 is practically 1 (Table 3). Quite expected, the less information contained in the smaller number of observations make the inverse problem tougher to solve, which is signalized by the less accurate values and the more iterations needed for convergence.

7 Conclusion

In the last decades there were massive losses of honeybee colonies which have raised concern. Although there is still no consensus on what causes the phenomenon Colony Collapse Disorder, the model proposed in the paper helps to delineate the population dynamics of honeybee colonies. It accounts for both healthy and extinction behaviour of the honey bees, modeling CCD as a contagious infection brought to the hive by the foragers, caused by metabolic stress, environmental pollution and other factors.

The described computational algorithm reveals how the parameters affect the system. It could support the beekeepers in managing the colonies in such a way to mitigate the severity or, hopefully, to completely eradicate the disorder's effect on the hives. A future work might include modification and improvement of the current model to produce more precise description of the dynamics and a deeper analysis to suggest methodology for prevention of such diseases, impacting both environmental balance and economics.

Acknowledgement. The authors would like to thank the reviewers for their constructive comments and suggestions, which significantly improved the quality of the paper.

The study of the first author is supported by contract 2020-FNI-01, funded by the Research Fund of the University of Ruse. The second and third author are supported by the Bulgarian National Science Fund under Project DN 12/4 "Advanced analytical and numerical methods for nonlinear differential equations with application in finance and environmental pollution" from 2017.

References

1. Amdam, G.V., Omholt, S.W.: The hive bee to forager transition in honeybee colonies: the double repressor hypothesis. J. Theoret. Biol. **223**(4), 451–464 (2003)
2. Bailey, L.: The Isle of Wight Desease. Central Association of Bee-Keepers (2002)
3. Banks, H.T., Davidian, M., Samuels, J.R.: An inverse problem statistical methodology summary. Mathematical and statistical estimation approaches in epidemiology, pp. 249–302 (2009). https://doi.org/10.1007/978-90-481-2313-1_11
4. Banks, H., Dediu, S., Ernstberger, S.: Sensitivity functions and their uses in inverse problems. J. Inverse Ill-Posed P. **15**(7), 683–708 (2008)
5. Becher, M.A., Osborne, J.L., Thorbek, P., Kennedy, P.J., Grimm, V.: Review: towards a systems approach for understanding honeybee decline: a stocktaking and synthesis of existing models. J. Appl. Ecol. **50**, 868–880 (2013)
6. Booton, R.D., Iwasa, Y., Marshall, J.A.R., Childs, D.Z.: Stress-mediated Alle effects can cause the sudden collapse of honey bee colonies. J. Theoret. Biol. **420**, 213–219 (2017)
7. Coleman, T.F., Li, Y.: An interior, trust region approach for nonlinear minimization subject to bounds. SIAM J. Opt. **6**, 418–445 (1996)
8. Cox-Foster, D.L., et al.: A metagenomic survey of microbes in honey bee Colony Collapse Disorder. Sci. **318**, 283–287 (2007)
9. Dormand, J.R., Prince, P.J.: A family of embedded Runge-Kutta formulae. J. Comp. Appl. Math. **6**, 19–26 (1980)
10. Dornberger, L., et al.: Death of the bees: a mathematical model of colony collapse disorder. Technical report 2012–12, Mathematics Preprint Series, University of Texas at Arlington Mathematics Department (2012)
11. Ellis, J.D., Evans, J.D., Pettis, J.: Colony losses, managed colony population decline, and colony collapse disorder in the United States. J. Apic. Res. **49**(1), 134–136 (2010)
12. Sheet, F.: The economic challange posed by declining pollinator populations. Office the Press Secretary, The White House (2015)
13. Finley, J., Camazine, S., Frazier, M.: The epidemic of honey bee colony losses during the 1995–1996 season. Am. Bee J. **136**, 805–808 (1996)
14. Hundsdorfer, W., Verwer, J.G.: Numerical Solution of Time-Dependent Advection-Diffusion-Reaction Equations. Springer, N.Y. (2003). https://doi.org/10.1007/978-3-662-09017-6
15. Khoury, D.S., Myerscough, M.R., Barron, A.B.: A quantitative model of honey bee colony population dynamics. PLoS ONE **6**(4), e18491 (2011)

16. Khoury, D.S., Barron, A.B., Meyerscough, M.R.: Modelling food and population dynamics honey bee colonies. PLoS ONE **8**(5), e0059084 (2013)

17. Kralj, J., Fuchs, S.: Parasitic mites influence flight duration and homing ability of infested Apis mellifera foragers. Apidologie **37**, 577–587 (2006)

18. Kribs-Zaleta, C.M., Mitchell, C.: Modeling Colony Collapse Disorder in honeybees as a contagion. Math. Biol. Eng. **11**(6), 1275–1294 (2014)

19. Kulincevic, J.M., Rothenbuhler, W.C., Rinderer, T.E.: Disappearing disease. Part 1 - effects of certain protein sources given to honey bee colonies in Florida. Am. Bee J. **122**, 189–191 (1982)

20. López-Incera, A., Nouvian, M., Ried, K., Müller, T., Briegel, H.J.: Collective defense of honeybee colonies: experimental results and theoretical modeling. arXiv:2010.07326 [q-bio.PE], pp. 1–20 (2020)

21. Ma, Z., Zhou, Y., Wu, J.: Modeling and Dynamics of Infectious Diseases. World Scientific Publishers, Singapore (2009)

22. MathWorks: solve nonlinear least squares (nonlinear data-fitting) problems. https://www.mathworks.com/help/optim/ug/lsqnonlin.html. Accessed 13 Sep 2020

23. MathWorks: solve nonstiff differential equations - medium order method. https://www.mathworks.com/help/matlab/ref/ode45.html. Accessed 13 Sep 2020

24. Oldroyd, B.P.: What's killing american honey bees? PLoS Biol. **5**(6), e168 1195–1199 (2007)

25. Paxton, R.J.: Does infection by Nosema ceranae cause "Colony Collapse Disorder" in honey bees (Apis mellifera)? J. Apic. Res. **49**(1), 80–84 (2010)

26. Ratti, V., Kevan, P.G., Eberl, H.J.: A mathematical model of forager loss in honeybee colonies infested with Varroa destructor and the acute bee paralysis virus. Bull. Math. Biol. **79**(6), 1218–1253 (2017)

27. Russel, S., Barron, A.B., Harris, D.: Dynamics modelling of honeybee (Apis mellifera) colony growth and failure. Ecolog. Model. **265**, 138–169 (2013)

28. Shampine, L.F., Reichelt, M.W.: The MATLAB ODE Suite. SIAM J. Sci. Comp. **18**, 1–22 (1997)

29. Silver, J.: Bee disease on the Isle of Wight. Irish Bee J. **7**, 10 (1907)

30. Switanek, M., Crailsheim, K., Truhetz, H., Brodschneider, R.: Modelling seasonal effects of temperature and precipitation on honey bee winter mortality in a temperate climate. Sci. Tot. Environ. **579**, 1581–1587 (2017)

31. Torres, D.J., Ricoy, V.M., Roybal, S.: Modelling honey bee populations. PLoS ONE **10**(7), e0130966 (2015)

32. Van der Zee, R., et al.: Managed honey bee colony losses in Canada, China, Europe, Israel and Turkey for the winters of 2008–2009 and 2009–2010. J. Appl. Res. **51**(1), 100–114 (2012)

33. van Engelsdorp, D., et al.: Colony collapse disorder: a descriptive study. PLoS ONE **4**(8), e6481 (2009)

34. Winston, W.L.: The Biology of the Honey Bee. Harvard University Press, Cambridge Mass (1991)

35. Yıldız, T.A.: A fractional dynamical model for honeybee colony population. Int. J. Biomath. **11**(5), 1850063 (2018)

Ontology Engineering

A Semantic Approach to Multi-parameter Personalisation of E-Learning Systems

Humam K. Majeed Al-Chalabi[1]([✉])(iD) and Ufuoma Chima Apoki[2](iD)

[1] Faculty of Automatics, Computer Science and Electronics, University of Craiova, Craiova, Romania
`hemoomajeed@gmail.com`
[2] Faculty of Computer Science, Alexandru Ioan Cuza University, Iasi, Romania
`apoki.ufuoma@info.uaic.ro`

Abstract. This paper presents a model framework for achieving semantic adaptation in e-learning systems using multiple parameters for personalisation. The proposed model, which utilises semantic technologies, aims to boost learning experiences and outcomes within the process of learning. This is often achieved through a mechanism that adapts the educational contents of a course in keeping with student's preferences expressed by multiple parameters (such as their learning styles, media preferences, level of data, language, etc.) The variation process involves real-time mapping of learning resources and student data, semantic annotation, metadata enrichment of learning resources, creation of student profile with relevant preferences, and personalisation of every course in step with the foremost suitable (or preferred) parameters. Achieving this entailed the creation of an ontology and several other modules that work in the background of the adaptive process.

Keywords: E-learning systems · Adaptive learning · Semantic web · Ontologies · Adaptive educational systems

1 Introduction

The purpose of personalised learning as one of the major needs of this century is the ability to recognise students' needs and preferences, and also their capabilities [1]. An online learning environment brings together learners possessing different learning capacities based on backgrounds and needs, and therefore requires different learning paths to achieve optimal satisfaction for each learner [2]. Individual characteristics and preferences of each learner (such as educational background, learning styles, learning objectives, motivation) are useful (if properly utilised) in the provision of optimal paths in the quest to accomplish individual learning outcomes.

Several platforms like Learning Management Systems (LMS) and Intelligent Tutoring Systems (ITS) exist for the delivery of e-learning content to students, as well as the administration and monitoring of student activities.

© Springer Nature Switzerland AG 2021
D. Simian and L. F. Stoica (Eds.): MDIS 2020, CCIS 1341, pp. 381–393, 2021.
https://doi.org/10.1007/978-3-030-68527-0_24

They offer the possibility of presenting enrolled students with a wide range of courses with highly customisable features. Learning could take different forms which include Computer Managed Learning, Computer Assisted Instruction, Synchronous/Asynchronous Learning, Fixed/Adaptive Learning, Linear/Interactive Learning, and Individual/Collaborative Learning [3].

Personalisation and adaptivity (interchangeably used, most times) have become key necessities in e-learning systems. While personalisation focuses on the customisation of learning content by an instructor, adaptivity refers to software/technology that can alter learning paths or course content in real-time from information that is gained from monitoring students and their interactions with the learning system. However, the majority of e-learning platforms that exist do not offer many options for personalisation or adaptivity; they are mostly achieved by customised learning platforms or by extending LMS (such as Moodle) through plugins or web services.

Personalisation and adaptivity can be achieved by creating different learning paths and/or experiences utilising different features of users. While most systems focus on adaptivity in general, others focus on adaptivity based on few parameters which make such systems course-specific and not easily customisable for other courses [4]. Adaptivity and personalisation can be achieved through adaptive content (which is widely used in implementation), adaptive instruction, and adaptive presentation [5].

There are multiple parameters in the literature used in the personalisation of learning scenarios. Criteria that can be used for personalisation include the learner's preferences, the status and history of the learner, the parameters of the learning medium, and other pedagogical and domain parameters. Level of knowledge and learning styles are popularly utilised in modeling and implementation. Personalisation parameters are covered in [5–7].

To achieve personalisation with multiple parameters, the authors in [7] outlined four main strategies:

- Applying all parameters to personalise each course,
- Applying a subset of parameters which represents only the preferences of the learners,
- Applying a subset based on standardisation of course materials,
- Applying a subset of parameters suggested by a domain expert supervising the course.

The drawback of the first two strategies is the high number of possible learning paths when the set of possible parameters are greater than two or three with different dimensions. This will involve a lot of tests and questionnaires and workload to ensure all dimensions of each personalisation parameters are represented for each concept to be studied. The third strategy takes advantage of metadata standards which already exist, while the last option utilises the expertise of the course instructor. To properly combine multiple parameters, it is imperative to explore the last two options.

Learning objects, which have been described as "any entity, digital or non-digital, which can be used, re-used, or referenced during technology-supported *learning"* [8] by the IEEE Learning Technology Standards Committee, have become fundamental in the development of educational resources in e-learning platforms. With constant developments in Information and Communications Technology (ICT) in the educational sector, the number and complexity of learning objects are on the rise. However, there is a lack of interoperability and compatibility of educational repositories, making it cumbersome in the design and maintenance of semantic education libraries and repositories, and the intelligent search of learning objects.

The semantic web, which is an extension of the current web, plays a huge part in the development of personalised learning in e-learning systems. The technologies of the semantic web, which include RDF, XML, and ontologies, can be useful for the intelligent discovery, annotation, semantic enrichment, and transformation of learning objects. Ontologies, which can be described as a *"specification of a conceptualisation"* [9], provide the advantage of solving the challenges of interoperability between the educational repositories of different e-learning systems. With well-defined ontologies, which can be extendable, personalised search and recommendation can be achieved, because computers are meaningfully able to process data due to the commonality of semantic meaning and relationships between terms [10].

As e-learning systems become more prevalent and hard to ignore in learning, there is a growing need for shared learning resources between already-existing learning systems for reusability and adaptability. A feasible approach requires a domain-independent, automatic, and unsupervised method to detect relevant features from heterogeneous learning resources, and associate them to concepts to be learnt which are modelled in a background ontology [11]. These learning resources need to be transformed through annotation into learning objects, which conform to metadata standards. These learning objects can then be set up hierarchically in an ontology. One challenge, though, with ontologies is providing semantic and structural uniformity. One way to solve this is the ontology mapping for the interoperability of learning resources presented in [12].

Within this framework, the rest of this paper is structured as follows: Sect. 2 gives a brief description of adaptivity in e-learning, and the opportunities of semantic web technologies in influencing adaptivity in online learning. Section 3 describes the architecture of the proposed model, and also details the technologies used in the design. The paper ends with Sect.4, which describes future work and improvements on the model and approach.

2 Literature Review

A major differentiating factor between online learning and traditional learning is the ability to redirect the focus of learning to a learner-centered environment. This involves the application of learning analytics tools, which include data collection from learners (log data, student characteristics, educational background, and academic performance) to manage and improve learning. Utilisation of these techniques produce adaptive and intelligent web-based educational systems.

Adaptivity in e-learning can involve various forms, which include adapting learning resources, support, display, and other instructional elements. In the literature, adapting learning resources is the most applied dimension in e-learning platforms, and they can be broadly grouped into content adaptation and link-level adaptation [10]. Content adaptation entails dynamically altering the contents of the learning resources such as fragments, segments, or pages, and having different forms of presentation of content. The other involves presenting the most suitable learning content in the right order based on the learner's needs and preferences. In both cases, it has become necessary to organise educational content as learning objects. The quality and efficiency of e-learning systems depend largely on the quality of suitably-selected learning objects, the relationships between the concepts they instruct, and the parameters for adaptation.

Learning resources used in e-learning platforms are mostly digital, allowing possibilities for modification to suit learners' preferences and needs. Learning content is abundant on (and off) the web in a variety of formats. The heterogeneity and amount of accessible learning resources are gradually becoming a challenge for learners and educational instructors/designers who design systems for e-learning purposes. Several metadata standards have been developed in a bid to solve the problems of non-uniformity [6]. However, many available learning resources do not fit these structures. To achieve intelligent and automatic searching and indexing of these diverse learning content, it is necessary to define a simpler metadata standard, which is independent of implementation [13]. Another method of achieving this is the research into the automatic annotation of diverse learning resources.

One major change which promises a huge evolution of learning processes is the transition from the traditional web to the semantic web. The semantic web (Web 3.0) involves the restructuring of data from relational databases to semantic graphs. The main technologies of Web 3.0 include eXtensible Markup Language (XML), Resource Description Framework (RDF), and Ontologies. XML allows for the arbitrary structuring of documents in a format, readable by both humans and machines, without explicitly stating what the structures mean. RDF, which was originally meant to be a metadata modelling language, is used to express data models referred to as objects or "web resources" and the relationships between them. With ontologies, concepts and relationships in a particular domain can be described. This facilitates the meaningful processing of data if there is a common understanding of the concepts and relationships between them [10].

The goal of this study is to create an ontology model and enrich it the components of a Relational Database (RDB) schema using classes provided by a semantic framework. With the semantic framework, it will be possible to make semantic web recommendations.

In this context, we will be using the D2RQ Framework [14] which enables the extraction and restructuring of data from RDBs in RDF graph format using the OWLready2 [15], which is a package for ontology-oriented programming in Python programming language. This makes the data also available for the semantic web. With the D2R server connection to the RDB, SPARQL queries [16] can be run to get semantic data.

Within ontologies and the semantic web, reasoners function to derive information from a knowledge base in an inference engine. The W3C (World wide web consortium) has a list of reasoners which include FaCT++, HermiT, and Pellet [17]. These reasoners can be used when they are interfaced to an ontology through an API. Pellet (developed in Java) is the first sound and complete OWL-DL reasoner with extensive support for reasoning with individuals, user-defined datatypes, and debugging support for ontologies. HermiT checks the consistency of an ontology, and can be used to identify subsumption relationships between ontology classes.

Semantic Web Rule Language (SWRL) integrates rules, concepts, and the relationships between concepts defined in Web Ontology Language (OWL), thereby extending their expressiveness. The possibility of generating new knowledge is achieved by creating rule sets in SWRL and it fundamentally serves as the inference engine [18].

3   Semantic Model for Recommending Learning Resources

Our research is different from related works in several regards. We provide personalisation to the students with a possibility of different parameters. This allows for flexibility and prevents the personalisation algorithm from being course-specific. Personalisation can be achieved by using parameters a course instructor considers necessary or according to the learning materials available for a course. Also, adaptive learning is achieved through different methodologies such as ontologies and inference rules. The proposed model is intended to maximise the learner's ability to learn and also provide individualised learning paths based on the learners' preferences. This is realised by obtaining the learner's initial abilities and preferences and using semantic reasoning and rule-based reasoning to predict the optimum learning path.

Figure 1 describes the architecture and main model elements of the proposed model. It includes the following: the learner model (which holds preferences and capabilities), learning objects metadata for annotation and semantic enrichment, ontologies describing learning objects and learners, and the personalisation parameters that will be used for adaptation. The learners and the course instructor access the LMS through the user interface. The learner model stores information about the learner in the Learner Data Repository. Information stored include learner characteristics and educational background. Learning resources are semantically enriched with information related to personalisation parameters, and subsequently stored in the Learning Object Data Repository. The information from the learner model and domain model are subsequently utilised in the mechanism for adaptation.

The path to personalised learning paths for the model will be described. The course instructor adds learning resources for each course according to different

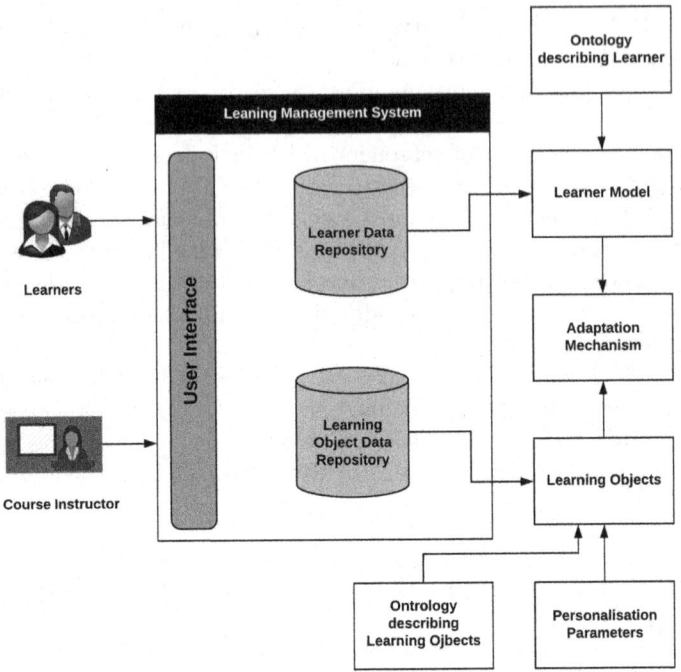

Fig. 1. The architecture of the model

dimensions of personalisation parameters with the user interface. The learning resources are subsequently transformed into learning objects by adding descriptions specific to the LOM (Learning Object Metadata) standard [19]. The course instructor then chooses the parameters that will be used in a certain course (for example, language, level of knowledge, and Honey-Mumford Learning Style). An alternative will be to select the most important parameters based on available learning objects in a course using an algorithm and a concept-parameter matrix for the learning objects in the course. This is achieved through the relationships between data elements and personalisation parameters in the domain ontology as shown in Fig. 2.

Table 1 shows the relationships between dimensions of personalisation parameters in the ontology and data elements of the LOM standard. The first column describes possible parameters that can be used to provide personalisation for students. These are usually specified by an instructor or an educational expert. The column, linguistic terms, details different dimensions of the parameters, which learners can be grouped into. Linguistic terms are better suited for characterisation because they provide more concrete demarcations than numerical scales [7]. The next 3 columns represent elements of the LOM standard which the personalisation parameters are mapped to, with the element name and value space mapped to parameters and linguistic terms, respectively.

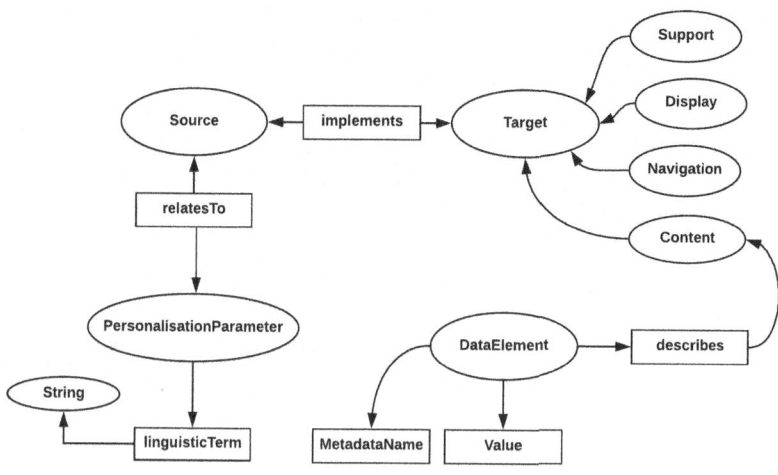

Fig. 2. Ontology mapping between data elements and personalisation parameters

The learner also accesses the LMS through the user interface, which serves as the communication component of the interactions between the learner and the learning system. The LMS was built with Laravel [20], which is an open-source PHP-web framework for the development of web applications. When the user signs up for a course, (s)he is required to take questionnaires or tests for the most important parameters in that course. If the user has previously used the system, the system will have a history of his previous (static) preferences, and (s)he won't have to go through those tests again.

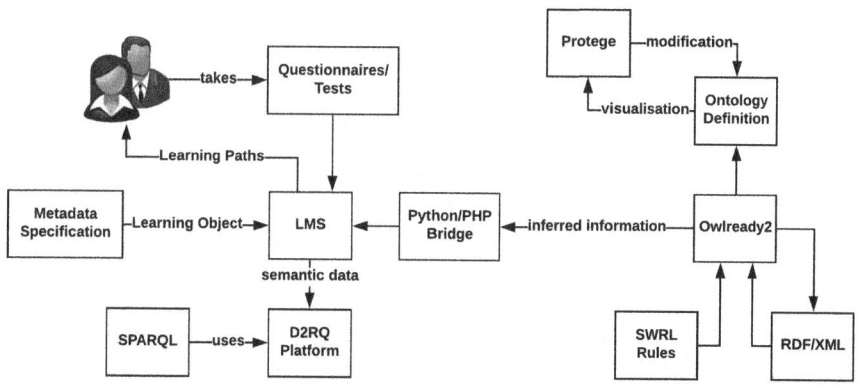

Fig. 3. Main elements of the model architecture

Table 1. Ontology mapping between personalisation parameters and LOM standard

Parameter	Linguistic terms	Nr.	Element name	Value space
Language	English	1.3	Language	en
	English	5.11	Language	en
Media Preference	Audio, text, video	4.1	Format	Audio, image, text, video
		5.2	Learning Resource type	diagram, figure, graph, slide, narrative text
Navigation Preference	Breadth-first, depth-first	1.7	Structure	Hierarchical
Level of Knowledge	Beginner, intermediate, advanced	5.8	Difficulty	Easy, medium, difficult
Bloom's Taxonomy of Learning Goals	Knowledge, comprehension, application	5.2	Learning Resource Type	Exercise, narrative text, exam, experiment, self-assessment, lecture
		9.1	Purpose	Concepts, theories, ideas, examples, exercises, tests
Felder-Silverman Learning Style	Active, reflective	5.1	Interactivity type	Active, expositive, mixed
		5.3	Interactivity level	Low, medium, high
	Sensory, intuitive	9.1	Purpose	Facts, details, principles, theories
	Sequential, global	1.7	Structure	Collection, networked, hierarchical, linear
	Visual, verbal	4.1	Format	Audio, image, text, video
		5.2	Learning Resource Type	diagram, figure, graph, index, narrative text, simulation, slide
Witkin Cognitive Style	Field-dependent, Field-independent	9.1	Purpose	Global approach, analytical approach

Figure 3 shows the main model elements of the architecture and the procedure for recommendation. The D2R server of the D2QR platform transforms the data stored in the LMS relational database to a semantic database, which is organised in tables. The next phase involves creating the ontology using the OWLready2 platform. The ontology is populated with classes, individuals, and properties using a SPARQL client and ontology-oriented programming (Python and OWLready2). SWRL rules defined in the ontology receive recommendations that determine how learning objects are related to personalisation parameters. With the inference engine, the recommended content is inferred for the learners. The results of these recommendations are provided to the LMS through a Python-PHP bridge. The LMS subsequently presents the personalised learning path to the students.

3.1   Semantic Mapping to the LMS

Mapping the relational database of the LMS to a semantic form can be achieved using the D2RQ platform, which is a system that allows for accessing relational databases as virtual, read-only RDF graphs, without having to recreate the RDB into an RDF store. The main components of interest of the D2RQ platform include a declarative mapping language, which describes relations between an ontology and a relational data model. Another is the D2R server, which uses an HTTP connection to provide a linked data view and HTML view for debugging, and a SPARQL Protocol endpoint over the database which can be accessed with a SPARQL client. These components provide the following functionalities:

- Querying the LMS database (which is non-RDF) using SPARQL,
- Accessing the contents of the LMS database as Linked Data over the Web,
- Creating custom dumps of the LMS database in RDF formats to load into an RDF store,
- Accessing information from the LMS database with a SPARQL client.

Figure 4 shows the D2RQ Engine for the LMS classes, which also specifies a SPARQL endpoint for the LMS dataset mapping. This provides possibilities for executing SPARQL queries to map classes and properties. The queries produce results which can be used in OWLready2 in JSON or XML syntax.

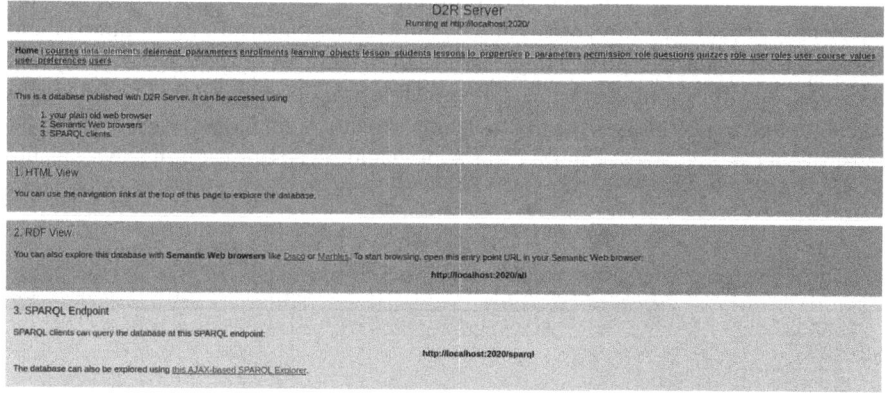

Fig. 4. Classes of the LMS mapped on the D2R Server

3.2 Creating a Domain Ontology in OWLready2

In designing ontologies, the first ideal step would be identifying the goal and scope of the ontology. When ontologies are designed properly, they can be used to accurately describe a domain. It is, however, imperative to balance the expressiveness and complexity of the design. For this research, we took advantage of existing standard vocabularies and ontologies, and applied their classes and properties when creating the mapping file that was used on the D2RQ platform. Learning Objects were modelled according to IEEE LOM ontology standards and the learners were modelled according to the basic FOAF (Friend of a Friend) [21] ontology standard as shown in Fig. 5.

```
<rdf:RDF xmlns:rdf="http://www.w3.org/1999/02/22-rdf-syntax-ns#"
         xmlns:xsd="http://www.w3.org/2001/XMLSchema#"
         xmlns:rdfs="http://www.w3.org/2000/01/rdf-schema#"
         xmlns:owl="http://www.w3.org/2002/07/owl#"
         xmlns:foaf="http://xmlns.com/foaf/0.1#"
         xml:base="http://test.org/lmsontology.owl"
         xmlns="http://test.org/lmsontology.owl#"
         xmlns:swrl="http://www.w3.org/2003/11/swrl#">
```

Fig. 5. Using existing standards in the ontology design

In the literature, there are three main strategies for accessing ontologies in a programming language [15]. The first strategy involves using a query language such as SPARQL. The second strategy involves the use of an Application Programming Interface such as OWL API and Jena. The third strategy, which was used, involves ontology-oriented programming (OWLready2 and Python, in this case). This utilises the advantage of the similarities between object models and ontologies, with classes, properties, and individuals in an ontology corresponding to classes, attributes, and instances, respectively, in object models. It also allows for the definition of classes and hierarchies, variables and restrictions, the relationships between classes. Figure 6 shows a visual representation of the resulting ontology using Protege.

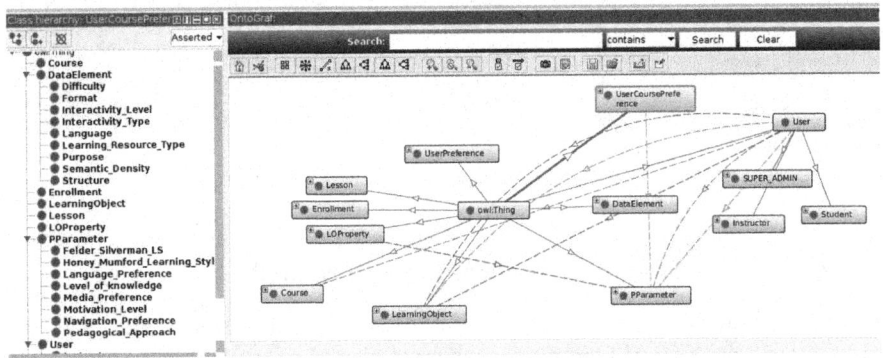

Fig. 6. Visualising the ontology with Protege

3.3   SWRL Rules

SWRL rules are basically used to integrate 'if ..., then ... ' associations in ontologies. SWRL rules were used to define the relationships between data elements of the LOM standard and different dimensions of a personalisation parameter. These rules are then used to identify the suitability of a learning object (based on its data properties) to a learner (based on the results of tests and questionnaires). Some examples of SWRL rules are described in Fig. 7. In Fig. 8, a list of

```
In [ ]:  with onto:
             rule3 = Imp() #Relationship between users and courses enrolled
             rule3.set_as_rule("""Enrollment(?e), User(?u), enrollment_user_id(?e, ?euid), user_id(?u, ?euid), Course(?c),\
             enrollment_course_id(?e, ?ecid), course_id(?c, ?ecid) -> isEnrolledIn(?u, ?c)""")

             rule5 = Imp() #Relationship between Personalisation Parameter and Data Element values
             rule5.set_as_rule("""PParameter(?pp), DataElement(?de), pp_de(?pp, ?id), de_pp(?de, ?id) -> relatesTo(?pp, ?de)"

             rule6 = Imp()#Relationship between Learning Objects and Course
             rule6.set_as_rule("""LearningObject(?lo), Lesson(?l), Course(?c), belongsToLesson (?lo, ?l),\
             belongsToCourse (?l, ?c)  -> loBelongsToCourse(?lo, ?c)""")

             rule25 = Imp() #Relationship between users and suitable learning objects
             rule25.set_as_rule("""User(?u), LearningObject(?lo), hasPreference(?u, ?up), isDescribedBy(?lo, ?lom),\
             relatesTo(?up, ?lom) -> isSuitedTo(?u, ?lo)""")

             rule26 = Imp() #Relationship between users and suitable learning objects
             rule26.set_as_rule("""User(?u), UserCoursePreference(?ucp), user_id(?u, ?uid), userInCourse(?ucp, ?uid),\
             Course(?c), courseForUser(?ucp, ?cid), course_id(?c, ?cid), LearningObject(?lo), loBelongsToCourse(?lo, ?c),\
             hasCoursePreference(?ucp, ?up), isDescribedBy(?lo, ?lom), relatesTo(?up, ?lom) -> isCourseSuitedTo(?u, ?lo)""")
```

Fig. 7. SWRL rules

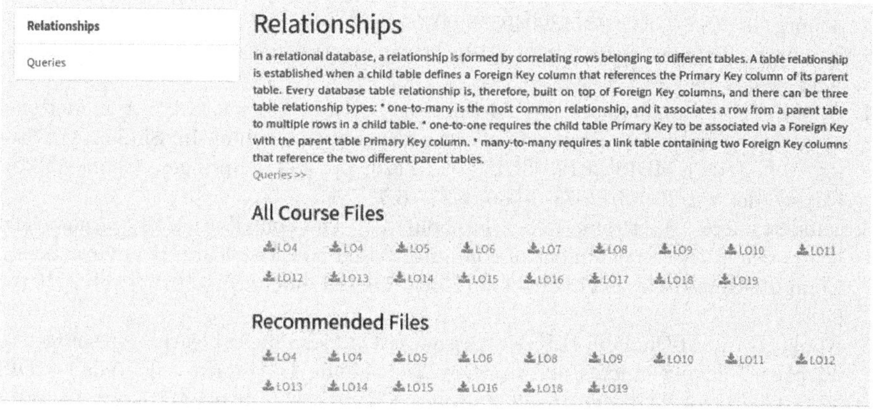

Fig. 8. Recommending learning objects

learning objects recommended for a user based on her preferences for the lesson on "Database Relationships" is shown.

The SWRL rules are executed sequentially in Pellet (which is embedded in OWLready2). The values that belong to the class of individuals and inferred values can be stored in the ontology.

4 Conclusion and Future Work

It has become a fact that incorporating e-learning into traditional learning curriculums is inevitable, and most educational institutions are using learning management systems to augment classroom activities. We have, in this research paper, proposed a semantic approach for adapting learning resources while incorporating multiple learning parameters.

This process involved using semantic technologies, ontologies, and adaptation rules. The advantages of this approach is the fact that the methodology is not restricted to specific courses and the course instructor can choose specific parameters based on experience, or parameters can be selected based on the learning resources in the course. For future work, we will focus on using an LMS such as Moodle, which has an expressive and comprehensive RDB. This is useful for incorporating dynamic adaptivity. Also, we will work on defining the algorithm to select parameters based on the course materials.

References

1. National Academy of Engineering: 14 grand challenges for engineering in the 21st century. http://www.engineeringchallenges.org/challenges.aspx. Accessed 20 Sep 2020
2. Brusilovsky, P.: Adaptive hypermedia for education and training (2012). https://doi.org/10.1017/CBO9781139049580.006
3. Tamm, S.: Types of E-learning (2019). https://e-student.org/types-of-e-learning/. Accessed 12 Sep 2020
4. Apoki, U.C., Ennouamani, S., Al-Chalabi, H.K.M., Crisan, G.C.: A model of a weighted agent system for personalised e-learning curriculum. In: Simian, D., Stoica, L.F. (eds.) MDIS 2019. CCIS, vol. 1126, pp. 3–17. Springer, Cham (2020). https://doi.org/10.1007/978-3-030-39237-6_1
5. Vandewaetere, M., Desmet, P., Clarebout, G.: The contribution of learner characteristics in the development of computer-based adaptive learning environments. Comput. Hum. Behav. **27**, 118–130 (2011). https://doi.org/10.1016/j.chb.2010.07.038
6. Apoki, U.C., Al-Chalabi, H.K.M., Crisan, G.C.: From digital learning resources to adaptive learning objects: an overview. In: Simian, D., Stoica, L.F. (eds.) MDIS 2019. CCIS, vol. 1126, pp. 18–32. Springer, Cham (2020). https://doi.org/10.1007/978-3-030-39237-6_2
7. Essalmi, F., Ayed, L.J.B., Jemni, M., Kinshuk, Graf, S.: A fully personalization strategy of E-learning scenarios. Comput. Hum. Behav. **26**, 581–591 (2010). https://doi.org/10.1016/j.chb.2009.12.010
8. Learning Technology Standards Committee (IEEE): The learning object metadata standard - ieee learning technology standards committee. https://www.ieeeltsc.org/working-groups/wg12LOM/lomDescription. Accessed 17 Sep 2020
9. Gruber, T.R.: A translation approach to portable ontology specifications. Knowl. Acquis. **5**, 199–220 (1993). https://doi.org/10.1006/knac.1993.1008
10. Kurilovas, E., Kubilinskiene, S., Dagiene, V.: Web 3.0 - based personalisation of learning objects in virtual learning environments. Comput. Hum. Behav. **30**, 654–662 (2014). https://doi.org/10.1016/j.chb.2013.07.039
11. Vicient, C., Sánchez, D., Moreno, A.: An automatic approach for ontology-based feature extraction from heterogeneous textualresources. Eng. Appl. Artif. Intell. **26**, 1092–1106 (2013). https://doi.org/10.1016/j.engappai.2012.08.002
12. Arch-int, N., Arch-int, S.: Semantic ontology mapping for interoperability of learning resource systems using a rule-based reasoning approach. Expert Syst. Appl. 40, 7428–7443 (2013). https://doi.org/10.1016/j.eswa.2013.07.027
13. Perisic, J., Bogdanovic, Z., Duric, I.: Semantic model for adaptive e-learning systems. In: Symposium proceedings-XV International symposium Symorg 2016: Reshaping the Future Through Sustainable Business Development and Entrepreneurship, Belgrade, Serbia, pp. 345–353 (2016)

14. D2RQ: Accessing Relational Databases as Virtual RDF Graphs. http://d2rq.org/. Accessed 22 Sep 2020
15. Lamy, J.B.: Owlready: ontology-oriented programming in Python with automatic classification and high level constructs for biomedical ontologies. Artif. Intell. Med. **80**, 11–28 (2017). https://doi.org/10.1016/j.artmed.2017.07.002
16. SPARQL Query Language for RDF, W3C. https://www.w3.org/TR/rdf-sparql-query/. Accessed 23 Sep 2020
17. OWL/Implementations.   https://www.w3.org/2001/sw/wiki/OWL/Implementa tions. Accessed 23 Sep 2020
18. Hassanpour, S., O'Connor, M.J., Das, A.K.: Visualizing logical dependencies in SWRL rule bases. Presented at the (2010). https://doi.org/10.1007/978-3-642-16289-3_22
19. Learning object standard - EduTech Wiki. http://edutechwiki.unige.ch/en/ Learning_object_standard. Accessed 25 Sep 2020
20. The PHP Framework for Web Artisans. https://laravel.com/. Accessed 25 Sep 2020
21. Brickley, D., Miller L.: FOAF vocabulary specification 0.91. Citeseer (2007)

Defining a Core Ontology for Medical Devices in Germany to Ensure Semantic Interoperability

Andreas E. Schütz$^{(\boxtimes)}$ ⑩, Tobias Fertig⑩, and Kristin Weber⑩

Faculty of Computer Science and Business Information Systems,
University of Applied Sciences Würzburg-Schweinfurt,
Sanderheinrichsleitenweg 20, 97074 Würzburg, Germany
{andreas.schuetz,kristin.weber}@fhws.de

Abstract. To manage the data of medical devices, systems have to fulfill high demands. These high demands are due to the legal requirements for medical devices and due to specific needs of users of these information. Furthermore, the lack of semantic interoperability prevents a centralized provision of high-quality data sets. The goal of our project was to develop an ontology for medical devices, and to derive a UML data model from our ontology. For this purpose, the approach "Ontology Development 101" was used. Based on the German law for medical devices and expert interviews, we could identify relevant domain topics and terms. Subsequently, we used the language OWL to define classes, properties and constraints for the ontology. Afterwards, the ontology was translated into a UML data model using an already defined set of rules. This led to the final hypothesis that semantic interoperability between hospital information systems can be achieved when using the ontology.

Keywords: Medical devices · Ontology · Semantic interoperability · Semantic web

1 Introduction

Digitization was still in its early stages when the German law for medical devices was enacted on 02.08.1994. Only 3% of the technological information capacity worldwide was digitized in 1993, according to estimations. However, already 94% were digitized in 2007 [9]. Nowadays, cross-sectoral digitization does not even stop at health care facilities. 89% of all hospitals use a hospital information system, and 87% of all hospitals believe that the benefits of digitization outweighs the risks [11]. Therefore, being able to manage medical devices digitally is important for healthcare facilities such as hospitals, practices or nursing homes. Moreover, it is important for the users as well as the operators of medical devices. Health institutions use software solutions for the administration of medical devices such as hospital information systems.

© Springer Nature Switzerland AG 2021
D. Simian and L. F. Stoica (Eds.): MDIS 2020, CCIS 1341, pp. 394–410, 2021.
https://doi.org/10.1007/978-3-030-68527-0_25

The solutions are either specifically designed or general industry solutions. Since medical devices are directly intended for use on humans, the requirements of legislators for those devices are very high and detailed. Mapping these complex regulations into data records is a challenge for many institutions. Therefore, it is not surprising that the data quality in the industry solutions is often not consistently implemented, because the requirements to be mapped are not exactly known [Appendix 1, 2].

Especially the specialized software solutions aim at fulfilling relevant parts of the Medical Devices Act and the associated guidelines [15]. However, software working within local environments can have issues when exchanging, merging or comparing data with other systems, even if the legal requirements are completely fulfilled. The data records created by different systems often differ in their structure [Appendix 1, 2]. Moreover, the data records are not standardized and cannot be imported into every system 1, 2]. This is due to different data suppliers, e.g. the manufacturers of medical devices. This problem of heterogenous data records can be solved if a semantic interoperability would be available for the systems [2].

The article Semantic Web by [1] recommends using an ontology to achieve semantic interoperability. This is also supported by [22], who sees ontologies as an interface between different languages and representations. An ontology is a formal description of real-world objects and their properties and relationships to each other [17]. According to Schönbein, the task of an ontology is to provide an integrated and uniform environment for the components and systems involved, and to offer a semantic basis for transformation between these different systems [22].

While ontologies for the Semantic Web were only established with the aforementioned articles by Berners-Lee, the medical sector has been working with ontologies for some time. The Semantic Web is an extension of the current web, which enriches information with well-defined meaning to enable computers or people to work together [1]. Musen describes ontologies as a possibility to share digital knowledge across software environments [17]. The deductive conclusion leads to the assumption that medical devices could also be represented in an ontology. Furthermore, such an ontology can be used to derive a data model for software environments.

The aim of our research project is to create a core ontology for the domain of medical devices, restricted to the essential terms. Therefore, we have to restrict the ontology to the essential terms. However, we are using scientific methods and tools for the definition, and afterwards, we will proof, that the ontology can be used for information systems. The derivation of the data model will prove that the ontology can be used for information systems. The ontology will form a basic framework for extensions at the current state. The relevant requirements are limited to those that are relevant for data records provided or used by a medical device. The ontology is based on the guidelines of [18]. At the end of this paper, an explorative approach is used to generate a hypothesis. The hypothesis allows statements about the suitability for establishing the semantic interoperability of medical devices.

We will start with the required basics in Section "Theoretical Background". In Section "Methodology", we will give an overview of the research approach for setting up the ontology. Afterwards, in Section "Orientation" we define the domain of the ontology as well as the relevant terminology. Furthermore, we summarize possible interfaces to already existing ontologies. Subsequently, in Section "Defining the Ontology" classes are derived from the identified terms and ordered hierarchically. The required properties of the classes are also defined. In Section "Deriving the Data Model" we derived a data model from our ontology. In the final Section "Conclusion", the findings are finally summarized and an outlook on further research is given.

2   Theoretical Background

In this section, we explain the Medical Devices Act as well as existing ontologies in more detail. This is intended to increase the comprehensibility of the paper as well as the vocabulary used.

2.1   Medical Devices

The first version of the Medical Devices Act (MPG) was published in 1994 with the aim of protecting patients, users and third parties during treatment by, operation of, and usage of medical devices [Appendix 2, 1]. The MPG implements the European Council (EC) Directives 90/385/EEC [Appendix 2, 2], 93/42/EEC [Appendix 2, 3] and 98/79/EC [Appendix 2, 4] into German national law. According to the MPG, medical devices are all "instruments, apparatus, devices, software, substances and preparations made of substances or other objects used individually or in combination with each other, including software specifically intended by the manufacturer to be used for diagnostic or therapeutic purposes and used for the proper functioning of the medical device" [Appendix 2, 1, §3(1)]. Medical devices are clearly distinguishable from drugs, which means that they do not achieve their main effect either pharmacologically, immunologically or metabolically [Appendix, 2, 1, §3(1)].

According to the subdivision of the individual directives, medical devices are divided into the following three categories in MPG: active implants, in vitro diagnostics (IVD), and other medical devices. The group of active implants includes pacemakers or insulin pumps. IVD devices include blood glucose meters or HIV tests. Other medical devices include X-ray machines or wheelchairs.

Due to the continuous development of medical devices, the MPG has been subject to a number of changes since its adoption, some of which were also triggered by European amending directives. In addition to the MPG, we considered the consolidated EC directives and the individual regulations to which the MPG refers in many sections.

In addition, discussions and interviews with representatives from industry helped to identify further requirements for medical devices. All requirements from representatives are trying to increase the usability of the different products. Since our ontology aims to also support the information systems of the industry, these requirements are given preferential treatment in design decisions as long as they do not violate requirements of the MPG.

2.2 Ontology

In computer science, the term ontology can be described as "an explicit formal specification of a common conceptualization" [5]. With the help of an ontology a domain is described by a standardized terminology. Relationships as well as derivation rules are established under the individual terms. For structuring, the ontology uses classes, relations, functions and axioms which are statements that represent truths, e.g. "A medical device has only one manufacturer" [8].

The Web Ontology Language (OWL) specified by the W3C can be used to define the ontology. In OWL classes and attributes (properties) can be used to represent complex relations [10]. The instantiations of the classes are called individuals. With OWL, classes can be represented in hierarchies, which means that subclasses can be defined. The subclasses inherit properties of the superclasses. The properties correspond to the properties of a class but are declared in OWL similar to classes. Each property consists of a domain and a range.

The domain is the class in which the property is stored, while the range represents the value range. This can be either a Datatype Property or an Object Property. Datatype properties connect individuals with a common data type like integer or string. Object Properties create a link between individuals and thus act as a description of relationships. To describe object properties, it may also be suitable to declare them as functional. The term functional means that their range can have a unique value or no value at all instead of several values. However, this value can be assigned to several domains. If the property is declared as inverse functional, the instance in the range may only be assigned exclusively to one domain. However, the domain itself can take several values.

3 Methodology

Ontologies aim at representing a domain formally and schematically [26]. As different domains are, as different ontologies can be in their appearance. Therefore, it is important to decide in advance which form of an ontology is suitable for the given domain.

The classification of our ontology is based on the classification of [20]. With regard to the criterion linguistic expressiveness and formality, the ontology can be assigned to the area of formal ontologies. Formal ontologies are characterized by their clear semantics and strict rules [20], which is a basis for achieving semantic interoperability. In contrast to a software ontology, the data in a formal ontology is not linked to a specific method, but should be retrievable [20]. This also applies to the purpose of our ontology for medical devices.

Our medical device ontology can be classified as "domain ontology". It focuses on a specific area, but is still abstract enough that it can be used for different applications [20]. There are also different methods and procedures for the creation of an ontology [4]. This work is based on the methodology "Ontology Development 101" of [18]. The procedure comprises the seven steps shown in Fig. 1. The paper divides these steps into the two phases orientation and definition, whereas the last step instantiation is not part of this paper. Instead a data model is derived.

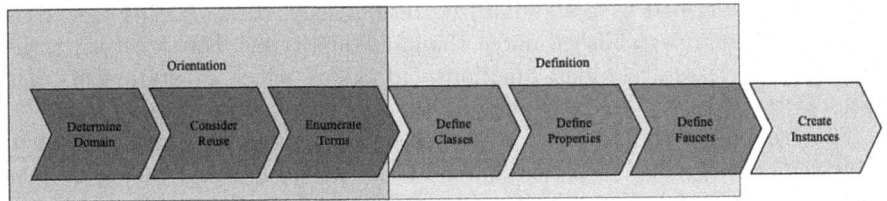

Fig. 1. Process of defining an ontology according to [18]

Ontology Development 101 is a process-oriented method that allows for setbacks and aims to achieve a continuous improvement of the ontology during multiple iterations. This characteristic is important for our ontology, which has to be extended in a continuous process due to its complexity.

The domain is subject to frequent changes due to the constant development. Therefore, it is recommended to define the ontology under aspects of the Open World Assumption (OWA) [10]. This establishes an ontology under the assumption that a knowledge base is always potentially incomplete [10]. We aim to create the core ontology, which can be extended later on.

4   Orientation

The phase orientation serves to open up the knowledge area which will be mapped in the ontology. Following the method Ontology Development 101, this section is divided into three subsections. First, the domain of the ontology is defined. Second, already existing ontologies are discussed, and last, the relevant terms of the domain are defined.

4.1   Determine Domain

In order to set a clear focus for the definition of an ontology, it is recommended to define the viewing area and the domain. A definition is provided by the following paragraph in combination with the subsequent definition of competence questions.

The domain of our ontology is the representation of medical devices. Thereby the ontology guarantees a semantic interoperability of different medical devices. The ontology helps to subdivide medical devices in a meaningful way and to represent the legal requirements. The ontology is maintained by manufacturers and used by different professional information systems.

In the methodology of [6], so-called competence questions are presented as a tool for defining the domain. These are used to evaluate the ontology by checking whether the result can answer the competence questions [6]. Questions from the expert interview as well as from research of medical law are taken into account.

CQ1: Which medical devices are intended for self-use?
CQ2: Which medical devices are subject to a specific legal regulation?
CQ3: What is the risk classification of a medical device?
CQ4: Is a certain instance heavy and large?
CQ5: Are there instances of a particular manufacturer?
CQ6: Which instances have expired?
CQ7: Does the instance of a medical device have to be sterile?
CQ8: Which instances were under a metrological test?
CQ9: Which selected institute monitors the medical device?

4.2 Consider Reuse

The reuse of ontologies in the same domain can save time in the ontology creation process. The following projects were found during the research for already existing ontologies in the domain of medical devices:

OntoVigilance: Ontology for post-market surveillance strategy. The project ran from 2013 to 2015 and resulted in a prototype software. This software collects Internet-based data from medical devices that are already on the market. The European Directive for Medical Devices 93/42/EEC [Appendix 2, 3], which requires a system of technical safety surveillance (93/42/EEC), is to be implemented. The ontology was further developed in the following project [23].

OntoPMS: follow-up project to OntoVigilance. The project, which ended in 2018, intended to complete and extend the status achieved in the previous project. Both, OntoVigilance and OntoPMS focus on adverse events of medical devices [23].

An Ontology Model for The Medical Device Development Process in Europe: scientific work representing the development process for medical devices [21]. The result is an informal ontology in the form of a flowchart diagram. The diagram represents chronologically the product development process. Therefore, we could not reuse the proposed ontology model within our own ontology.

To sum up, the research did not uncover any reusable ontology.

4.3 Enumerate Terms

The terms relevant for our ontology could be extracted from the MPG. In addition, technical literature was consulted for supplementary research. Some very relevant terms will be explained for a better understanding.

– Medical device: Medical devices in general have already been defined in Section "Theoretical Background". They are the main domain of this work and thus represent a central term. Medical devices are divided into active implants, in-vitro diagnostics (IVD) and other medical devices according to the European Directives RL 90/385/EEC [Appendix 2, 2], RL 98/79/EC [Appendix 2, 4] and RL 93/42/EEC [Appendix 2, 3]. All medical devices, including the subgroups, must meet the definition of a medical device. In practice, however, for administrative reasons, a distinction is made between medical devices and consumables [Appendix 1, 1].

– Active implants: Active implants are a subgroup of medical devices. They are called legally active implantable medical devices and describe "any active medical device designed to be wholly or partially inserted, surgically or medically, into the human body or by medical intervention into a natural orifice and intended to remain there after surgery" [Appendix 2, 2]. Active implants are not divided into risk classes because they generally have a very high risk [19].

– In Vitro Diagnostics (IVD): do not act directly on the human body but on samples taken from the human body [7]. This includes sample containers or smear materials. A detailed definition can be found in [Appendix 2, 1, §3(4)]. IVDs are classified in different lists according to Annex II of the European Council Directive 98/79/EC [Appendix 2, 4,]. List A contains IVDs that are particularly high-risk (e.g. HIV tests). IVDs with a high risk are on list B (e.g. blood glucose meters). All other IVDs are assigned to the general class.

– Other Medical Devices: medical devices, which do not fall under the definition of active implants or IVDs [Appendix 2, 3, §1(5)]. Other medical devices are assigned to different risk classes on the basis of the classification rules specified in Directive 93 42 EEC [Appendix 2, 3]. In addition to the main risk classes I, II and III, there are various subclasses.

– Medical Assets: In the economic classification in practice, medical assets and consumer goods are differentiated [Appendix 1, 1]. This also has an effect on data storage, as the two forms have different requirements [3]. They are characterized by the fact that they can be used several times in the performance process [13]. The term covers both: medical devices and IVDs [Appendix 1, 1].

– Medical Consumables: Classification as a medical consumable product is also preferred in practice for economic reasons [Appendix 1, 1]. They are "consumed or unusable by their intended use or they are used exclusively by a patient and remain with him" [13]. This classification includes other medical devices, IVDs and active implants [Appendix 1, 1].

– CE marking: The CE marking must be affixed to the product by the responsible person before it is released on the market. If the medical device meets various requirements with regard to risk classification [Appendix 2, 1, §9(3)], a conformity assessment procedure and the submission of a declaration of conformity is required in advance [7]. In this case, the product must bear the number of the selected institution involved in the conformity procedure on the CE marking [Appendix 2, 1, §9(3)].

- Manufacturer: In principle, the manufacturer is responsible for the design, manufacture, packaging and labelling of a medical device for its first release on the market. Since the manufacturer must observe legal obligations, they are an important player in medical device act [Appendix 2, 1, §3].
- Operator: The respective medical device is in operation at the operator's premises. The operator also has responsibilities such as maintenance [Appendix 2, 1, §3] or safety inspections [Appendix 2, 1, §6].
- Selected institution: the selected institution carries out the conformity assessment procedure for a medical device. In this case the identification number of the selected institution is noted on the CE marking of the product [Appendix 2, 1, §9(3)].
- Legal procedures: MPG requires numerous tests and procedures. These include clinical trials, metrological tests and biological safety tests.
- Nomenclature: The nomenclature required in the DIMDI Regulation of 4 December 2002 as amended on 25 July 2014 (Federal Law Gazette I p. 4456 §3) has a similar purpose as our ontology: Support of the international exchange of information [7]. The uniform naming is used for the registration of the medical device by the responsible person. The Universal Medical Device Nomenclature System (UMDNS) is used for medical devices and the European Diagnostic Market Statistics (EDMS) for IVDs. In the future, however, these will be replaced by the Global Medical Device Nomenclature (GMDN).

5   Defining the Ontology

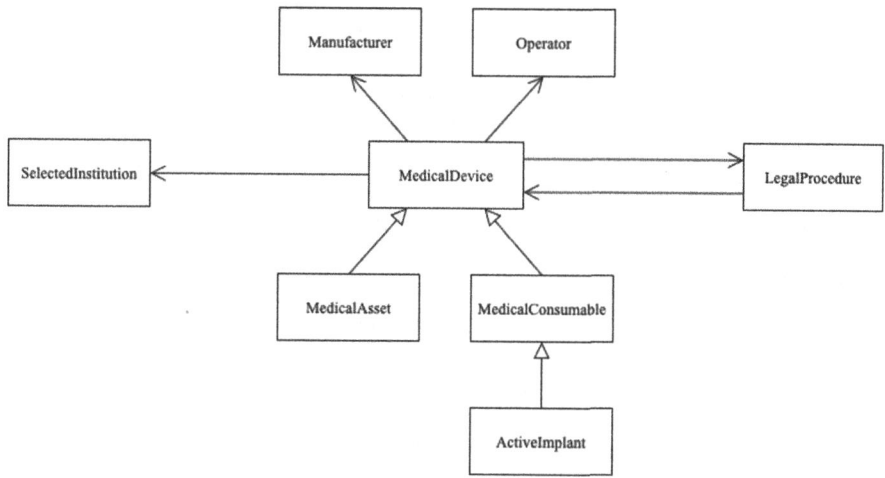

Fig. 2. UML class diagram of the ontology for medical devices

After defining the domain and the relevant terms, this section aims to structure the information and to develop the framework for an ontology. Based on the method Ontology Development 101 the classes, the properties as well as the constraints are defined. We used a combination of the top-down and the bottom-up approach for the arrangement of the classes. At first, the classes are defined and afterwards they get generalized and specified [18]. Moreover, we derive classes from the terms described in the previous section. However, only the suitable terms are used for the derivation into classes. The remaining terms serve as properties, i.e. datatype properties of the derived classes. It should be noted that the focus of this paper is on the classes and their relationships (Object Properties). As an example, some relevant datatype properties identified in the MPG are created for the individual classes. However, these do not claim to be legally complete.

The visualized ontology is shown in Fig. 2 as UML class diagram. Moreover, Fig. 2 shows the arrangement and linking of the individual classes. The generalization arrows represent generalization and specialization of the individual classes. The black, unidirectional arrows represent an object property, with the arrowhead pointing to the range of the object property. We will introduce the individual classes and their associated properties in this section. In contrast to the provided tables, each class does have a datatype property for its unique identification number. Moreover, we will discuss design decisions.

`MedicalDevice`: The central class of the ontology is the class `MedicalDevice`. This class can be further subdivided by specialization into the subclasses `ActiveImplant`, `InVitroDiagnostic`, and `OtherMedicalDevice`. The CE marking is not defined as a separate class, as it does not exist in its own, but rather represents a characteristic of the medical device. A `MedicalDevice` consists of the properties shown in Table 1.

`MedicalAsset`: This class is shown in Table 2 and is a subclass of `MedicalDevice`. Therefore, it inherits the properties of `MedicalDevice`. Due to the many different properties of assets and consumable goods, the classification of products for the core ontology was chosen and not the classifaction based on legal aspects (Active Implants, IVD, Other Medical Devices). Moreover, the chosen classifaction was prefered by experts. This also corresponds to the economic use of information systems. The legal subdivision can later be used additionally in an extended ontology.

`MedicalConsumable`: As a subclass of `MedicalDevice`, this class inherits their properties. As already explained for the class `MedicalDevice`, the inheritance was designed based on their relevance in practice and their different properties. The properties are shown in Table 3.

`ActiveImplant`: This class inherits all properties of Consumable. Due to the transitivity, the properties of the class `MedicalDevice` are also inherited. In addition, some data properties but no additional object properties are defined. The properties are shown in Table 4.

`Operator`: This class stores the data of the natural or legal person who operates the respective medical device. In our core ontology, the class is only used by the class `MedicalDevice` as a range. However, the class has the potential

Table 1. Properties of the class `MedicalDevice`

Name	Range	Description
Datatype Properties:		
description	String	Description of medical device
nomenclatureCode	String	Code of nomenclature
riskClassification	String	Each medical device has its own classification of risk
manufacturingDate	Date	Date of the manufacturing
refId	Integer	Id required to order the medical device
ceMarking	Boolean	Flag whether an ce marking has to be on the medical device
quantityUnit	String	Describes the units of quantity used for the medical device
microbiologicalStandard	String	Which standard has to be adhered to
Object Properties:		
isOperatedBy	Operator	The association to the operator of the medical device. The property is functional
manufacturedBy	Manufacturer	Association to the manufacturer. The property is functional
verifiedBy	SelectedInstitution	Sometimes, the id of the selected institution must be attached to the medical device. This property is functional
underlies	LegalProcedure	A medical device is subject to certain legal procedures

to be related to other classes in an extended ontology. The properties are shown in Table 5.

Manufacturer: This class describes various maintenance tasks that may be necessary for a medical device. These include sterilization or maintenance, for example. In order to record individual maintenance activities, a class `MaintenanceInProgress` should be added to an extended ontology. This class will be related to the classes Operator and Maintenance. Since documentation of the performed maintenance is not explicitly required by law, it is not considered in our core ontology. The properties are shown in Table 6.

Table 2. Properties of the class `MedicalAsset`

Name	Range	Description
Datatype Properties:		
purchasingDate	Date	The purchasing date is required for economic processes, for example amortization
price	Double	The prize of the medical asset is required for economic processes
weight	Integer	The weight of the medical asset can be used for planning transportations
dimensions	Integer	The dimensions are required for the final location
accessories	String	This property stores the information about accessories
lastMaintenance	Date	The date of the latest maintenance
interval	Integer	The interval for maintenances

Table 3. Properties of the class `MedicalConsumable`

Name	Range	Description
Datatype Properties:		
expirationDate	Date	The date of expiration
personalUsage	Boolean	Flag whether the consumable is suitable for personal use

Table 4. Properties of the class `ActiveImplant`

Name	Range	Description
Datatype Properties:		
implantationDate	Date	The date of implantation must be stored in the data set
physician	String	The name of the physician who performed the implantation
patient	String	The name of the patient who carries the implant
customized	Boolean	Indicates if the product is a custom-made product

Table 5. Properties of the class `Operator`

Name	Range	Description
Datatype Properties:		
name	String	The name or designation of the natural or legal person operating a medical device
address	Address	Contact Information of the operator

Table 6. Properties of the class `Manufacturer`

Name	Range	Description
Datatype Properties:		
name	String	Name of the manufacturer
address	Address	Address of the manufacturer

`SelectedInstitution`: This class is mainly involved in legal proceedings. It comes into contact with the medical device itself, as the identification number often has to be affixed in combination with the CE marking. The properties are shown in Table 7.

Table 7. Properties of the class `SelectedInstitution`

Name	Range	Description
Datatype Properties:		
temporary	Date	The appointment selected institution is limited in time. This property indicates the date of renewal
task	String	The task of the selected institution regarding the associated medical device

`LegalProcedure`: This class describes the execution of a legal procedure. It encapsulates the complexity of the different legal procedures. By extending the ontology, the class can be further specialized in subclasses, which are then associated to the class `MedicalDevice` or its subclasses. The properties are shown in Table 8.

Table 8. Properties of the class `LegalProcedure`

Name	Range	Description
Datatype Properties:		
name	String	The name of the legal procedure
legalBasis	String	The underlying legal basis of the procedure
result	String	Result of the legal procedure, for example a classification
Object Properties:		
medicalDevice	MedicalDevice	Association to the medical device on which a procedure was performed. Property is functional.

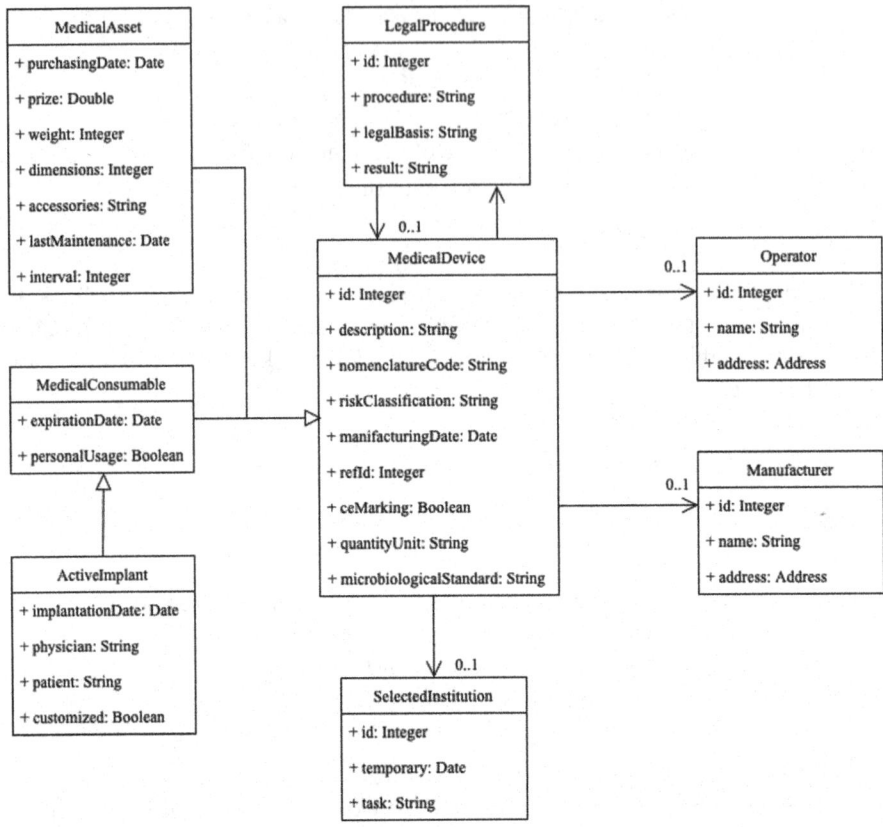

Fig. 3. The derived datamodel of the ontology for medical devices

6   Deriving the Data Model

In order to check whether the designed core ontology is suitable for use in an information system, a data model will be derived in this section. The Unified Modeling Language (UML) is accepted as an international standard for object-oriented analysis and design, which is characterized by its semantic possibilities [16]. Especially the class diagram is suitable for the representation of relationships. The versatile UML can also ease the derivation of a relational model [16]. Thus, UML can serve as an exemplary target model for derivation.

Kiko and Atkinson note that a direct translation from OWL to UML always leads to a loss of information since common language constructs are used differently [14]. However, the amount of information loss can be reduced by the consistent application of translation rules. A collection of some examples of such rules can be found in the work of [24,27]. According to Zedlitz, problems in the translation from OWL to UML occur with OWL-specific functions such as automatic classification, nested class expressions or special property properties like reflectivity, symmetry or transivity [27]. However, these were not used in our core ontology. Therefore, we won't have any loss of information during the transformation of our ontology. The ontology has already been developed with regard to information systems, and therefore, basic concepts for data storage and processing, such as a unique identifier for each instance, have already been considered. Nevertheless, the used element types are checked for their suitability before transformation.

The class `MedicalDevice` has an generalization relationship to its subclasses. Since both superclass and subclass can be transformed into a UML class, the transformation of the relationship is easy [27].

Data properties and object properties are used in OWL and also in the created ontology. In UML, data properties are represented as class-dependent attributes [27]. Object properties represent relations between two classes. In UML, they can be represented as class-dependent attributes or as associations [27]. Also, the inverse relationship between the classes `MedicalDevice` and `LegalProcedure` can be represented in UML. UML defines the two ends of a binary, bidirectional association as two inverse relations.

Another requirement is to be able to translate the declared functional properties. These are actually only syntactic identifiers for certain cardinality restrictions. Cardinalities can be represented in UML. Therefore, a 0..1 multiplicity must be created for the respective association. Figure 3 shows the derived data model after the translation [27]. The translation of an ontology into UML class diagrams can also be automated. The OWL2UML plugin for the Protégé ontology development environment, for example, is ideal for this purpose [12].

7   Conclusion

In this article we proved that the essential terms of the domain of medical devices can be represented in a core ontology. The MPG was examined, and the most relevant terms were identified. The knowledge gained was enriched with terms and information from industry representatives who also have an influence on medical devices. The enumerated terms were then transformed into classes and properties and linked semantically in a core ontology. The derivation into a UML data model shows that the ontology can basically be used in an information system. With the knowledge that semantic interoperability of information systems can be achieved with ontologies, this leads to the following hypothesis: If the core ontology is used by information systems, then semantic interoperability between these information systems can be achieved.

With the help of this semantic interoperability, medical institutions can finally succeed in improving their data quality. However, to make the ontology usable, it should be available publicly. The open source software WebProtégé, which can be hosted on a dedicated server, is a good example. Ontologies can be created in OWL format and maintained within the development environment. The ontology can also be downloaded in various formats [25].

The development of the ontology is application-oriented and primarily oriented towards the requirements of practical applications. Since data from interviews are mainly used, the ontology should, however, be validated with further scientific methods in the future.

The defined ontology is a core ontology that can be extended at will. Reality is subject to constant change and so the ontology should also participate in this change as a reflection of the real world. In future work the classes for legal foundations could be specified. Furthermore, the ontology project identified in Section "Consider Reuse" could be connected in the future. In contrast to the presented ontology, which all have a very specific use case, our ontology focuses on the basic concepts of medical devices. With these features it could serve as a mediator that allows the specialized ontologies to communicate with each other.

A   Appendix 1 Interviews

The following two interviews were conducted to determine the ontology requirements:

1. Interview with Nick Dähnhart at 05. July 2016: . Nick Dähnhart is Management Consultant at the INBEX Institute for Business and Solution Excellence GmbH. In his professional career he has gained a lot of experience in the management of medical devices in IT.
2. Interview with Christian Fürber at 29. March 2016: . Christian Fürber is Senior Data Strategy Consultant and CEO of the Information Quality Institute GmbH. During his many years of service at the German Federal Armed Forces, he developed and implemented the forces' data quality strategy. A large part of this work dealt with medical devices.

B   Appendix 2 Laws

1. Bundesgesetzblatt Nr. 657/1996: Gesetz über Medizinprodukte (Medizinproduktegesetz - MPG), version from 08. September 2015. Link to the original German Version: https://www.gesetze-im-internet.de/mpg/MPG.pdf Link to the English version: https://www.bundesgesundheitsministerium.de/fileadmin/Dateien/3_Downloads/Gesetze_und_Verordnungen/GuV/M/MPG_englisch.pdf
2. Council Directive 90/385/EEC of 20. June 1990 on the approximation of the laws of the Member States relating to active implantable medical devices. Link: https://eur-lex.europa.eu/legal-content/EN/TXT/?uri=CELEX%3A01990L0385-20071011
3. Council Directive 93/42/EEC of 14. June 1993 concerning medical devices. Link: https://eur-lex.europa.eu/legal-content/DE/TXT/?uri=CELEX%3A-31993L0042
4. Council Directive 98/79/EC of the European Parliament and of the Council of 27. October 1998 on in vitro diagnostic medical devices. Link: https://eur-lex.europa.eu/legal-content/EN/ALL/?uri=CELEX%3A31998L0079

References

1. Berners-Lee, T., Hendler, J., Lassila, O.: The Semantic Web - A new form of Web content that is meaningful to computers will unleash a revolution of new possibilites. Sci. Am. **284**, 34–43 (2001)
2. Bointner, K., Duftschmid, G.: HL7 template model and EN/ISO 13606 archetype object model - a comparison. Stud. Health Technol. Inf. **150**, 249 (2009)
3. Eichhorn, S.: Handbuch Krankenhaus-Rechnungswesen: Grundlagen-Verfahren-Anwendungen. Gabler Verlag (1988) https://books.google.de/books?id=atOZPA-AACAAJ
4. Fernández-López, M., Gómez-Pérez, A.: Overview and analysis of methodologies for building ontologies. Knowl. Eng. Rev. **17**(2), 129–156 (2002). https://doi.org/10.1017/S0269888902000462
5. Gruber, T.R.: A translation approach to portable ontology specifications. Knowledge Acquisition **5**(2), 199–220 (1993). https://doi.org/10.1006/knac.1993.1008, http://www.sciencedirect.com/science/article/pii/S1042814383710083
6. Grüninger, M., Fox, M.S.: Methodology for the design and evaluation of ontologies. In: Workshop on Basic Ontological Issues in Knowledge Sharing. Montreal, August 1995
7. Harer, J.: Anforderungen an Medizinprodukte: Praxisleitfaden für Hersteller und Zulieferer. Carl Hanser Verlag GmbH & Company KG (2014)
8. Hesse, W.: Ontologie(n), July 2005 https://gi.de/informatiklexikon/ontologien. Accessed 07 Dec 2020
9. Hilbert, M., López, P.: The world's technological capacity to store, communicate, and compute information. Science **332**(6025), 60–65 (2011). https://doi.org/10.1126/science.1200970
10. Hitzler, P., Krötzsch, M., Rudolph, S., Sure, Y.: Semantic Web: Grundlagen. eXamen.press, Springer, Heidelberg (2007). https://books.google.de/books?id=hBEkBAAAQBAJ
11. Inverto, A.G.: Studie zur Digitalisierung und technologischen Vernetzung in deutschen Krankenhäusern. Technical report, Inverto AG (2015)
12. Istochnick, I.: OWL2UML - Protege Wiki, May 2009. https://protegewiki.stanford.edu/wiki/OWL2UML. Accessed 07 Dec 2020
13. Kerres, A., Seeberger, B., Mühlbauer, B.: Lehrbuch Pflegemanagement III. Lehrbuch Pflegemanagement, Springer (2003). https://books.google.de/books?id=USY6Tf4O-XMC
14. Kiko, K., Atkinson, C.: A Detailed Comparison of UML and OWL. Ph.D. thesis, Uni Mannheim, June 2008. http://ub-madoc.bib.uni-mannheim.de/1898. Accessed 07 Dec 2020
15. Loy & Hutz: MED Software zur optimalen Verwaltung medizintechnischer Geräte (2016). https://www.loyhutz.de/individuelle-cafm-software-loesungen/pakete/med-medizintechnikverwaltung/. Accessed 07 Dec 2020
16. Matthiessen, G., Unterstein, M.: Relationale Datenbanken und Standard-SQL: Konzepte der Entwicklung und Anwendung. Pearson Deutschland (2008)
17. Musen, M.A.: Dimensions of knowledge sharing and reuse. Comput. Biomed. Res. Int. J. **25**(5), 435–467 (1992). https://doi.org/10.1016/0010-4809(92)90003-s
18. Noy, N.F., McGuinness, D.L.: Ontology Development 101: A Guide to Creating Your First Ontology. Technical Report KSL-01-05, Standorf Knowledge Systems Laboratory and Standorf Medical Informatics, March 2001. http://www.ksl.stanford.edu/people/dlm/papers/ontology-tutorial-noy-mcguinness-abstract.html. A.ccessed 07 Dec 2020

19. Prinz, T.: Zulassung und Konformitätserklärung von Medizinprodukten. In: Expertenbericht Biomedizinische Technik. DGBMT im VDE e.V., December 2015. https://www.vde.com/de/dgbmt/publikationen/expertenbericht. Accessed 07 Dec 2020

20. Roussey, C., Pinet, F., Kang, M.A., Corcho, O.: An introduction to ontologies and ontology engineering. In: Ontologies in Urban Development Projects, vol. 1, pp. 9–38. Springer, London (2011)

21. Santos, I.C.T., Gazelle, G.S., Rocha, L.A., Tavares, J.M.R.: An ontology model for the medical device development process in Europe. In: IPC - Resumos alargados em actas de encontros científicos internacionais com arbitragem. University of Brescia (2012)

22. Schönbein, R.: Wissensrepräsentation mittels Ontologien. visIT Wissensrepräsentation **2**(2006), 4–5 (2006)

23. Uciteli, A., et al.: Ontology-based specification and generation of search queries for post-market surveillance. J. Biomed. Semant. **10**(9), May 2019. https://doi.org/10.1186/s13326-019-0203-7. https://jbiomedsem.biomedcentral.com/articles/10.1186/s13326-019-0203-7

24. Vo, M.H.L., Hoang, Q.: Transformation of UML class diagram into OWL Ontology. J. Inf. Telecommun. **4**(1), 1–16 (2020). https://doi.org/10.1080/24751839.2019.1686681

25. WebProtégé: WebProtégé - Protégé Wiki. https://protegewiki.stanford.edu/wiki/WebProtege. Accessed 07 Dec 2020

26. Weller, K.: Ontologien. In: Grundlagen der praktischen Information und Dokumentation: Handbuch zur Einführung in die Informationswissenschaft und -praxis, pp. 207–218. De Gruyter Saur, 6 edn., March 2013

27. Zedlitz, J., Luttenberger, N.: Conceptual modelling in UML and OWL-2. Int. J. Adv. Softw. **7**(1&2), 15 (2014)

Author Index

Printed in the United States
By Bookmasters